Lecture Notes in Earth Sciences

72

Editors:
S. Bhattacharji, Brooklyn
G. M. Friedman, Brooklyn and Troy
H. J. Neugebauer, Bonn
A. Seilacher, Tuebingen and Yale

W0107230

Springer-Verlag
Berlin Heidelberg GmbH

Julio C. Wasserman
Emmanoel V. Silva-Filho
Roberto Villas-Boas (Eds.)

Environmental Geochemistry in the Tropics

With 103 Figures and 52 Tables

 Springer

Editors

Prof. Dr. Julio C. Wasserman
Laboratoire de Chimie Bio-Inorganique et Environnement
CNRS – EP 132, Centre Helioparc
2, av. du Président Angot, F-64000 Pau, France

Prof. Dr. Emmanoel V. Silva-Filho
Dept. de Geoquímica – UFF
Outeiro de São-João Batista s/n°
Centro, Niterói, RJ, 24020-150, Brazil

Prof. Dr. Roberto Villas-Boas
Centro de Tecnologia Mineral, CETEM/CNPq
Rua 4, quadra D, Cidade Universitária – Ilha do Fundão
Rio de Janeiro, RJ, 21941-590, Brazil

"For all Lecture Notes in Earth Sciences published till now please see final pages of the book"

Cataloging-in-Publication data applied for

Die Deutsche Bibliothek - CIP-Einheitsaufnahme

Environmental geochemistry in the tropics / Julio C. Wasserman ... (ed.).

(Lecture notes in earth sciences ; 72)
ISBN 978-3-540-63730-1   ISBN 978-3-540-69638-4 (eBook)
DOI 10.1007/978-3-540-69638-4

ISSN 0930-0317
ISBN 978-3-540-63730-1

Typesetting: Camera ready by author
SPIN: 10656748   32/3142-543210 - Printed on acid-free paper

Preface

This book is a compilation of contributions for the enhancement of environmental geochemistry in the tropics. Some of the papers began as extended abstracts and presentations at the first *International Symposium on Environmental Geochemistry in the Tropical Countries* (ISEGTC), held in Niterói in 1993, and have now been expanded and up-dated. The book includes a range of aspects of tropical research where little has been done to date.

New fields of tropical research are emerging all the time. Although tropical research has a tendency to imitate methods already applied to temperate environments, it is not always easy to obtain data in tropical environments and to explain these data in light of what has already been done in colder climates. Reaction velocity, biological activity, especially microbial, and the lack of well defined seasons are examples of some of the parameters that distinguish tropical and temperate environments. Iron provides a classic example of the importance of climate on geochemical behavior. While the amount of iron in surface sediments and soils is very low in temperate locations, this metal is a significant percentage of Brazilian soils. The presence of iron oxides is the result of more drastic weathering, typical of tropical environments. The geochemical implications of iron are enormous, since its presence can change the behavior of a number of other important variables such as organic matter, metals and micro-organic pollutants. Another example is presented in Chapter 13, where the authors establish a relationship between the geochemistry of organic matter and mercury. Presently, it is hypothesized that mercury is not associated with organic matter but in an organic matter form, as methyl compounds. The implications of mercury geochemistry for risk assessment studies in the tropics are of grave concern.

Chapter 1 is a transcription of the discussions during the closing session of the ISEGTC, summarized by Dr. Egbert K. Duursma. The Symposium had been the forum for discussions on the present state of and new directions for environmental research in the tropics.

Chapter 2 is an example of studies being conducted in Brazil on the paleoclimates of the humid forest environment, using carbon isotopes to identify C-3 and C-4 plant debris. Chapter 3 is a significant attempt to establish transfer factors for artificial radionuclides in cultivated plants. Chapters 4 and 5 discuss models of soil (and mineral) leaching of elements and metals, a major process in tropical weathering. Chapters 6 and 7, although focusing on very different subjects, are important tools for environmental research. Remote sensing has already proven to be a powerful tool in the development of mass balance studies, necessary, but very rare in tropical areas. Chapter 6 is an example of the application of satellite image analysis to biomass studies in tropical settings. Chapter 7 introduces concepts, e.g. the reference plant and reference freshwater, resulting from the concentrations measured in a large number of samples, using multi-element analysis techniques. Chapters 8, 9, 10 and 11 are classical geochemistry studies from sub-tropical and tropical coastal environments, but they are of importance because of the use of different geochemical tools. Chapters 12 and 13 focus on the dynamics of heavy metals (cadmium, zinc and mercury) in coastal environments. The next three chapters focus on environmental problems caused by gold mining. Although Chapter 16 furnishes a temperate example of the environmental problems associated with gold mining, it is a particularly important contribution which emphasizes that mercury behaves similarly in tropical and in temperate climates subject to this kind of activity. Chapters 14 and 15 are examples of the application of techniques for the exploitation of gold in lesser developed countries in the tropics, where serious environmental problems occur as a result of mercury release.

Groundwater chemistry is poorly known in the tropics. Due to the deficiency of sanitary facilities in the developing countries, groundwater is frequently used not only for water supply but also for disposal of used waters. As a consequence, the water table is reduced and groundwater quality degraded. Chapter 17 presents the study of a small drainage basin where over-exploitation of groundwaters is becoming an insoluble problem.

The next chapter presents the results of a large scale air monitoring program in the tropics. In this chapter, the influence of forest burnings on air quality is assessed.

The final chapter is concerned with environmental modeling, and particularly budgets of materials in tropical areas. The simple calculations that are involved in the budgets can be improved considerably when dynamic processes are included and more realistic validation data have been obtained. Nevertheless, the apparent ease of making a mass balance hides the amount of information needed and the degree of knowledge of processes. I think that this contribution is an appropriate finale, since it does not provide information, but uses information, for a better understanding of the environment.

I would like to acknowledge all contributors and thank them for their patience. I also wish to thank Mr. Reginaldo Machado Filho for his work on many of the figures that appear in this book.

Rio de Janeiro   Julio Cesar Wasserman
March, 1998

Contents

CONTRIBUTORS

Aravena, R., University of Waterloo, Waterloo, Canada

Artaxo, P., Instituto de Física, Universidade de São Paulo, São Paulo, Caixa Postal 66318, 05389-970, São Paulo, Brazil

Baisch, P. R., Departamento de Oceanografia Geológica, Fundação Universidade de Rio Grande, Rio Grande, Caixa Postal 474, 96200-000, Rio Grande, Brazil

Barcellos, C., Departamento de Geoquímica, Universidade Federal Fluminense, Outeiro de São João Batista, s/nº, 24020-150, Niterói -Brazil (Present address: DIS, Fundação Oswaldo Cruz, Rua Leopoldo Bulhões 1480, 21041-210 Rio de Janeiro, Brazil)

Barral, A. O., Instituto Nacional de Investigación y Desarrollo Pesquero (INIDEP), C.C. 175, 7600, Mar del Plata, Argentina.

Barrocas, P. R. G., CESTEH, Fundação Oswaldo Cruz, Rua Leopoldo Bulhões 1480, 21041-210 Rio de Janeiro, Brazil

Bidone, E. D., Departamento de Geoquímica, Universidade Federal Fluminense, Outeiro de São João Batista, s/nº, 24020-150, Niterói -Brazil

Boulet, R., ORSTOM, Institute of Geosciences, NUPEGEL, France.

Carvalho, I. G., Instituto de Geociência, UFBA, Rua Caetano Moura, 123, Salvador, 40170-290, Salvador, Brazil

Ceradini, S., Divisione Ambiente, CISE, Via Reggio Emilia 39, MI-20090, Milan, Italy

Duursma, E. K. (retired), Res. les Marguerites, appt. 15, 1305, Chemin des Revoirs, La Turbie, F-06320, France

Ferrer, L. D., Lab. de Química Marina, Instituto Argentino de Oceanografía (IADO), Av. Alem, 53, 8000, Bahía Blanca, Argentina

Gerab, F., Instituto de Pesquisas Energéticas e Nucleares/CNEN, São Paulo, Brazil

Gonzalez, M. L., INGEOMINAS, Diagonal 53 # 34-53, Bogotá, Colombia

Laybauer, L., Curso de Pós-Graduação em Geociências, Universidade Federal do Rio Grande do Sul, Caixa Postal 15065, 91501-970, Porto Alegre, RS, Brazil

Lacerda, L. D., Departamento de Geoquímica, Universidade Federal Fluminense, Outeiro de São João Batista, s/n°, 24020-150, Niterói -Brazil

Lechler, P. J., Nevada Bureau of Mines and Geology, University of Nevada, Reno, Nevada, 89557, USA

Maddock, J. E. L., Departamento de Geoquímica, Universidade Federal Fluminense, Outeiro de São João Batista, s/n°, 24020-150, Niterói -Brazil

Marcovecchio, J. E., Lab. de Química Marina, Instituto Argentino de Oceanografía (IADO), Av. Alem, 53, 8000, Bahía Blanca, Argentina

Markert, B., Zittau International Hoschschul, Markt 23, 027A3, Zittau, Germany

Marques Jr., A. N. M., Programa de Pós-Graduação em Biologia Marinha, Universidade Federal Fluminense, Outeiro de João João Batista s/n°, 24020-150, Niterói, Brazil

Melamed, R., Centro de Tecnologia Mineral/CNPq, Rua 4, Quadra D, Cidade Universitária, Ilha do Fundão, 21941-590, Rio de Janeiro, Brazil

Miller, J. R., Department of Geology, Indiana University/ Purdue University, 723W Michigan Street, Indianapolis, Indiana, 46202-5132, USA

Mogollón, J. L., Reservoir Department, INTEVEP S. A., Aptdo. 76343, Caracas 1070-A, Venezuela

Mosser, C., Instituto de Geociência, UFBA, Rua Caetano Moura, 123, Salvador, 40170-290, Salvador, Brazil

Paredes, J. F., Instituto de Geociência, UFBA, Rua Caetano Moura, 123, Salvador, 40170-290, Salvador, Brazil

Pessenda, L. C. R., Center for Nuclear Energy in Agriculture (CENA/USP), Caixa Postal 96, 13400-970, Piracicaba, Brazil

Prieto, G. R., INGEOMINAS, Diagonal 53 # 34-53, Bogotá, Colombia

Pucci, A. E., Lab. de Química Marina, Instituto Argentino de Oceanografía (IADO), Av. Alem 53, 8000, Bahía Blanca, Argentina

Queiroz, A. F. S., Instituto de Geociência, UFBA, Rua Caetano Moura, 123, Salvador, 40170-290, Salvador, Brazil

Ramos, M. A. S. B., Instituto de Geociência, UFBA, Rua Caetano Moura, 123, Salvador, 40170-290, Salvador, Brazil

Rodrigues Filho, S., Centro de Tecnologia Mineral/CNPq, Rua 4, Quadra D, Cidade Universitária, Ilha do Fundão, 21941-590, Rio de Janeiro, Brazil

Santos, A. L. F., Instituto de Geociência, UFBA, Rua Caetano Moura, 123, Salvador, 40170-290, Salvador, Brazil

Scagliola, M. O., Obras Sanitarias Mar del Plata Sociedad de Estado (OSSE), Roca 1213, 7600, Mar del Plata, Argentina

Silva-Filho, E. V., Departamento de Geoquímica, Universidade Federal Fluminense, Outeiro de São João Batista, s/n°, 24020-150, Niterói -Brazil

Telles, E. C. C., Center for Nuclear Energy in Agriculture (CENA/USP), Caixa Postal 96, 13400-970, Piracicaba, Brazil

Touré, I., LGGA/IMG, Université de Nice-Sophia-Antipolis, Parc Valrose, Nice, 06108, France

Tubbs, D., Departamento de Geociências, Universidade Federal Rural do Rio de Janeiro, Seropédica, RJ, Brazil

Valencia, E. P. E., Instituto Boliviano de Ciencia y Tecnologia Nuclear, La Paz, Bolivia

Wasserman, J. C., Departamento de Geoquímica, Universidade Federal Fluminense, Outeiro de São João Batista, s/n°, 24020-150, Niterói -Brazil (Present address: Laboratoire de Chimie Bio-Inorganique et Environnement, CNRS E.P. 132. Helioparc, 2, av. du Président Angot F-64000, Pau, France)

Wasserman, M. A., Instituto de Radioproteção e Dosimetria/CNEN, Av. Salvador Allende, s/n°, 22780-160, Rio de Janeiro, Brazil (Present address: Laboratoire de Chimie Bio-Inorganique et Environnement, CNRS E.P. 132. Helioparc, 2, av. du Président Angot F-64000, Pau, France)

Yamasoe, M. A., Instituto de Física, Universidade de São Paulo São Paulo, Caixa Postal 66318, 05389-970, São Paulo, Brazil

1 Trends in Environmental Geochemistry in Tropical Countries[1]

Egbert K. Duursma
Res. les Marguerites, app. 15, 1305, Chemin des Revoirs, La Turbie, F-06320, France

General

1. The participation of 80 scientists, coming from 15 different countries, presenting 80 papers and 32 posters, has shown that wide interest existed in this first international tropical environmental geochemistry symposium. It has also demonstrated the necessity of such a meeting in which the perspectives for tropical environmental problems can be presented and discussed.
2. Science has a great moral responsibility in detecting and forecasting global and regional changes in environments, whether caused by human activities or due to natural changes on a global or regional scale. Science should express concern and present solutions on environmental issues to public authorities that are legally responsible for regional and national sustainable development, considering both human and natural resources. Science should equally inform the public at large.
3. The symposium focused on eight different themes:

   - Geochemistry of aqueous environments
   - Geochemistry of soils and its relationships with organisms
   - Dynamics of geochemical processes; application to pollution control
   - Atmospheric quality
   - Organic matter and its importance in tropical geochemical cycles
   - Analytical methods applied to environmental geochemistry; adaptations for samples from tropical environments
   - Programmes of environmental quality and monitoring in tropical countries
   - Modelling, geochemical balances and environmental diagnosis in the tropics

   A special workshop and forum discussion was also held on heavy metals in tropical countries.
4. The symposium took an environmental geochemical approach which brought together a number of well-known and promising scientists working in very

---

[1] This chapter was based on the discussions of the closing session of the First International Symposium on Environmental Geochemistry in the Tropical Countries, held in Niterói (Brazil), on December, 3, 1993.

different fields, such as atmosphere, terrestrial systems including vegetation, and tropical river and estuarine systems. Among the participants were also engineers and public health scientists.

5. As a whole, the symposium was a great success. It brought into discussion a great many current problems on which detailed and well-documented results were presented. These ranged from local problems, such as contamination from mining operations at the mining sites, to climatic changes over the South-American continent as related to Pacific and Atlantic ocean and meteorological circulation patterns.

6. The tropics play an essential role in global physico-chemical processes where atmospheric transport of natural and anthropogenically produced substances is concerned.

7. Mining operations, such as of uranium, thorium, tin, copper, coal and gold, may have small to serious environmental side effects. In particular where metal mercury is applied for gold mining, the amount of mercury which is released into the environment is of great concern. This mercury creates serious toxic effects to living species and to humans who consume these contaminated species. Due to the fact that tons of mercury now present in the environment, are transformed by sedimentary bacteria into the very toxic methyl-mercury which accumulates in fish, scientists speak of a "Chemical Time-Bomb", which will sooner or later endanger large parts of, e.g., the Amazon and some Asian systems. The key point is that it is still unknown when this "Time-Bomb" will create these serious problems, which may occur in large areas and last for a long time.

8. Tropical countries will endure contamination problems which temperate and developed countries have experienced for a long time, and which they are now trying to repair at great expense the national public budgets. The pressing question for the tropics is how to avoid a repetition of history by controlling sources and pathways of contaminants to the environment, or whether pollution control has to be understood the "hard way", which usually is the common human reaction, first awaiting disasters.

9. Although environmental legislation may exist in a number of cases, its implementation is of great concern. Environmental legislation should be revised permanently and obtain the support of scientific commissions (such as the Environmental Protection Agency in the USA) in order to incorporate up-to-dated scientific information.

Results and Opinions

10. Apart from climatic conditions, the environmental problems in tropical countries differ from those of temperate countries, in land use, which in some tropical countries is rapidly expanding, while that of temperate countries is usually much less changing.

11. Due to elevated temperatures and year-round growth and mineralization, the cycling rates of substances in tropical countries are much higher than those in temperate countries. In this context a number of studies were presented on the availability of micro-nutrients as metals to terrestrial plants.

12. Several "recent" geological records may be used for extrapolation of geological and climatic events from the past to the future, in particular those related to the cycling climatic conditions which determine the oceanic and meteorological situation at the South-American Atlantic coast.

13. In this context it is suggested that the climate of the South-American continent may be very sensitive to global climatic changes, e.g., due to greenhouse effects, and thus trigger abrupt changes from dry to wet climate and vice versa, as has been observed for the last 8000 years, and which are of an "El Niño" type. These changes will affect almost all regions of Brazil and all equatorial countries of South America.

14. Pollution (contamination) studies in tropical countries have shown the existence of environmental problems, but there is still a lack of sensitive equipment to arrive at quantitative results at different levels of all kinds of inorganic and organic pollutants. Improvement in the analytical capacity would make it possible to study the dynamics of the contamination in different compartments of the environment, such as water, suspended matter, sediments and biota, in much more detail. Thus it will be possible to determine with greater accuracy the bio-availability of contaminants occurring in soil and water, which probably is different from that in temperate countries.

15. A number of environmental problems have been studied little or not at all, such as those which concern pesticides and PCBs in natural ecosystems, groundwater for, e.g., phosphate speciation and geochemical studies linked to ecosystem studies concerning pathways and transfer through food chains (not necessarily accumulation by the food chain).

16. Since biotic processes are also key controlling ones of contaminant pathways, more inter-disciplinary research has to be carried out, with e.g., biologists, for the prediction of the toxicity of pollutants in ecosystems and to man as a principal consumer. In goldmining regions this is particularly urgent with respect to mercury contamination. In this context geochemical models should be developed to understand and predict the kinetics of the processes involved in order to set up a better risk assessment and to limit the contamination. Collaboration with engineers and public authorities is therefore also required to change the mining procedures which cause that pollution.

17. Proper chemical research should be based on regular standardization of analytical methods, the use of reference samples and participation in inter-calibration exercises. Since the matrixes of samples from tropical countries may differ from those supplied by organizations such as the US National Institute of Standards and Technologies, the IAEA Laboratory in Monaco and the EC Institute in Ispra (Italy), it is suggested that institutes in the tropics take

their own initiatives in this direction, if possible jointly with one of the above-mentioned organizations.

18. As an alternative to traditional maximum permissible standards for contaminants in soils, water, fish and agricultural products, scientific research should be applied in risk assessment, taking into account the tropical dynamics and pathways of these contaminants. It is not justified to apply the available standards as determined in temperate countries because of the differences in the dynamics and pathways. This requires adapted research which should if possible be carried out with relevant international organizations such as WHO, FAO and IAEA.

19. Most of the geochemical contributions to the symposium dealt with terrestrial, river and estuarine systems. Since a number of problems extend to continental shelves and adjacent oceans, these systems should be part of the overall studies.

20. Contamination of bays, lagoons, rivers and estuaries with urban sewage is a burning problem for many tropical countries, due to the absence of proper sewage-treatment plants. This causes not only undesired eutrophication but is a consistent health problem for man due to the presence of pathogenic micro-organisms in water and fish. It is stressed that this very urgent problem should receive high priority and action by all responsible authorities. Research should be enhanced to find at least some interim solutions, depending on cost-benefit analysis, such as marsh or water hyacinth growth techniques, prior to the construction of fully operative sewage treatment plants.

21. Savannah and "deforestation" fires represent a great concern for atmospheric quality. The fires in tropical countries contribute 80% of the world's fires, excluding the higher fossil-fuel burning emissions and re-growth of vegetation. These fires contribute also to metal fallout on a global scale, while additionally they have a regional influence on ozone concentrations in the troposphere.

22. Tropical countries are undergoing rapid changes due to growing and migrating population causing increased exploitation of the natural resources that often occurs in an unsustainable manner. Therefore environmental geochemists in tropical countries need to develop a solution-oriented approach in addition to the usual problem-oriented approach. This means that geochemists must work with and educate farmers, miners and fishermen as well as regional and national government officials, engineers and businessmen.

23. A capacity building in environmental science is required in which the environmental geochemists take initiatives among other solution-oriented scientists, including socio-economic scientists and managers.

24. Environmental modelling is one of the tools to combine results from different disciplines and should be encouraged for the local and global problems in tropical countries. The results should be applicable to assessment and management strategies.

25. After the 1992 UNCED conference in Rio there has been a strong international development towards environment management coupled with sustainable development. Impact assessment and expert opinion modelling programmes have been and are being developed in order to couple science, management and public and private activities. New studies should take advantage of these models for which demos are freely available.

2 Paleoclimate Studies in Brazil Using Carbon Isotopes in Soils

Luiz C.R. Pessenda[1], E.P.E. Valencia[2], R. Aravena[3], E.C.C. Telles[1] and R. Boulet[4]

[1] Center for Nuclear Energy in Agriculture (CENA), Piracicaba, Brazil
[2] Instituto Boliviano de Ciencia y Tecnologia Nuclear, La Paz, Bolivia
[3] University of Waterloo, Waterloo, Ontario, Canada
[4] ORSTOM, Institute of Geosciences, NUPEGEL, France

Abstract

This chapter presents carbon isotope data of soil organic matter (SOM), collected in natural forest ecosystems in different sites from Brazil. The studied areas are located in Londrina (Southern part of the country), Piracicaba (Southeast), Salitre (Central) and Altamira (Northern). This study is part of the research program on tropical and sub-tropical soils in Brazil, of which the main objective is to use carbon isotopes to provide information on vegetation changes and their relationships with climatic changes during the Holocene. ^{14}C data of charcoal samples, the humin fraction and soil organic matter (SOM) contents, indicate that the organic matter of this area is at least of Holocene age. ^{13}C data in SOM indicate that C_4 plants were the dominant vegetation in Londrina and Piracicaba during the early and middle Holocene, while C_3 plants were the dominant vegetation in Altamira and probably a mixture of C_3 and C_4 plants occurred in Salitre during the Holocene.

Introduction

Paleo-reconstruction of vegetation changes and their relation to climate in tropical and sub-tropical forest is essential for the understanding of the response of these ecosystems to future climatic change. Different approaches that include geomorphological (Ab'Saber, 1977; 1982; Servant et al. 1981; Bigarella and Andrade Lima 1982), biological, botanical (Hafter 1969; Prance 1973; Gentry 1982), palynological (Absy et al. 1991; Ledru 1993) and isotope studies (Pessenda et al. in press a and b; Martinelli et al., in press) have been used to infer past climatic changes in the Amazonia, Central, Southeast and South regions of Brazil.

Paleo-ecological and geomorphological studies suggest the occurrence of severe climatic changes in the South American continent. It has been hypothesized that there were drier periods during the Pleistocene and Holocene than the present, when the tropical forest was replaced by savannah vegetation with predominance of grasses (Van der Hammen, 1974; Absy and Van der Hammen, 1976; Absy,

1980; Ab'Saber, 1982; Leyden, 1985; Bush and Colinvaux, 1990; Markgraf 1991). An understanding of the degree to which these changes affected the composition of the soil organic matter (SOM) would improve our ability to understand changes in the future.

The stable carbon isotope composition ($^{13}C/^{12}C$, or $\delta^{13}C$) of SOM contains information regarding the presence/absence of C_3 and C_4 plant species in past plant communities, and their relative contribution to community net primary production (Throughton et al., 1974; Stout et al., 1975). This information has been utilized to document vegetation change (Hendy et al., 1972; Dzurec et al., 1985), to infer climate change (Hendy et al., 1972) and to estimate rates of SOM turnover (Cerri et al., 1985).

The $\delta^{13}C$ values of C_3 plant species range from approximately -32‰ to -20‰ PDB, with a mean of -27‰, while, in contrast, the $\delta^{13}C$ values of C_4 species range from -17‰ to -9‰, with a mean of -13‰. Thus, C_3 and C_4 plant species have distinct $\delta^{13}C$ values and differ from each other by approximately 14‰ (Boutton, 1991).

In this chapter changes in the SOM of Brazilian soils and their relationships with past climatic changes are analyzed. The aim of the project developed in the Radiocarbon Laboratory of CENA was to correlate radiocarbon datings to the ^{13}C composition of SOM, with the objective of studying the evolution of local vegetation in the past. ^{14}C dating was used to estimate soil organic matter chronology and $\delta^{13}C$ was used as an indicator of the vegetation types (C_3 vs. C_4) in the local environment.

Material and Methods

The soil samples were collected from natural forests of Londrina, 51°10'W, 23°18'S, Paraná State, in the south of Brazil; Piracicaba, 47°38'W, 22°43'S, São Paulo State, in the southeast; Salitre de Minas or Salitre, 46°46'W, 19°00'S, central Brazil; and Altamira, 3°30'S, 52°58'W, Pará State, in the north (Fig. 1).

The natural forest at Londrina, Piracicaba and Salitre is a Mesophitic semi-deciduous type forest and at Altamira is part of the Amazon forest. The soils of Londrina and Altamira are "Terra Roxa Estruturada", according to the Brazilian soil classification, Alfisol in the American soil classification. The soil sample from Piracicaba is a "Latossolo Vermelho Escuro" (Dark Red Latosol), according to Brazilian soil classification, Oxisol in the American soil classification, and the soil from Salitre is a "Latossolo Vermelho Amarelo" (Yellow Red Latosol), Oxisol in the American soil classification.

Fig. 1: Map showing the sites studied

Soils at all sites were sampled by collecting up to 10 kg of material in 10 cm increments from the surface to 180 cm depth. 1 kg of the sampled material was dried at 60°C to constant weight and root fragments were discarded by hand-picking. Any remaining plant debris was removed by flotation in hydrochloric acid 0.01 M, dried again to constant weight and sieved (0.105 mm for δ^{13}C and 0.210 mm for ^{14}C).

The humin fraction was extracted from the 0.210 mm fraction (2.5 kg), using conventional methods (Dabin, 1971; Goh, 1978; Anderson and Paul, 1984): a) acid digestion in hydrochloric acid 0.5 M at 70°C - 80°C for 4 hours and washing with distilled water until pH reaches 3 to 4; b) reaction of the solid residue with at least 30 liters (10 liters per extraction) of sodium pyrophosphate - sodium hydroxide 0.10 M for about 36 hours (12 hours per extraction) and washing with distilled water until pH reaches 3 - 4; c) hydrolysis of residue with 4 liters of hydrochloric acid 3 M at 100°C for 12 hours, followed by washing with distilled water until pH reaches 3 to 4; d) the solid residue was dried at 40°C for 48 hours and sieved (<0.210 mm).

The charcoal samples were hand separated from 10 kg soil samples from Salitre, oven dried at 90°C, weighed and treated using the conventional AAA (acid-alkaline-acid) treatment. From the deepest samples (2.20 m to 3.45 m) charcoal was hand separated from small amounts of soil with the aid of a 5 mm sieve. These samples were used only for ^{13}C analyses.

The ^{14}C analyses on charcoal and humin samples were carried out at the Radiocarbon Laboratory of the Center for Nuclear Energy in Agriculture (CENA), using the benzene method and liquid scintillation counting (Pessenda and Camargo, 1991). Benzene samples were counted for at least 48 hours in a low level Packard 1550 liquid scintillation counter. Radiocarbon ages are expressed in years BP and percent modern carbon (pmC) relative to 95% of the activity of oxalic acid standard in 1950 and normalized to a $\delta^{13}C$ of -25‰ PDB (Stuiver and Polach, 1977).

The carbon isotopic ratios of SOM were determined by mass spectrometry of CO_2 from sample combustion in an atmosphere of pure oxygen at 900°C. Results were expressed as $\delta^{13}C$ with respect to PDB standard in the conventional δ (‰) notations.

Carbon contents of soil samples were determined in 1 to 5 g of <0.105 mm fraction by combustion in a C,H auto-analyzer and values expressed as weight percent of dry sample.

Results and Discussion

The total organic carbon contents of soils are shown in Fig. 2. The total carbon concentration decreased from the surface to the deeper depths for all soils. The highest concentrations in all soil profiles were observed in Salitre, probably due to the presence of charcoal found in the soil.

The values of $\delta^{13}C$ in SOM are shown in Fig. 3. The values obtained between the surface and the 40-50 cm interval are representative of C_3 plants, reflecting the current local vegetation (forest) in all soil environments. These values remained almost constant in the case of Altamira, characterizing the predominance of C_3 vegetation during the last 9,810 yrs. BP. In Salitre soil, the values ranged from -26.0 ‰ to -21.0‰ and the isotopic trend observed in these data could be due to isotope fractionation effect, occurring during microbial respiration, resulting in a gradual increase in $\delta^{13}C$ of SOM with time (Stout et al., 1975; Becker-Heidmann and Scharpenseel, 1992) and/or a change from a mixed C_3 and C_4 vegetation to predominantly C_3 vegetation in the period from 8,790 yrs BP to 1,820 years BP. For Londrina and Piracicaba soils, the values show a significant change from -21.6‰ to -15.0‰, probably indicating predominance of C_4 vegetation in both sites in the periods from 10,800 yrs BP to 3,090 yrs BP and about 6,000 yrs BP to 3,440 yrs BP, respectively. These results suggest that in the case of tropical regions, a) the savannah vegetation was dominated by C_3 grasses; b) woody vegetation instead of grasses dominated the vegetation changes during these phases; c) the replacement of the forest by savannah vegetation was restricted to small areas.

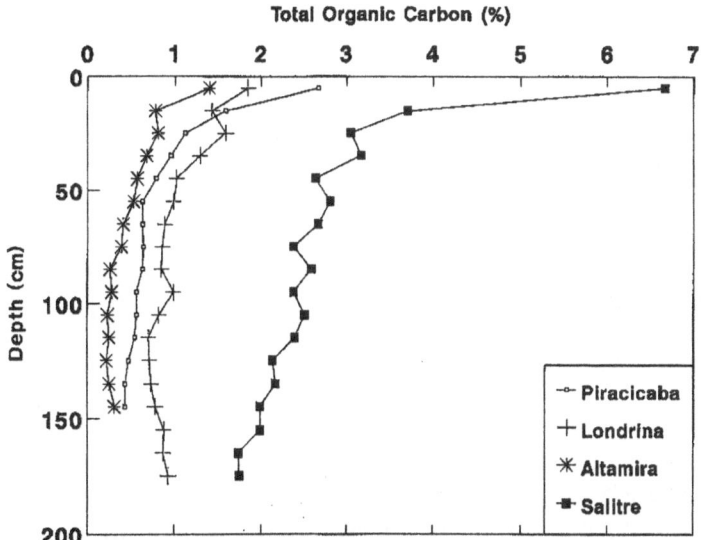

Fig. 2: Total organic carbon of soil profiles

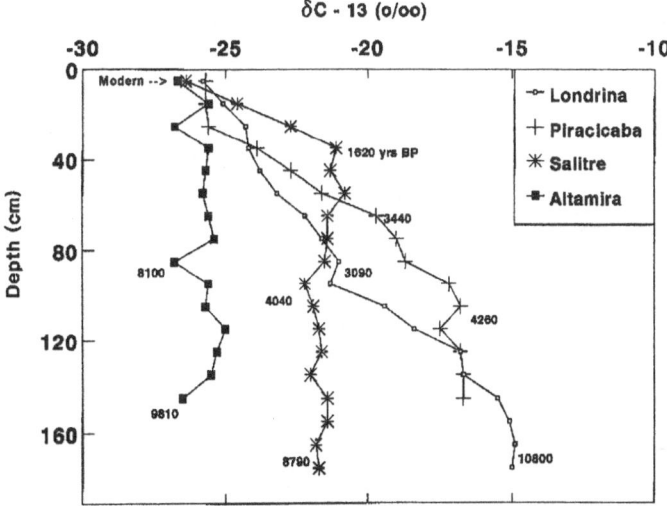

Fig. 3: ^{14}C age and δ^{13}C variation as function of soil depth

In Central Brazil, in the Salitre area, Ledru (1993) described changes during the last 32,000 yrs BP based on pollen analyses. She postulated two major periods of forest retreat and predominance of savannah vegetation (grasses), probably associated with very dry climatic conditions. These were from 11,000 to 10,000 yrs BP and 6,000 to 4,500 yrs BP. Dry periods have also been reported in the Central Amazon Basin and other areas of South America during the Holocene (Absy, 1982; Van der Hammen, 1982). The most significant dry phases recorded in these areas and Southern Brazil are 7,500 to 6,000 yrs BP; 4,200 to 3,500 yrs BP; 2,700 to 2,000 yrs BP; 1,500 to 1,200 yrs BP and 700 to 400 yrs BP (Bigarella, 1971; Fairbridge, 1976; Absy et al. 1991).

It seems that the dry phases registered by Ledru (1993) in the Salitre area and by several authors in other regions are in very good agreement with the data obtained in this paper for Central and Southern parts of Brazil, probably indicating that the sub-tropical region was much drier than the tropical region during climate changes in the early and middle Holocene.

Charcoal is present along the entire soil profiles at the three sampling locations in Salitre (Fig. 4). Some peak values are observed at certain soil depths, without any clear correlation between the three locations. The presence of charcoal in this soil is a clear indication that this area has been affected by forest fires, probably during most of its history. The extremely elevated charcoal contents in some soil horizons indicate that these events were more severe during some periods, probably indicating drier conditions. In the upper Rio Negro (Colombia and Venezuela), the presence of charcoal was reported, indicating the occurrence of frequent and wide-spread fires in the Amazon Basin, possibly associated with extremely dry periods and/or human disturbances (Saldarriaga and West, 1986). These events range from 6,000 yrs BP to the present and several events coincide with dry phases recorded during the Holocene (Absy, et al., 1982; Van der Hammen, 1982).

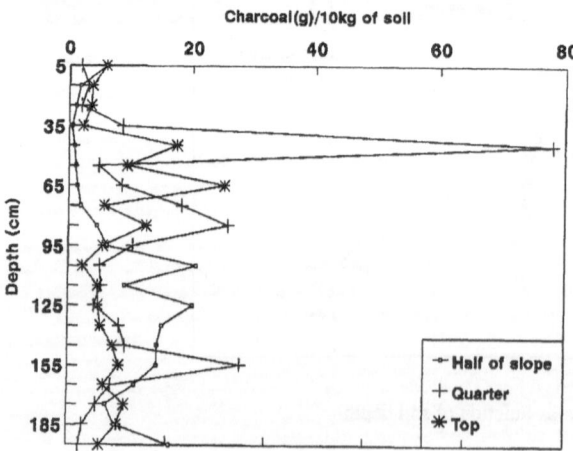

Fig. 4: Charcoal distribution as function of soil depth in Salitre

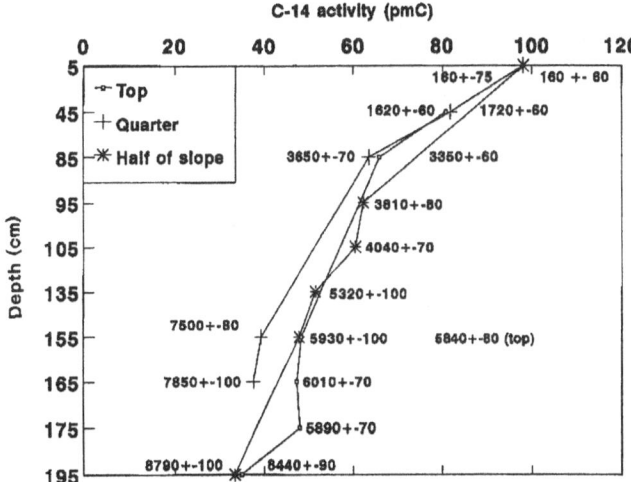

Fig. 5: ^{14}C dating of charcoal samples from Salitre as function of soil depth

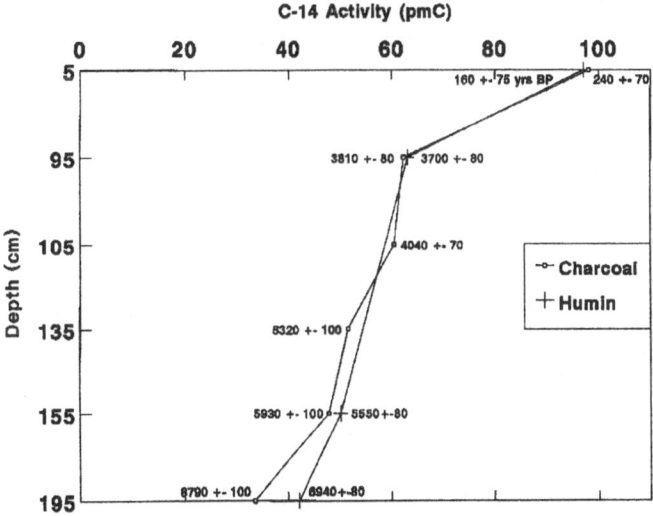

Fig. 6: ^{14}C dating of charcoal and humin samples in Salitre

Charcoal ^{14}C ages range from 160 yrs BP close to the soil surface to 8,790 yrs BP at 2.0 m depth (Fig. 5). Considering the age profiles, it is possible to expect that charcoal at the deepest locations should have a radiocarbon age of at least 12,000 yrs BP. These data show that the ages at the three locations were similar in some horizons. However the charcoal samples from the fourth location seem to be older than in the other locations. At the above location, between 150 and 180 cm, the ages were very similar, suggesting a mixing of charcoal at this depth interval.

Figure 6 shows the ^{14}C dates of charcoal and humin samples. No significant age differences were obtained until a depth of 155 cm, demonstrating that the humin can be a useful fraction to date SOM. At 2.0 m depth, the humin was about 2,000 years younger than the charcoal age, indicating that the input of younger carbon by translocation significantly affected the humin in the deepest parts of the soil profile.

Conclusions

Carbon isotope data in SOM (humin fraction) collected in soil profiles from Altamira, in the Amazon region, Salitre de Minas, central Brazil, and Piracicaba and Londrina, in the southern region of Brazil, indicate that the organic matter in these soils is at least Holocene in age. The presence of significant amounts of charcoal in the Salitre area suggests that forest fires were a significant process during the Holocene. δ^{13}C data in SOM indicate that C_3 plants were the dominant vegetation in the Altamira region and probably a mixture of C_3 and C_4 plants was the dominant vegetation in Salitre during the period 8,790 yrs BP to 1,820 yrs BP. For Piracicaba and Londrina the δ^{13}C data indicate a predominance of C_4 plants during the early and middle Holocene, probably indicating that the southern region of Brazil was much drier than the tropical region during climate changes in the past.

Acknowledgements

We gratefully acknowledge the financial support of FAPESP (São Paulo Foundation for Research), grants nos. 90/3312-0 and 91/1868-1. Most of the laboratory work was done by Maria Valéria L.Cruz, Paulo Ferreira, Cláudio Sérgio Lisi, Gláucia Pessin and Márcio Arruda, to whom the authors are grateful.

References

Ab'Saber, A.N. (1977). Espaços ocupados pela expansão dos climas secos na América do Sul, por ocasião dos períodos glaciais quaternários. *Paleoclimas* (São Paulo), 3:1-20.

Ab'Saber, A.N. (1982). The paleoclimate and paleoecology of Brazilian Amazonia. In Prance, G.T., ed., *Biological Diversification in the Tropics*, New York, Columbia University Press, pp:41-59.

Absy, M.L., Van der Hammen (1976). Some paleo-ecological data from Rondônia, southern part of Amazonian Basin. *Acta Amazônica*, 10:929-932.

Absy, M.L. (1980). Dados sobre as mudanças do clima e da vegetação da Amazônia durante o Quaternário. *Acta Amazônica*, 10:929-932.

Absy, M.L. (1982). Quaternary palynological studies in the Amazon basin . In Prance, G.T. (ed.) *Biological Diversification in the Tropics*, New York, Columbia University Press, pp:67-73.

Absy, M.L., Cleef, A., Fournier, M. , Martin, L. , Servant, M., Sifeddine, A. , Ferreira da Silva, M. , Soubies, F., Suguio, K., Turcq, B., Van der Hammen, T. (1991). Mise en évidence de quatre phases d'ouverture de la forêt dense dans le sud-est de l'Amazonie au cours des 60.000 dernières années. Première comparaison avec d'autres régions tropicales. *C.R. Acad Sci Paris*, t. 312, Sér II 312:673-678

Anderson, D.W., Paul, E.A. (1984). Organo-mineral complexes and their study by radiocarbon dating. *Soil Science Society American Journal*, 48:298-301.

Becker-Heidmann, P., Scharpenseel, H.W. (1992). The use of natural ^{14}C and ^{13}C in soils for studies on global climate change. In Long, A. and Kra, R.S. (eds.), *Proceedings of the 14th International ^{14}C Conference. radiocarbon*, 31(3):535-540.

Bigarella, J.J. (1971). Variações climáticas no Quaternário Superior do Brasil e sua datação radiométrica pelo método do carbono 14. Instituto de Geografia - Universidade de São Paulo. *Paleoclimas*, 1:1-22.

Bigarella, J.J., Andrade-Lima, D. (1982). Paleoenvironmental changes in Brazil. In: Prance GT (ed) *Biological Diversification in the Tropics*. Columbia Univ. Press, New York, pp:27-40.

Boutton, T.W. (1991). Stable carbon isotope ratios of natural materials. II. Atmospheric, terrestrial, marine and freshwater environments. In Coleman, D.C. and Fry, B. (eds.), *Carbon Isotope Techniques*, New York, Academic Press, pp:173-185.

Bush , M., Colinvaux, P.A. (1990). A pollen record of a complete glacial cycle from lowland Panama. *J Vegetat Sci*, 1:105-118.

Cerri, C.C., Feller, C., Balesdent, J., Victoria, R., Plenecassagne, A. (1985). Application du traçage isotopique naturel en ^{13}C, a l'étude de la dinamique de la matière organique dans les sols. *C.R. Acad. Sci.*, Paris, 300, Série II, 9:423-428.

Dabin, B. (1971). Etude d'une méthode d 'extraction de la matière humique du sol. *Science du Sol*, 1:47-63.

Dzurec, R.S., Boutton, T.W., Caldwell, M.M., Smith, B.N. (1985). Carbon isotope ratios of soil organic matter and their use in assessing community composition changes in Curlew Valley, Utah. *Oecologia*, 66:17-24.

Fairbridge, R.W. (1976). Shellfish-eating preceramic Indians in coastal Brazil. *Science*, 191: 353-359.

Gentry, A.H. (1982). Phytogeography patterns as evidence for a Chocó refuge. In Prance GT (ed) *Biological Diversification in the Tropics*. Columbia Univ. Press, New York, pp:112-135.

Goh, K.M., Molloy, B.P.J. (1978) Radiocarbon dating of paleosols using organic matter components. *J. Soil Sci*, 29(4):567-573.

Haffer, J. (1969). Speciation in Amazonian forest birds. *Science* 165: 131-137.

Hendy, C.H. 1972 , Rafter, T.A., MacIntoshi, N.W.G. (1972). The formation of carbonate nodules in the soils of the Darling Downs, Queensland, Australia and the dating of the Talgai cranium. In Rafter, T.A. and Grant-Taylor, T. (eds.), *Proceedings 8th International Conference*, Lower Hutt, New Zealand, Wellington, Royal Society of New Zealand, pp:D106-D126.

Ledru, M.P. (1993). Late quaternary environmental and climatic changes in central Brazil. *Quaternary Research*, 39:90-98.

Leyden, B.W. (1985) Late quaternary aridity and Holocene moisture fluctuations in the Lake Valencia basin, Venezuela. *Ecology*, 66:1279-1295.

Markgraf , V. (1991). Younger Dryas in southern South America. *Boreas*, 20:63-69.

Martinelli, L., Pessenda, L.C.R., Valencia, E.P.E., Camargo, P.B., Telles, E.C.C., Cerri, C.C., Victória, R.L., Aravena, R., Richey, J., Trumbore, S (1997). Carbon -13 and carbon-14 depth variation in soil profiles of sub-tropical and tropical regions of Brazil and relations with climate changes during the Quaternary. *Oecologia*. (in press)

Pessenda, L.C.R., Camargo, P.B. (1991). Datação radiocarbônica de amostras de interesse arqueológico e geológico por espectrometria de cintilação líquida de baixo nível de radiação de fundo. *Química Nova*, 14(2):98-103.

Pessenda, L.C.R., Valencia, E.P.E., Camargo , P.B., Telles, E.C.C., Martinelli, L.A., Cerri, C.C., Aravena, R., Rozanski, K. (in press a). Radiocarbon measurements in Brazilian soils developed on basic rocks. *Radiocarbon*, in press.

Pessenda, L.C.R., Aravena, R., Melfi, A.J., Telles, E.C.C., Boulet, R., Valencia, E.P.E., Tomazello, M. (in press b). The use of carbon isotopes (^{13}C, ^{14}C) in soil to evaluate vegetation changes during the Holocene in central Brazil. *Radiocarbon*, in press.

Prance, G.T. (1973). Phytogeographic support for the theory of Pleistocene forest refuges in the Amazon basin, based on evidence from distribution patterns in Caryocaraceae, Chrysbonaceae, Dichapetalaceae and Lecythidaceae. *Acta Amazonica*, 3(3):5-28.

Saldarriaga, J.G., West, P. (1986). Holocene fires in the northern Amazon basin. *Quaternary Research*, 26:358-366.

Servant, M., Fontes, J.C., Rieu, M., Saliége, X. (1981). Phases climatiques arides holocènes dans le sud-ouest de l'Amazonie (Bolivie). *C.R. Acad. Sci*, II, 292:1295-1297.

Stout, J.D., Rafter, T.A., Throughton, J.H. (1975). The possible significance of isotopic ratios in paleoecology. *In*: Suggate, R.P. and Cresswell, M.M. (eds). *Quaternary Studies*. Wellington, Royal Society of New Zealand, 279-286.

Stuiver , M., Polach, H. A. (1977) Discussion: Reporting ^{14}C data. *Radiocarbon*, 19:355-363.

Throughton, J.H., Stout, J.D., Rafter, T. (1974). Long-term stability of plant communities. *Carnegie Institute of Washington Yearbook*, 73:838-845.

Van der Hammen, T. (1974). The Pleistocene changes of vegetation and climate in tropical South-America. *J.Biogeogr.*, 1:3-26.

Van der Hammen, T. (1982). Paleoecology of tropical South America. In Prance G.T. (ed.). *Biological Diversification in the Tropics*. New York, Columbia, Univ. Press, pp:60-65

3 Soil-to-Plant Transfer of ^{137}Cs Related to its Geochemical Partitioning in Oxisols of Tropical Areas

Maria Angélica M. Wasserman

Instituto de Radioproteção e Dosimetria / CNEN, Av. Salvador Allende s/n°, Rio de Janeiro, Brazil, CEP: 22780-160

Abstract

Differences in ^{137}Cs soil-to-plant transfer for two types of soils under tropical climatic conditions are discussed with reference to pedology and geochemical partitioning. Using acid oxisol soils with low exchangeable K contents, transfer factor (Tf) values ranged from 0.18 to 0.41 for black beans. Basic oxisol with normal exchangeable K contents presented lower Tf values: 0.06 to 0.11. These values were higher than mean values reported by IUR for ^{137}Cs for beans under temperate climate: 0.03. Results of sequential extraction showed ^{137}Cs weakly bound to soil components and underline the importance of Fe oxides in the control of ^{137}Cs availability.

Introduction

Transfer of Radionuclides from soil to plants and through the food chain is one of the paths by wich radioactivity harm can reach man. In agricultural systems, the parameter that best describes the soil-plant interactions is the transfer factor (Tf). This parameter has been widely calculated using International Union of Radioecologists (IUR) recommendations (International Union of Radioecologists, 1989), wich relate the total amount of radionuclide present in the plant and in the soil:

$$Tf = \frac{Ap}{As}$$

Ap = Activity in the edible part of the plant (Bq/kg dry weight).
As = Activity in the soil (Bq/kg dry weight).

The differences found between different soils and types of vegetation engender a great variety of measured transfer factors (Yasuda and Uchida, 1994). This picture is further complicated by the use of fertilizers and other agricultural practices that modify root uptake by directly influencing the concentration of stable elements, pH and other physico-chemical properties of the soil that modify the availability of elements (Aarkrog, 1979; Cawse and Turner, 1982; Simmonds, 1985; Claus et al.,1990).

In this work, the differences in soil/plant transfer of ^{137}Cs in two types of soils under tropical climatic conditions will be discussed with reference to pedology and geochemical partitioning.

Material and Methods

The chosen culture for transfer factor determination was black bean (*Phaseolus vulgaris*). The plants were cultivated for two consecutive years in masonry lysimeters following the agricultural recommendations.

The lysimeters measuring 1 m^2 in area and 1 m in depth, were installed in a plot 10.4 x 2.7 m, built in a restricted area of the Institute for Radioprotection and Dosimetry (IRD/CNEN/Brazil). The lysimeters were filled with 15 cm of coarse material (sand and gravel), 30 cm of uncontaminated soil and 40 cm of contaminated soil. Two types of soils were used in the experiments. One of them was a yellow-red oxisol, artificially contaminated by adding 40 ml of a solution containing approximately 60 µCi of ^{137}Cs in a liter of water for every 2 cm layer (lysimeters 1 to 6). The other soil type was an urban one collected in Goiânia, at a site where a radiological accident occurred in 1987 (lysimeters 9 and 10).

Harvesting and sampling were done when the beans were ready for consumption. Direct measurements of the activity of ^{137}Cs in beans and soil were done by gamma spectrometry using a NaI detector.

Soil analyses were performed by the Brazilian Enterprise for Agricultural Research (EMBRAPA) in order to characterize the soil and to determine soil needs for fertilizers and corrections.

Sequential extractions were performed in four soil samples, collected in lysimeters 1, 2, 9 and 10. The method used was that of Meguellati (1982), drawing five geochemical phases in the following order: Exchangeable, carbonatic, reducible, oxidizable and residual. All of the extracts were analyzed by gamma spectrometry using a Ge detector, for determination of ^{137}Cs content.

Factors Affecting Soil-to-Plant Transfer

Characteristics of the Radionuclides

The chemical form and physico-chemical characteristics of the radionuclides in the rooting zone (e.g. half-life, oxidation state, chemical similarity to essential nutrients) determines the degree of root uptake for each radionuclide. Some elements such as Cs and Sr are metabolically well assimilated by plants, but differences in their root uptake have been attributed to their availability in soils (in general Cs tends to be fixed in clay structures). However, plants can be more tolerant to a high concentration of Sr than Cs. Other elements such as Ru are less well metabolized than Cs and Sr, and elements such as Pu, Am or Np are not metabolized at all and can be extremely toxic for plants (Cawse and Turner, 1982). Elements as Pu, that present several oxidation states, have different uptake rates for each chemical form:

Pu(II) > Pu(IV) >Pu(VI) (Jacobson and Overstreet, 1948, cited in Cawse and Turner,1982).

In Table 1, where some examples of ^{137}Cs and ^{226}Ra Tf values in tropical soils are presented, some differences can be observed in the radionuclide uptake in the same crop. From this table it can be seen that transfer factors obtained in tropical soils are generally one order of magnitude higher than transfer factors obtained in temperate environments. The most accepted hypothesis for the absorption mechanism is the one postulating that a carrier that reacts selectively with ions present in the soil solution, forming an intermediary product (in a similar way to catalyzing enzymes in biochemical reactions), that enters the cell membrane (Sutcliffe and Baker, 1989). The biochemical similarities between ^{137}Cs and K and between ^{90}Sr and Ca lead us to expect that their biological cycle should be similar to that of natural K and Ca respectively (Brown and Bell, 1995; Andersen, 1973; Claus et al., 1990). Therefore, soils containing low concentrations of stable elements in the soil solution should be favorable to root uptake of radionuclides with a chemical behavior analogous to the stable nutrients, through competition of carriers.

Table 1: Transfer factor (Tf) values for ^{137}Cs and ^{226}Ra in tropical soils and transfer factor values from IUR data for temperate climates

Crop group	Crop	Radio-nuclide	Tf	IUR data*	Reference
Pods	Beans (grain)	^{137}Cs	2.6×10^{-1} $(6.0\times10^{-2} - 4.1\times10^{-1})$ (n=16)	2.9 E-2 $(1.0\times10^{-3} - 1.6\times10^{-1})$ (n=90)	This work
Pods	Beans (grain)	^{137}Cs	2.4×10^{-1} (n=15)	2.9×10^{-2} $(1.0\times10^{-3} - 1.6\times10^{-1})$ (n=90)	Scardino and Helene (1990)
Pods	Beans (grain)	^{226}Ra	6.6×10^{-2} $(1.2\times10^{-2} - 9.1\times10^{-2})$ (n = 4)	6.0×10^{-3} $(6.3\times10^{-3} - 1.4\times10^{-2})$ (n=6)	Lima (1988)
Pods	Peanuts (grain)	^{137}Cs	1.0	3.0×10^{-2}** (n = 84)	Delmas et al. (1987)
Roots	Radish (root)	^{137}Cs	1.6 $(4.6\times10^{-1} - 3.5)$ (n = 5)	5.5×10^{-2} $(1.0\times10^{-3} - 1.3\times10^{-1})$ (n = 8)	Wasserman and Belém (1996)
Roots	Carrot (root)	^{137}Cs	1.0 $(8.5\times10^{-2} - 1.5)$ (n =5)	1.5×10^{-1} $(2.6\times10^{-2} - 2.6\times10^{-1})$ (n = 4)	Wasserman and Belém (1996)
Roots	Carrot (root)	^{226}Ra	9.2×10^{-2} $(3.0\times10^{-2} - 1.5\times10^{-1})$ (n = 4)	4.1×10^{-3} $(2.5\times10^{-3} - 5.7\times10^{-3})$ (n=4)	Lima (1988)

*IUR/RIVM(1989)   **IUR/RIVM(1987)

In the agricultural systems, ^{137}Cs and ^{90}Sr are of special interest for root uptake because once in the soil they can cycle in the biosphere as long as their long radioactive half-life allows, and, furthermore, they enter in the system in a chemical form readily available to plants.

Plant Species

Plant species have all been shown to influence transfer. The metabolic needs for the elements in plants is typical according to plant species and accumulation rates tend to vary. Values presented in Table 1 show that significant differences were found in ^{137}Cs uptake between crops of the same group: radish and carrots, peanuts and beans. Evans and Dekker (1968) noticed that ^{137}Cs absorption in some plant cultures like beet was higher than in forage, and these in turn absorbed more than cereal.

Some authors have reported that radionuclide concentrations in grains are lower due to protection by husks. The distribution of ^{137}Cs in different parts of black beans presented in Fig. 1 shows that grains presented the lowest levels of ^{137}Cs (21.2%) at harvest time. ^{137}Cs was found mainly in pods (40.4 %) and leaves (38.4 %). These results are very similar to those found by Scardino and Helene (1990) for ^{137}Cs distribution in black beans growing on artificially contaminated Brazilian soil: 23% in beans, 34% in pods and 43% in leaves.

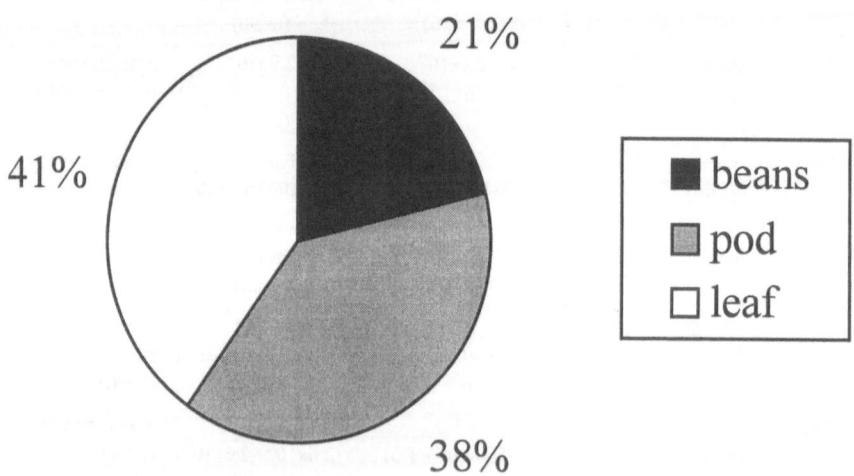

Fig. 1: Mean values for ^{137}Cs distribution in different parts of black beans (n = 6; sd = ±10%)

Pedological Characteristics

In addition to the chemical form and physico-chemical characteristics of the radionuclides available in the root zone and the differences in metabolism of different species, plant absorption of radionuclides is related to soil properties, texture, mineralogical composition, organic matter contents, pH and nutrient status (Cawse and Turner, 1982; Coughtrey and Thorne,1983; Delmas et al., 1987; Claus et al., 1990). An increased uptake of radioactive Cs has been observed in plants growing in peat or sandy soils (Bunz and Krake, 1989). In tropical soils, high transfer factor values have been reported compared with IUR mean values for the same culture and radionuclide as shown in Table 1.

The results of pedological analyses, presented in Table 2, show in lysimeters L1 to L6 an acid soil with average texture (sandy clay loam), high contents of organic matter, poor in clay minerals and nutritive elements, and low cation exchange capacity. The values of ΔpH and Ki are normal for soils rich in Fe and Al oxides. The ensemble of these results is common for most Brazilian oxisols and suggests a need for correction in a number of cultures (Malavolta, 1976). The soil was fertilized with urea, potassium chloride and calcium phosphate. The fertilization was applied to supply the nutritional needs for healthy black beans.

The soil of lysimeters L9 and L10 also showed an average texture (sandy clay loam), but they are basic soils, rich in nutrient elements, and therefore no fertilization was applied in these lysimeters. Contents of organic matter and cation exchange capacity were typical for tropical soils, and values of ΔpH and Ki were common for soils rich in Fe and Al oxides compared to other clay minerals.

Table 2: Pedological analysis of artificially and accidentally contaminated oxisols

	Artificially contaminated L1 to L6	Accidentally contaminated L9 and L10		Artificially contaminated L1 to L6	Accidentally contaminated L9 and L10
pH (H$_2$O)	4.8	7.8	SiO2 %	3.7	7.7
pH (KCl)	4.3	7.7	Al2O3 %	10.9	13.9
Δ pH	0.5	0.1	Ki	0.58	0.94
sand %	70	64	Fe$_2$O$_3$ %	3.9	5.3
clay %	24	20	TiO$_2$ %	1.15	1.19
silt %	6	16	CaCO$_3$ %	ND	0.92
CEC	5.0	10.7	Ca (meq/100g)	ND	9.9
C %	1.15	1.32	Na (meq/100g)	0.10	0.17
N %	0.07	0.09	Mg (meq/100g)	0.50	0.54
P (ppm)	1.00	50	Al^{+++}(meq/100g)	0.50	0.0
K(meq/100g)	0.09	0.17	H$^+$(meq/100g)	3.8	0.1

The values for ^{137}Cs activity in both contaminated soils and the measured transfer factors are presented in Table 3. No significant variation in root absorption of ^{137}Cs was observed for the different years of cultivation. However, substantial differences were observed for the two types of soils: Tf values for L1 and L2 were higher than those measured on L9 and L10. These values ranged from 0.06 to 0.41 and were higher than those reported in the literature for ^{137}Cs in beans in temperate climates: Tf values reported in the IUR data bank ranged from 0.001 to 0.159. Slight differences were also observed between transfer factors in soils with the same treatment: Tf for L1 was lower than Tf for L2. Low loads of exchangeable K in the soils of lysimeter 2 (L2) seem to increase ^{137}Cs absorption, engendering Tf values as high as 0.40 and 0.41 for the years 1994 and 1995 respectively. Lysimeters with higher loads of exchangeable K presented lower Tf for ^{137}Cs. Root uptake of ^{137}Cs in this study seems to be related to the presence of exchangeable K in the soil. These preliminary results agree with many other studies that have demonstrated that addition of K (and to a lesser extent, ammonium) can displace some Cs to soil solution due to competition for cation exchange sites increasing the ^{137}Cs uptake (Shaw and Bell, 1991; Brown and Bell, 1995). Sutcliffe (1957), studying root uptake of chemically similar elements, proposed that soil variables seem to be more important for Cs uptake than carrier competition effect. No significant difference in the K contents in beans was observed in different lysimeters, showing that plant variables as competition effect for carrier are not enough to explain root uptake of Cs.

The results of Tf values for Cs in beans as a function of pH values in the soil are presented in Fig. 2. These results suggest that soil pH can also affect Cs availability for plant, but more data are necessary to confirm this.

Table 3: Potassium contents (%) in beans pulses, total activity of ^{137}Cs in soils and transfer factor (Tf) of ^{137}Cs for beans, cultivated in 1994 and 1995

Lysimeter	K in crop (%)	^{137}Cs (Bq/kg dry soil)	K$^+$ (meq/100g)	Tf (1994)	Tf(1995)
L1	2.9	8485	0.07	0.21	0.18
L2	2.7	7037	0.02	0.40	0.41
L9	2.7	1339 ± 15	0.17	0.08	0.06
L10	2.7	1443 ± 3	0.14	0.11	0.09

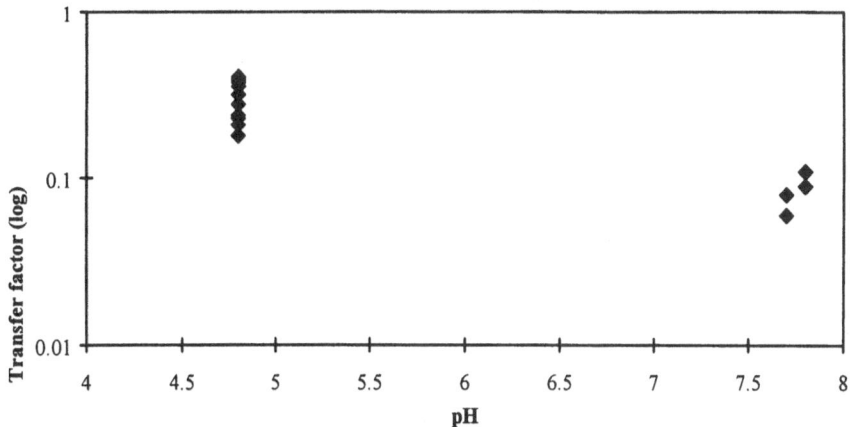

Fig. 2: Log of [137] Cs transfer factor for black beans as a function of pH in oxisols

Geochemical Distribution of ^{137}Cs in Oxisols

Only a small fraction of the total loads of elements present in the soil, whether of cultural or natural origin, is considered ready bioavailable: those present in the soil solution, adsorbed to clays and to organic matter (Malavolta, 1983; Polar and Bajülgen, 1991; Simkiss et al.,1993); thus, exchange processes have an important influence upon radionuclide availability (Andersen, 1973; Shaw and Bell, 1991; Brown and Bell, 1995). Stable Cs can be fixed to the crystalline lattice of some minerals present in soils, mainly 2:1 clay types, or weakly adsorbed by negative charges present in humic substances and clays (Evans and Dekker, 1966; Bunz and Shultz, 1985). Nonetheless, if the specific sites for sorption became saturated by stable Cs, NH_4^+ or K, the bioavailability of ^{137}Cs can increase. Soils showing a low content of exchangeable K and high concentrations of kaolinite (clay 1:1), as found in tropical areas, are expected to show a high Cs transfer to plants.

The geochemical partitioning of the radioelements furnishes a first approach to investigate their potential bioavailability. It also provides estimates of the percentage of the element bound to the lattice of primary or secondary minerals, in which destruction and subsequent release of the element to the environment is not observable on a human time scale (Meguellati, 1982; Oughton, 1990; Oughton et al., 1992). However, some soil compounds are unstable in some environmental conditions and elements can be released during modifications of physico-chemical conditions, with consequent incorporation by living organisms (Table 4). Physical, chemical and biological processes (e.g., leaching of radionuclides to deeper layers of the soil, rapid mineralization of organic matter with consequent release of the associated radionuclide to the environment and rupture of binding by weathering) may influence the radionuclide's mobility and its transfer to the trophic chain.

The elements present in the residual phase are referred to as those trapped in the structure of primary and secondary minerals, constituting the background levels for the region. Elements originating from human activities are distributed among the other physico-chemical phases, respective to their chemical form and behavior. Low soil to plant transfer for ^{137}Cs in temperate climates, has been reported as due to its strong association with soil components. Some ^{137}Cs found in the residual phase suggest that this radionuclide can be progressively fixed in clay structures (2:1 clay types) a few years after contamination (Oughton, 1990; Riise et al., 1990).

The results for geochemical partitioning of ^{137}Cs in oxisols presented in Fig. 3 show that this element is potentially available for root uptake since no ^{137}Cs bound to the lattice of primary or secondary minerals was detectable, despite the long period of contamination in both soils: contamination in lysimeters L1 and L2 occurred 3 years before sampling and the accidental contamination of the soil of lysimeters L9 and L10 occurred 8 years before. In oxisols, the predominant clays are 1:1 type, represented by kaolinites, which unlike 2:1 clays have a low cation fixation capacity, due to their molecular structure and low cation exchange capacity.

Table 4: Physico-chemical conditions necessary for the release of elements associated with the different geochemical fractions of the soils (after Meguellati, 1982)

Geochemical fractions	Conditions for release
Exchangeable	Increasing ionic strength
Carbonatic	Reduction of pH
Reducible	Reduction of Eh
Oxidizable	Destruction of organic matter or increasing redox potential (Eh)
Residual	Not observable in the human time scale

Since Cs and K are present in the soil predominantly in the ionic form, it was to be expected that a large percentage of these elements should be present in the exchangeable phase. Nevertheless, very little ^{137}Cs was detected in this phase (<10%). In the soils with a long period after contamination (L9 and L10) no exchangeable ^{137}Cs was detectable.

The concentrations of ^{137}Cs associated with the carbonatic phase are said to be hardly soluble under high pH conditions. The soil pH can affect the availability of the elements present in soils in different ways (Claus et al., 1990). Under high pH conditions, insoluble precipitates of phosphates, carbonates or sulfates can be formed, decreasing considerably the availability of the elements to plants. Under acid conditions, as simulated by the sequential procedure in this phase, H^+ ions can displace some adsorbed cations to solution or redissolve some precipitates. Mello et al. (1983) noticed that fixed K on soil compounds becomes exchangeable when acid conditions prevail.

The same mechanism is proposed for [137]Cs. It is important to note that soils from lysimeters 1 and 2 are acid soils; to reach the exchangeable phase, the pH must rise from 4.8 to 7, and this can mean that some exchangeable ^{137}Cs are retained in the soil compounds. Care must be taken to understand carbonatic phase results, since the sorption mechanism due to pH can affect exchangeable ^{137}Cs measurements. The Tf values for ^{137}Cs in beans versus pH values in soil presented in Fig. 2 seem to corroborate this point.

The ^{137}Cs in the oxisols studied showed low affinity to organic compounds (oxidizable phase - Fig. 3). In living tissues, the K is mostly found dissolved in the cell sap in ionic form, and no organic compound has been identified in which K constitute an essential element. This implies the rapid release of this element in the early stages of decomposition of the plants (Carmo, 1985; Sauras et al., 1994). Since the presence of ^{137}Cs in the plant has been shown to be related to its availability in the soil and to the process of competitive inhibition with K, similar behavior of the two elements in the oxidizable phase is expected.

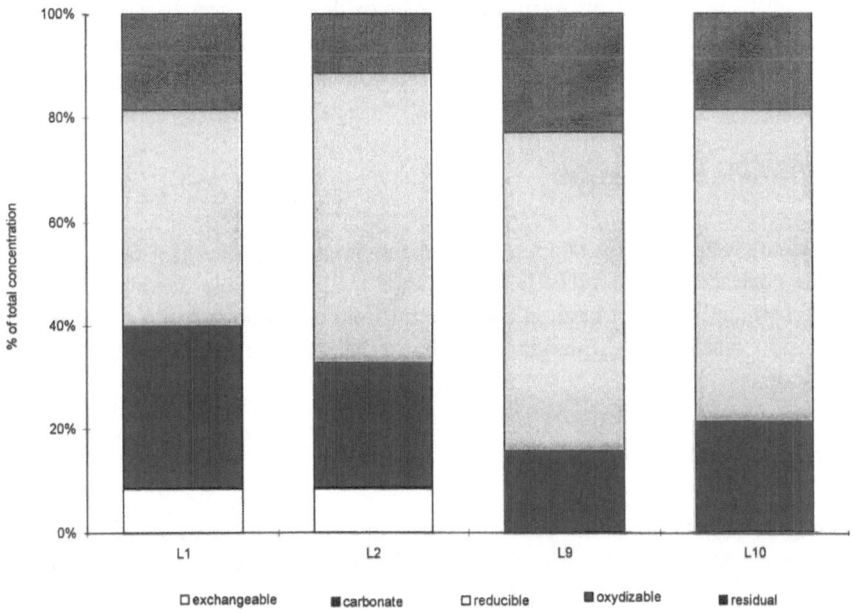

Fig. 3: Geochemical partitioning of ^{137}Cs in lysimeters 1, 2, 9 and 10

Finally, ^{137}Cs has a higher affinity for Fe and Mn oxide, the main compounds of the analyzed soils (reducible phase - Fig. 3). Tropical soils have kaolinite and Fe and Al oxides as main clay types (Mello et al., 1983). Lima (1988) has already noticed the importance of this fraction for ^{226} Ra distribution in oxisols. Since Fe and Al oxides can act as cement between clay minerals and organic matter to form soil aggregates (Kiehl,1979), reduced conditions, simulated by these phase reagents, can reduce Fe (III) and alter the stability of these aggregates, releasing elements sorbed on clays and organic matter. Probably, with the time, the exogenous radionuclides in exchange positions tend to be maintained in the soil aggregates in oxisols: 8 years after Goiânia accident 60% of the ^{137}Cs is associated to Fe oxide structures (L9 and L10), against 40% 3 years after artificial contamination (L1 and L2). This process may be responsible for decreases in the amount of ^{137}Cs in exchangeable fraction and consequently for reductions in root uptake in oxisols with time.

Conclusions

The dystrophic characteristics of Brazilian oxisols would seem to determine a high transfer of ^{137}Cs from soil to edible plants, compared to soils from temperate climates, once exchangeable K contents and the pH of soils are taken to be the main factors affecting ^{137}Cs root uptake. Based on sequential extraction results, ^{137}Cs in oxisols is weakly bound to soil components. The main geochemical phase controlling ^{137}Cs availability in oxisols seems to be the Fe oxides, rather than 2:1 clay minerals.

Concerning ^{137}Cs in beans, it can be concluded that pulses show lower concentration than leaves and pods, corroborating the results reported in the literature.

Acknowledgements

This work was partially supported by the International Atomic Energy Agency under contract no. 7468/RI/RB. The author thanks Virginia Medeiros, Heliane Zilberberg and Ernani Marchon from the Instituto de Radioproteção e Dosimetria/ Comissão Nacional de Energia Nuclear for help in the experimental area and analyses.

References

Aarkog, A.(1979). Environmental Studies on Radiological Sensitivity and Variability with Special Emphasis on the Fallout Nuclides Sr-90 and Cs-137. Danish Atomic Energy Commission Research Establishment. Roskilde. Denmark. *Report Risö*, R437.

Andersen, A.J. (1973). Plant Accumulation of Radioactive Strontium with Special Reference to the Strontium-Calcium Relationships as influenced by Nitrogen. Danish Atomic Energy Commission Research Establishment. Roskilde. Denmark. *Report Risö*, R278.

Brown, S.L. and Bell, N.B. (1995). Earthworms and Radionuclides, with Experimental Investigations on The Uptake and Exchangeability of Radiocaesium. *Environ. Pollution*, 88:27-39.

Bunz, K. and Schultz, W. (1985). Distribution Coefficients of 137 Cs and 85 Sr by mixtures of clay and humic material. *J. Radioanal. Nucl. Chem.* 90:230-237.

Bunz, K. and Kracke, W (1989). Seasonal Variation of Soil-to-Plant Transfer of K and Fallout [134,137] Cs in Peatland Vegetation. *Health Physics*, 57 (4):593-600.

Carmo, M.A.M. (1985). *O Papel de Eleocharis subarticulata (Nees) Boeckler (Cyperaceae) na Ciclagem de Nutrientes em um Brejo da Restinga de Maricá, RJ*. Master's thesis in Geosciences, Universidade Federal Fluminense, Niterói. 99p

Claus, B., B. Grahmann, V. Hormann, S. Keuneke, M. Leder, H. Müller, E. Peters, E.M. Rieger, I. Schmitz-Feuerhake and F. Wagschal (1990). Sr-90 Transfer Factors for Rye in Podzolic Soils: Dependence on Soil Parameters. *Radiat. Environ. Biophys.*, 29:241-245.

Cawse, P.A. and Turner, G.S. (1982). *The Uptake of Radionuclides by Plants: A review of Recent Literature*. Environmental and Medical Science Division. AERE-R-9887. 47p.

Coughtrey, P. J. and Thorne, M.C. (1983). *Radionuclide Distribution and Transport in terrestrial and Aquatic Ecosystems*. Balkema, Rotterdam, Vol.1, Chap. 7.

Delmas, J.; Grauby, A.; Paulin, R.; Rinauldi, C.; Ndoye, R. and Sekgassama, S. (1987). Etude Expérimentale du Transfert du Sr-90 et du Cs-137 d'un Sol de la Région du Cap Vert à l'Arachide. *Radioprotection*, 22 (4):334-340.

Evans, E.J. and Dekker, A.J. (1966). Plant Uptake of Caesium-137 from Nine Canadian Soils. *Can. J. Soil Sci.*, 46:167-176.

International Union of Radioecologists (1987). *V[th] Report of the Working Group Soil-to-Plant Transfer Factors*. IUR Report prepared by RIVM, Bilthoven, Netherlands

International Union of Radioecologists (1989). *VIth Report of the Working Group Soil-to-Plant Transfer Factors*. IUR Report prepared by RIVM, Bilthoven, Netherlands.

Kiehl, E.J.(1979). *Manual de Edafologia: Relações Solo-Planta*. Editora Agronômica Ceres:262p.

Lima, V.S.T.G. (1988). *Avaliação Comparativa da Partição e Incorporação radicular de [226]Ra Endógeno e Exógeno em solos agrícolas de Poços de Caldas: Formas de Ocorrência e Disponibilidade Vegetal*. Master's thesis, Universidade Federal do Rio de Janeiro, Brasil.183p.

Malavolta, E. (1976). *Manual de Química Agrícola*. Editora Agronômica Ceres: 528p.

Meguellati, N. (1982). *Mise au Point d'un Schèma d'Extractions Sélectives des Polluants Métalliques Associés aux Diverses Phases Constitutives des Sédiments*. Thèse 3ème Cycle. Université de Pau et de Pays de L'Adour. 134p.

Mello, F.A.F.; Sobrinho, M.O.C.B.; Arzolla, S.; Silveira, R.I.; Netto, A.C. and Kiehl, J.C. (1983). *Fertilidade do Solo*. Editora Nobel. 399p.

Oughton, D.H.(1990). Radiocaesium Association with Soil Components: The Application of a Sequential Extraction Technique. *In*: *VIIth Report of Working Group Soil-to-Plant Factors*. IUR Workshop.Uppsala, Sweden. pp:182-188.

Oughton, D.H., B. Salbu, G. Riise, H. Lien, G. Ostby and A. Noren (1992). Radionuclide Mobility and Bioavailability in Norwegian and Soviet Soils. *Analyst*, 117:481-486.

Polar, E. and Bajülgen, N. (1991). Differences in the Availabilities of Cesium-134,137 and Ruthenium-106 from a Chernobyl-Contaminated Soil to a Water Plant, Duckweed, and to the Terrestrial Plants, Bean and Lettuce. *J. Environ. Radioactivity*, 13:251-259.

Riise, G.; Bjørnstad, H.E.; Lien, H.N.; Oughton, D.H. and Salbu, B. (1990). A Study on Radionuclide Association with Soil Components Using a Sequential Extraction Procedure. *J. Radioanal. Nucl. Chem.* 142(2):531-538.

Sauras, T.; Roca, M.C.; Tent, J.; Llurado, M.; Vidal, M.; Rauret, G. and Vallejo, V.R. (1994). Migration Study of Radionuclides in a Mediterranean Forest Soil Using Synthetic Aerosols. *The Sci.Tot. Environ.*, 157:231-238.

Scardino, A.M.S. and Helene, O.A.M. (1990). Mecanismos de Difusão e Transporte de ^{137}Cs no meio Ambiente. *Anais da 13ª Reunião de Trabalho sobre Física Nuclear no Brasil*. Caxambu, MG (Basil):327

Shaw, G. and Bell, J.N.B. (1991). Competitive effects of Potassium and Ammonium on Caesium uptake Kinetics in Wheat. *J. Environ. Radioactivity*, 13: 238-296.

Simkiss *ET ALLI*. (1993). Radiocaesium in Natural Systems - A UK Coordinated Study. *J. Environ. Radioactivity*, 18(2):133-149.

Simmonds, J.R. (1985). The Influence of the Season of the Year on the Transfer of Radionuclides to Terrestrial Food Following an Accidental Release to Atmosphere. *NRPB-M-121*

Sutcliffe, J.F. (1957). The Selective Uptake of Alkali Cations by Red Beet Root Tissues. *J. Expt. Bot.*, 8:36-49.

Sutcliffe, J.F. and Baker, D.A. (1989). *As Plantas e os Sais Minerais*. Editora Pedagógica e Universitária Ltda. Vol.33: 80p.

Wasserman, M.A.M and Belém, L.M.J. (1996). Valores de Transferência do ^{137}Cs de Latossolos para Plantas comestíveis. *Anais do VI Congresso Geral de Energia Nuclear* (cd-rom).

Yasuda, H. and Uchida, S. (1994). Statistical Analyses of Soil-to-Plant Transfer Factors: Strontium and Cesium. *J. Nuclear Science and Technology*, 31(12):1308-1313.

4 Speciation, Reactivity and Mobility of Toxic Elements in Soil Systems

Ricardo Melamed
CETEM/CNPq - Centre for Mineral Technology, Rua 4, Quadra D,
Cidade Universitária, Ilha do Fundão, Rio de Janeiro, Brazil, 21941-590,

Abstract

The quality of groundwater, as well as surface waters, is of primary concern in waste and tailings disposal technology. However, due to natural processes at the disposal site, dissolution and mobilization of toxic compounds may occur. In this case, the sorption phenomena in soils play a key role, which determines the transport of toxic compounds to receiving waters, and consequently their fate. This chapter describes the importance of sorption mechanisms, as a result of the type of surface-chemical interaction, on the mobility of toxic chemicals. Specific oxyanion adsorption, metal adsorption and anion exclusion are highlighted. The effects of different physico-chemical parameters on the mobility of arsenic, mercury and nitrate are emphasized.

Introduction

The exploration and processing of minerals generate wastes and tailings that require appropriate management for the protection of the environment. This, in turn, requires an understanding of the processes involved within the substrate itself, as well as the interactions with the surroundings, recognizing the possible pathways of toxic elements from a mine or waste site to the atmosphere, to the soil, to water courses and to the groundwater. Because water is the major carrier of pollutants during leachate flow from disposal sites to water courses, the speciation of any toxic element, as a consequence of the water chemistry, affects the transport character of the chemical. Also, in considering the mobility of any compound through waste piles and soils, or seepage from tailing ponds, and ultimately the potential hazard to groundwater quality, the character of the surface with which toxic elements interact is important. In this regard, the sorption phenomena, as a result of the compound-mineral surface interaction, play a key role in the transport, reactions, bio-transformations and ultimately the fate of chemical constituents in aqueous systems. In addition, competition between chemical species for sorption sites on surfaces can be of major significance in the determination of the mobility of any potentially sorbing species.

This chapter is not intended to fully review retention mechanisms of chemicals, but rather to highlight some important reactions that may affect the mobility and fate of toxic elements, such as arsenic, mercury and nitrate generated at some point of the mining process.

Speciation in Solution

The presence of ions in solution influences the structure of water. In the region nearest the ion, water molecules form a dense mass, the primary solvation shell. In the next region outwards, water molecules forming the secondary solvation shell interact weakly with the ion. The properties of these shells vary with the concentration, valence and radius of the ion (Sposito, 1984).

In dilute solutions, major dissolved species can be seen as free ions. However, as the ionic strength increases, differences in ion activity become larger, due to short range interactions between adjacent ions forming complexes. In general, metal ions have vacant orbitals and can accept electrons, while ligands have at least one pair of electrons not shared in a covalent bond, and as such, can donate electrons. Metal ions may contain electrons in shared or unshared pair fashion. Similarly, ligands may be of low or high electronegativity.

Two types of soluble complexes are formed between metals and ligands: outer-sphere or ion pairs and inner-sphere complexes. The attractive force involved in an ion pair is coulombic, i.e., a relatively weak electrostatic association exists between a hydrated cation and a ligand, with preservation of the hydration shell. Inner-sphere complexes are strong associations in which a covalent bond exists between metals and ligands. No matter what type of complex is formed, the associations between metals and ligands can produce complexes with neutral, positive or negative charge. Many elements hydrolyze in water to form hydroxy-complexes. Mercury, for instance, may form hydroxy-complexes in the absence of the Cl^- ligand. However, as Cl^- ligand concentration increases, a strong inorganic complex is formed. In general, metals that form hydroxy-complexes are expected to form strong complexes with organic ligands. This is an important characteristic of speciation because dissolved ligands may compete with surface ligands for association with metals. If the dissolved ligand succeeds in complexing with all sites of the coordination shell of the metal, then specific adsorption is prevented.

The Electrified Surface of Minerals

The surfaces of colloids are electrically charged, due to an excess or a deficit of electrons. The nature of this charge development is either from isomorphic substitution in minerals, or from the complexation of ions, in the pore water solution, with surface functional groups (Sposito, 1984). As isomorphic substitution occurs in the crystal lattice of clay minerals, due to replacement of cations of higher valence by cations with lower valence, the resulting negative charge imbalance is permanent and is not influenced by the composition of the soil solution. On the other hand, at the edges of clay minerals as well as metal oxides, the constituent metal ions are unable to complete the coordination pattern existing in the bulk of the crystal. The result is unsatisfied bonds. In this case, the metal ion coordinates with OH groups, producing an hydroxylated surface which may accept protons and become positive or donate protons and become negative, depending on the pH of the solution.

Ions of the opposite charge to the surface will accumulate in the liquid phase near the charged surface. However, due to the concentration gradient created, diffusion and thermal forces tend to draw back these ions into the liquid phase. By the same token, ions with charge of equal sign to the surface are repelled from the vicinity of the surface with the diffusion forces acting to equalize the concentration of co-ions in solution. The general distribution of ions results in a diffuse ionic atmosphere surrounding a charged particle, which is the basis for the diffuse double layer theory (Gouy, 1910; Chapman, 1913).

The Gouy/Chapman model does not take into account the actual size of counterions. Ions are treated as point charges. The Stern modification (Stern, 1924) attempts to correct null counterion size of the diffuse double layer theory by allowing the ions to approach the surface within a certain minimum distance, producing a compact layer adjacent to the surface. Stern (1924) also introduced the possibility that ions might be adsorbed into the compact layer (the Stern layer) by forces other than purely electrostatic. Grahame (1947) postulated that ions are specifically adsorbed (inner-sphere surface complex) into the Stern layer, when they lose some of their water of hydration, whereas the hydrated ions are only electrostatically attracted to the surface (outer-sphere surface complex).

Arsenic Sorption

The processing of gold from refractory minerals, such as arsenopyrite (FeAsS), requires that arsenic (As) is removed and stabilized before disposal in landfills, as tailings or dumps. The pyrometallurgical process involves the crushing of the ore, flotation of the sulphides, sulphide roasting, washing and leaching. The roasting step to oxidize arsenopyrite results in the transformation of As to As(III). In this process, arsenic is transformed into a finely divided powder of arsenious oxide, which is leached in the presence of cyanide. This later stage solubilizes most of the arsenic. Further oxidation of the leached liquors to As(V) is usually carried out for the precipitation of As, as calcium or ferric arsenate compound, for disposal.

The stability of the metal arsenate is important for safe disposal, with no pollution threat to the water and to the air. However, even relatively protected, arsenical residues may be leached, eroded and volatilized, which were identified as the pathways for the arsenic spill from a zinc silicate ore disposal site in the Sepetiba Bay, Rio de Janeiro (Barcellos et al., 1992). Metal arsenates with low solubility may have their stability altered depending on the physical chemistry of the waste-site disposal system. Calcium arsenate has its solubility enhanced when in contact with atmospheric CO_2. In this reaction calcium arsenate is decomposed to $CaCO_3$, liberating contained arsenic (Robins and Tozawa, 1982). Solubility tests of various Fe-As compounds formed as products of high pressure leaching of arsenopyrite (Carageorgos, 1993) showed that the dissolution of Fe and As was incongruent, and that As dissolves preferentially over Fe in all solids studied. The products were identified as $Fe(OH)_3$, $H_2AsO_4^-$, and a concomitant drop in pH, during dissolution, was expressed as increased concentration of solution H^+. In general, all compounds formed as leached residues in the high pressure leaching process released As into solution, which varied from about 1 to 130 mg As L^{-1} in 98 days reaction time, depending on Fe/As ratio and temperature. The pH control is important in preserving the stability of arsenical wastes. At relatively high pH values, where the solubility of calcium arsenate is at its minimum, the decomposition of this compound to the hydroxide may still occur, liberating the free arsenate oxyanion. At pH values of about 4, ferric arsenate decomposes to ferric hydroxide (Carageorgos, 1993; Nishimura and Tozawa, 1981). During arsenopyrite processing, there is the possibility of oxidation of sulphides into sulphuric acid, with a resulting acid pH, and consequently, a delayed redissolution of metal arsenates. If redissolution of arseniferous wastes occurs, and arsenic compounds are free to move downward into the soil matrix, the selection of the soil type which will serve as disposal site is critical because the sorption process in soils may play a key role in preventing groundwater contamination. Figure 1 (Melamed et al., 1995a) shows the breakthrough curves (BTCs) of arsenate on two very distinct soil systems. The mobility of As through the Panoche soil (fine-loamy, calcareous) was considerably greater than through the Oxisol soil (kaolinite, quartz, goethite). This large difference in retention capacities between the two soils is explained by their differences in mineralogy (Jacobs et al., 1970; Fordham and Norrish, 1979), and by the differences in equilibrium pH between the two systems (Goldberg, 1986).

The capacity of the Oxisol soil to retain As (Fig. 1) is outstanding. About 30 pore volumes of As solution were displaced through the Oxisol before As was detected in the effluent, and 160 pore volumes were displaced before the highest effluent As concentration was reached. With 160 pore volumes displaced, effluent As concentration approached but did not reach the input concentration. The tailing and reduced maximum of the BTC suggest that surface nucleation of As with oxides was the major sorption mechanism involved at the latest stage of the As-BTC (Willet et al., 1988; van Riemsdijk and Lyklema, 1980). In contrast, As-BTC on the Panoche soil shows that As was detected in the effluent at about 5 pore volumes and did reach the input concentration at 15 pore volumes. These results reflect the importance of the selection of the soil used as a waste disposal site.

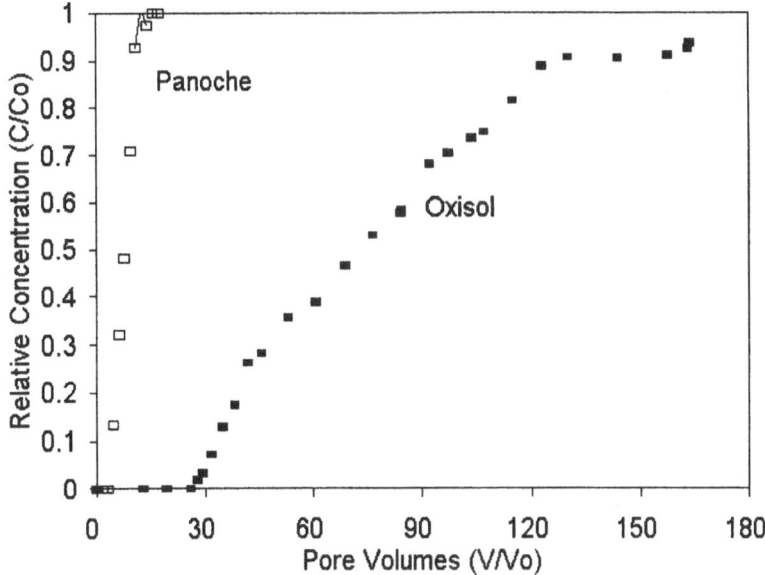

Fig. 1: As-breakthrough curves on Panoche and Oxisol soils

Fordham and Norrish (1979) studied arsenate uptake by the components of several acidic soils by means of statistical analysis of the data. In all the soils in which they were present, micron-size iron oxide-hydroxide pellets were outstanding in their ability to take up arsenate. Goethite and hematite pellets appeared to react equally well with arsenate. In one of the soils, where many particles were in fact microaggregates containing fine-grained units of gibbsite, hematite and kaolin, the arsenate uptake was dependent mostly on the proportion of hematite present. Titanium dioxide pellets, with surface deposits containing finely divided iron oxide, were able to take up more arsenate than expected from their ferric oxide contents. Both rutile and anatase were identified by X-ray diffraction, but no differences could be discerned between the two mineral forms in their ability to take up arsenate. Gibbsite, even though present in fine state, had little influence on arsenate uptake. Quartz was found to have no reactivity with arsenate.

Figure 2 (Melamed et al., 1995a) shows changes in effluent pH during As displacement through the Oxisol soil column. The initial effluent pH was 5.45, which was the pH of the background electrolyte solution. This pH is below the pK_2 (pH 6.7) of the arsenic acid. Thus, the monovalent anion ($H_2AsO_4^-$) was the predominant species in solution. The initial effluent pH value was also below the point of zero charge (PZC) (pH 6.0) of the soil which indicated that the density of surface functional aquo-groups (OH_2^+) was high, resulting in a positive surface potential that favoured the adsorption of As.

Arsenic appeared in the effluent at $V/V_0=20$. The constant value (5.45) of the effluent pH indicated that the primary ligand exchange mechanism, during phase 1 of the As-BTC, involved the substitution of As for labile OH_2^+ groups (Rajan, 1976). This ligand exchange reaction is represented by:

$$[Me(OH_2)_2]^+ + H_2AsO_4^- = [Me(OH_2)O_4AsH_2]^0 + H_2O \qquad (1)$$

where Me is the coordinating metal atom at the oxide surface.

The initial reaction of As with the oxide component of the soil can also be explained by the formation of a binuclear complex, involving the coordination of two oxygen atoms of the As molecule with two metal atoms at the oxide surface, as follows:

$$[Me_2(OH)(OH_2)_3]^+ + H_2AsO_4^- = [Me_2(OH_2)_2O_4AsH]^0 + 2H_2O \qquad (2)$$

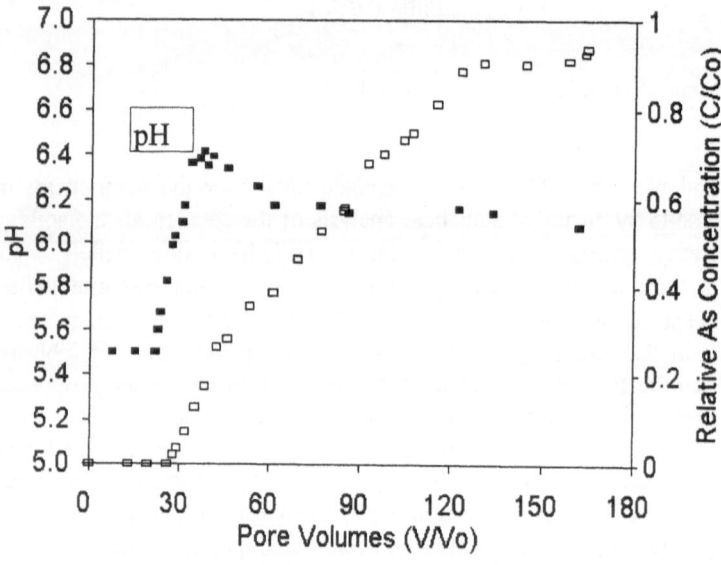

Fig. 2: Effluent pH during arsenate movement through the Oxisol soil

In reactions 1 and 2, the pH is constant during ligand exchange. Simultaneously with the appearance of As at $V/V_0=20$, the effluent pH increased sharply to a maximum of 6.4, at $V/V_0=40$. The increase in pH suggests that the primary ligand exchange mechanism, during phase 2, was between $H_2AsO_4^-$ and surface OH groups. The exchange with OH groups is favoured by the depletion of surface OH_2^+, and by the decrease in PZC of the soil associated with the inner-sphere complexation of As (Anderson and Malotky, 1979; Hingston et al., 1972). The exchange mechanism in phase 2 was written (Melamed et al., 1995a):

$$[Me(OH)(OH_2)]^0 + H_2AsO_4^- = [Me(OH_2)O_4AsH_2]^0 + OH^- \qquad (3)$$

As the pH increased to 6.4, the pK_2 of the arsenic acid (pH 6.7) was approached, which increased the proportion of soluble divalent $HAsO_4^{2-}$ species in solution. The increase in divalent As species, combined with the decrease in surface potential resulting from As complexation, increased the mobility of As (Barrow, 1985). The presence of complexed As can also be regarded as a self competitive mechanism in the movement of As (Barrow, 1974; Camargo et al., 1979; Logan and McLean, 1973).

The four plane model (Bowden et al., 1977) was discussed by Barrow (1985; 1987). In this model ions that replace hydroxyls or water molecules occupy a plane close to the surface and ions that form outer-sphere surface complexes are allocated to a different plane. The positions of the planes on which adsorption occurs are not fixed. Ions differ not only in their affinity to the surface but also in their mean position in the diffuse double layer (DDL) when they are adsorbed. Surface speciation is avoided. Instead it is assumed in the model that the surface activity of an ion is proportional to the ratio of occupied sites to vacant sites. The expression derived by Bowden et al. (1977) has the form of a competitive Langmuir equation:

$$A_i = [N_t k_i a_i \exp(-z \psi_a F/RT)] / [1 + k_i a_i \exp(-z \psi_a F / RT)] \qquad (4)$$

where A_i is adsorption of ion i, N_t the maximum adsorption, k_i a binding constant, a_i the activity in solution, z is the ion valence, ψ_a the surface electrostatic potential, F is the Faraday constant, R the gas constant and T the temperature.

In the case of arsenate sorption, z is negative and because the electrostatic potential decreases as pH increases, the value of the exponential term decreases. This effect tends to decrease adsorption. It is opposed by increased dissociation of the acid. If, for instance, the dissociated species to be adsorbed has a z value of -1, then at pH values below the pK, the concentration of this anion, though low, increases ten-fold for each unit increase in pH. For monovalent anions, this increase suffices to more than counterbalance the decrease in potential, and adsorption increases. Once the pK is reached, however, there is little further scope for increases in the concentration of the monovalent anion, so the effect of the electrostatic potential predominates and adsorption decreases. The effect is slightly different when a divalent species adsorbs because the z term has a value of -2, and hence the exponential term has a greater effect, i.e., the effect of changing the electrostatic potential is greater. The result is that adsorption decreases with increasing pH, with a slight increase in slope near the pK_2 of the acid. Hence, the model involves a choice between known species in solution rather than proposing surface species.

Phase 3 (Fig. 1) started at $V/V_0 = 40$. The tailing and reduced maximum As concentration displayed by the As-BTC of the control column, coupled with the decrease in effluent pH to a constant pH=6.1, appeared to involve the following mechanisms: 1) a decrease in ligand exchange involving As and surface OH groups, and sorption dominated by surface nucleation of As (Willett et al. 1988; Van Riemsdijk and Lyklema, 1980); 2) sorption of protons due to the increase in negative charge associated with the inner-sphere surface complexation of As, with a tendency to buffer the system to the PZC, which is the point of minimum solubility (Parks and de Bruin, 1962). The four plane model also accounts for the slow process that follows adsorption. Barrow (1985) defends the mechanism of solid state diffusion in which the anion penetrates the particles after a period of time has elapsed. The immediate source of the diffusive process is the surface concentration of adsorbate, not the solution concentration. The slow step observed in titration data was attributed to diffusion of protons into, or out of, the surface. If an adsorbed anion diffuses into the interior of the adsorbing material, a surface site will be vacated. However, the electrostatic potential of the surface will be less favourable for adsorption than it was originally (Barrow, 1985). A solid state diffusion mechanism was also used by Van Riemsdijk et al. (1984) to describe the effects of time on the reaction of phosphate with soils.

For further stabilization of arsenical wastes, Laguitton (1976) suggested the use of phosphate in combination with lime, to take advantage of the stability of a combination of calcium phosphate and calcium arsenate, which are known to form a solid solution. The suggestion that phosphate in the lime system would further stabilize As was found to be correct in terms of thermodynamic stability. However, if redissolution occurs, the presence of phosphate may contribute to an increase in As mobility in soils through competition for adsorption sites (Roy et al., 1986), thus preventing arsenate from forming inner-sphere surface complexes with oxides (Barrow, 1989; Goldberg, 1986). The leftward shifts of the As-BTCs (Fig. 3) indicate that As mobility through the Oxisol soil increased with P levels. At the highest P level, As appeared in the effluent after 10 pore volumes of solution were displaced (20 pore volumes less than in the control), and the greatest concentration of As in the effluent occurred when 60 pore volumes of solution moved through the column (60 pore volumes less than in the control). These data confirm the competitive effectiveness of P with As for sorption on the oxide surface (Miller et al., 1989; Roy et al., 1986; Goldberg, 1986; Hingston, 1981; Barrow, 1974).

Frost and Griffin (1977) demonstrated a much lower retention capacity on clay minerals (montmorillonite and kaolinite) of arsenite as compared to arsenate. In additon to being more toxic, As(III) is more mobile. Because the pK of arsenious acid (H_3AsO_3) is about 9.2, adsorption is low in the acidic range and weakly basic solutions, and increases slightly as pH 9.2 is approached.

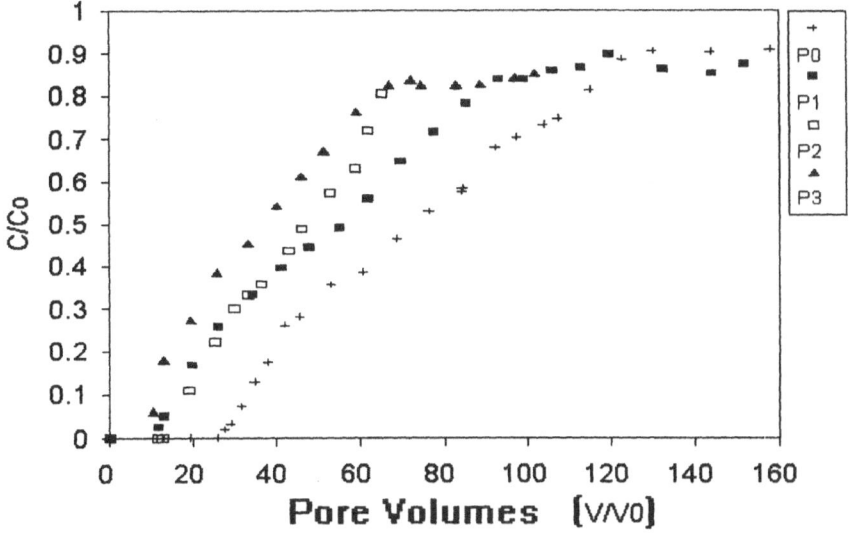

Fig. 3: Effect of phosphate on arsenate movement through Oxisol soil

Mercury Chemistry

The use of metallic mercury (Hg^0) for gold recovery is of concern because it has been shown that, after discharge into the environment, especially by burning the Au-Hg amalgam, volatilized Hg^0 can be transformed by ozone to ionic mercury (Hg(II)) (Iverfeldt and Lindqvist, 1986; Munthe and McElroy, 1992), which, in turn, may be transferred to rivers and soils through wet deposition (Padberg and Stoeppler, 1991). Once in the ionic form, mercury can be transformed to methyl mercury (CH_3Hg^+), the most toxic form of the metal. Methyl mercury accumulates and biomagnifies in the food chain (Huckabee et al., 1979), causing damage to the central nervous system of humans (Wood et al., 1978). The biotic transformation of mercury to methyl mercury was considered to be a detoxification mechanism, in which micro-organisms transfer a methyl carbanion (CH_3^-) from methylcobalamin to a Hg(II) species in either aquated, ionic, or complexed form (Craig and Moreton, 1985):

$$CH_3CoB_{12} + Hg^{2+} + H_2O = CH_3Hg^+ + H_2OCoB_{12}^+ \qquad (5)$$

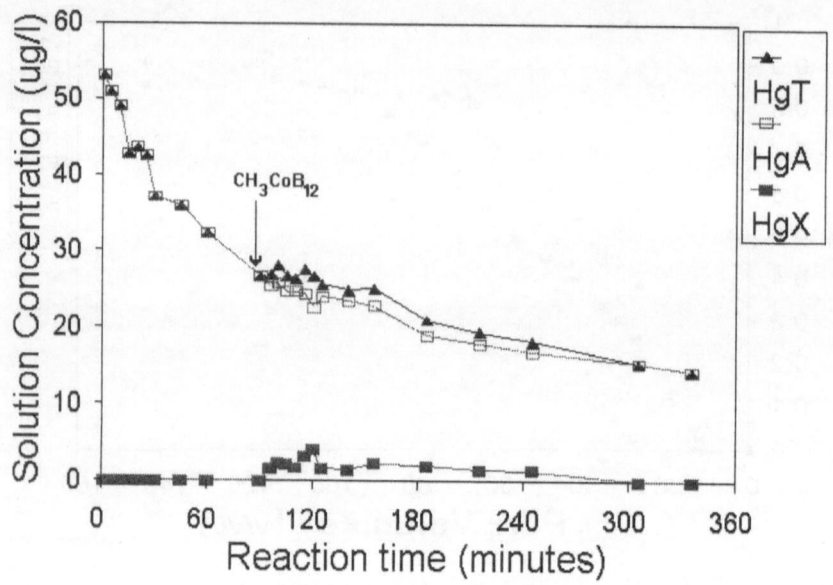

Fig. 4: Adsorption kinetics of Hg on river sediments. Arrow indicates addition of methylcobalamin. Hg^T: Total mercury; Hg^A: ionic mercury; Hg^X: methyl mercury

In addition of being very toxic, methylation of ionic mercury was also shown (Melamed et al., 1995b) to enhance the mobility of Hg through the aquatic environment. Figure 4 shows Hg(II) adsorption kinetics and the effect of the addition of methylcobalamin (CH_3CoB_{12}), a methylating agent, at 90 min. reaction time.

Initially, the rate of adsorption was relatively fast, and about 50% of Hg(II) in solution was taken up by the river sediments. Then, upon addition of the methylating agent, a temporary decrease in adsorption rate was observed, reflected by a relative stabilization of Hg^T and Hg^A solution concentrations, as well as a concomitant increase in Hg^X. This decrease in Hg adsorption rate after the addition of CH_3CoB_{12} is attributed to the relative lower affinity of Hg^X (methyl mercury) for sorption at the sediment/solution interface as compared to $HgCl_2^0$. This change in affinity upon methylation arises primarily from the fact that methyl groups have neither empty orbitals nor non-bonding electrons available for intermolecular interactions (Thayer, 1989).

In alluvium type gold mining, in which placer ores are dredged and processed on river barges, Hg^0 is introduced directly into the rivers, and the metal can be frequently observed at the bottom of the water column. Although Hg^0 is relatively immobile in the aquatic environment, and its solubility is low ($25\mu g\ L^{-1}$) in water (Hem, 1970), Hg contamination in people living upstream and downstream from gold recovery operations has been reported (Cleary et al., 1994). The relevance of dissolved organic acids in enhancing the solubility of metallic Hg was demonstrated by Veiga (1993) and Melamed et al. (1995b). Although the role of complexation with dissolved organic ligands in the Hg biogeochemical cycle is not clear (Verta et al., 1986; Driscoll et al., 1994), the complex formed is more mobile, with a relative lower affinity for adsorption at the surface of sediments (Melamed et al., 1995b). Melamed et al. (1996a) also showed that a decrease in pH decreases the degree of complexation, which was attributed to the competitive behaviour of H^+ with Hg(II) for complexation at the humic acid chain (Langford and Cook, 1995; Fletcher and Beckett, 1987). Fletcher and Beckett (1987) found that ion exchange with protons is a principal component in metal binding by soluble organic matter, which displays two distinguishable groups of exchange sites.

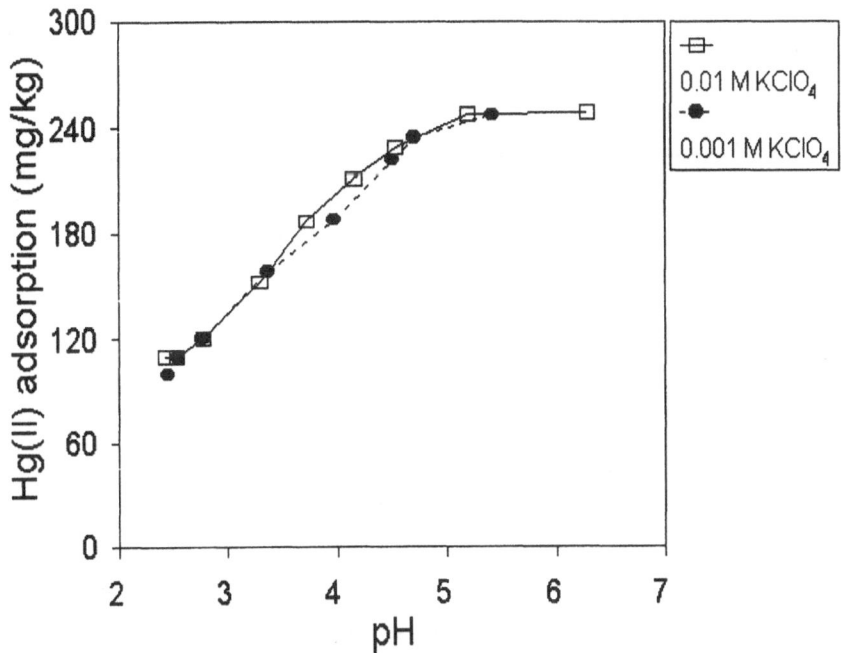

Fig. 5: The "adsorption edge" of Hg(II) on Oxisol at 0.01 M and 0.001 M KClO$_4$ background electrolyte concentrations

In gold mining activities where the ore is explored in quartz veins, Hg^0 can reach levels of 200 mg kg^{-1} in the tailings (Farid et al., 1991), which are left exposed over the soil surface, susceptible to oxidation and leaching. The leaching of Hg from the tailings to the soil underneath and the transport through the soil profile is strongly dependent on the pH of the soil solution. Figure 5 (Melamed et al., 1996b) shows the effect of pH and ionic strength on the retention of Hg(II) onto an Oxisol. The "adsorption edge" indicates a relatively low adsorption of Hg(II) below pH 3. Adsorption increases in the range of pH 3 to 5 and stabilizes. Hg(II) adsorption follows the same pattern of specific adsorption of other hydrolysable cations (MacNaughton and James, 1974) such as Pb, Cu and Zn, reflecting the effective competitive behaviour of protons with Hg(II) for adsorption at the mineral surface of this soil, and the presence of $Hg(OH)_2^0$ as the predominant species interacting at the surface as pH increases (Bolan and Barrow, 1984). Although there was practically no effect of ionic strength on Hg adsorption, the data in Fig. 5 indicate a slight tendency towards increased Hg adsorption as the ionic strength increases, in the range of pH values from 3.5 to 4.7. This pH range is below the point of zero charge (PZC) of the Oxisol (pH 4.7), where the surface is net positively charged. An increase in electrolyte concentration may enhance the adsorption of Hg(II) due to the favourable surface potential, promoted by increased concentration in electrolyte anions near the surface (Bolan and Barrow, 1984).

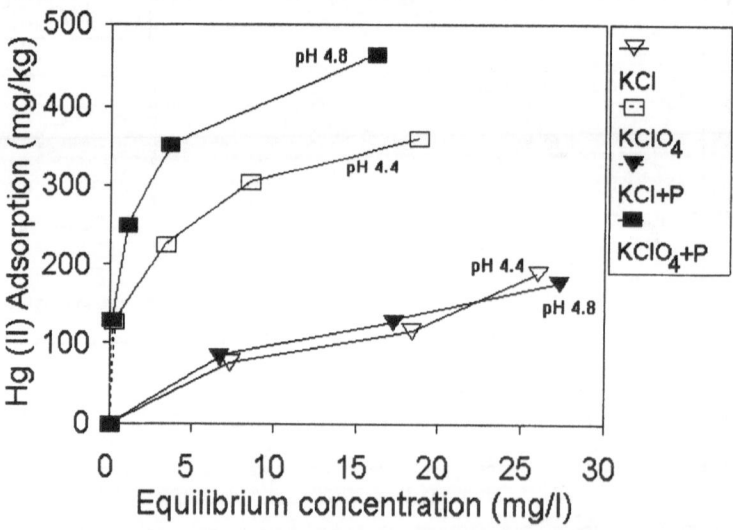

Fig. 6: Effect of P on Hg(II) adsorption isotherms on Oxisol under different electrolyte systems

The complexity of Hg chemistry in the environment reflects the various possible Hg species and complexes with different physical-chemical properties (Luoma, 1983). These complexes depend upon the types and concentration of the ligands involved in the system. Thus, mercury speciation and complexation, as well as physico-chemical parameters, determine the partition of the metal between solid and liquid phases and control Hg mobility and availability. The effect of the nature of the electrolyte on Hg adsorption and transport is demonstrated in Figs. 6 and 7 (Melamed et al., 1996b). A much higher adsorption of Hg resulted in the $KClO_4$ system as compared to the KCl system (Fig. 6). These results are in agreement with other studies of Hg(II) interaction with pure SiO_2 (MacNaughton and James, 1974). The formation of mercury chloride complex in solution promotes a much lower interaction of Hg at the soil solution interface, due to an outer-sphere surface complex interaction. The presence of phosphate (P) has practically no effect on Hg(II) adsorption in the KCl system. However, P enhances the adsorption of Hg(II) in the $KClO_4$ system considerably, due to the favourable surface potential and negative charge created.

As a result, as demonstrated in Fig. 7 (Melamed et al., 1996b), the retention of Hg(II) in the upper half of the soil column is much higher in the $KClO_4$ system than in the KCl system. Mercury moves deeper in the soil column when chloride is present, due to the formation of the mercury chloride complex. The mobility of Hg(II) in the presence of P is decreased more effectively in the $KClO_4$ system.

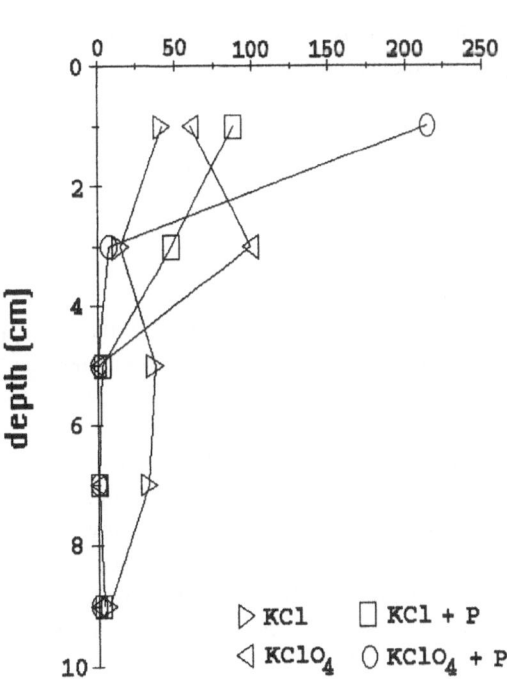

Fig. 7: Hg(II) transport through Oxisol soil column in KCl and $KClO_4$ systems in the presence of phosphate (P)

Nitrate Exclusion

Nitrate contamination to surface and groundwaters, due to the use of explosives in mining operations, is of concern in many parts of the world. To predict the transport and fate of nitrate in natural systems, an understanding of anion interaction with mineral surfaces is of fundamental importance.

Although anions were generally assumed to be non-reactive during miscible displacement, negatively charged ions are actually repelled from a region near negatively charged surfaces (Wagenet, 1983). The phenomenon gives rise to a local deficit of anions in the vicinity of negatively charged particles and an excess of anions in the bulk solution (De Haan and Bolt, 1963). Many studies have shown that negative adsorption of anions at a negatively charged surface results in anion exclusion during solute transport (Smith, 1972; Kissel et al, 1973). The importance of anion exclusion is that the anions are removed from the relatively immobile water associated with the DDL and positioned in the faster moving pore water. The result is that the average rate of transport of the anion in the soil is greater than the average pore water velocity. Thus, for a given duration of displacement, anions will move further in a porous material in the presence of anion exclusion than in its absence because the volume of water immediately adjacent to the clay surfaces does not hold anions, and as such, does not participate in the leaching reaction. So, the larger the anion exclusion volume, the greater the mobility of salt at a given soil water content and water application (Thomas and Swoboda, 1970).

The lack of complete mixing with increasing non-equilibrium for solute transfer by diffusion between inter-and intra-soil aggregates at high pore water velocities was studied by many investigators (Van Genuchten and Wierenga, 1976; Gaudet et al., 1977, Nkedi-Kizza et al., 1982, 1984). Van Genuchten and Wierenga (1976) described a model in which the liquid phase is partitioned into mobile (macroporosity-convection controlled) and immobile (microporosity-diffusion controlled) regions. They suggested that only a certain fraction of sorption sites, i.e., those located around the larger pores, participated actively in the adsorption process. This would have the same effect as a reduction in adsorption, leading to a relatively faster movement of the chemical and earlier breakthrough curve. At the same time, a considerable portion of the chemical must diffuse to remaining sites located inside aggregates. The slow diffusion process would continuously remove material from the larger saturated pores resulting in extensive tailing. James and Rubin (1986), on the other hand, demonstrated that at sufficiently low flow rates the diffusive mixing between mobile and immobile water is sufficiently fast, and that the anion exclusion function developed by Bresler (1973) would improve predictions of the convection-dispersion equation. Smith (1972) suggested that when the flow rate is fast, spatial variations become important and the anion simply may not have enough time to associate with all the soil water before passing through the soil. However, when the flow rate is slow, even though the anion has time to associate with all the soil water, electrical repulsion can prevent its association both with water near negatively charged soil surfaces and with water in any expanded clay interlayers.

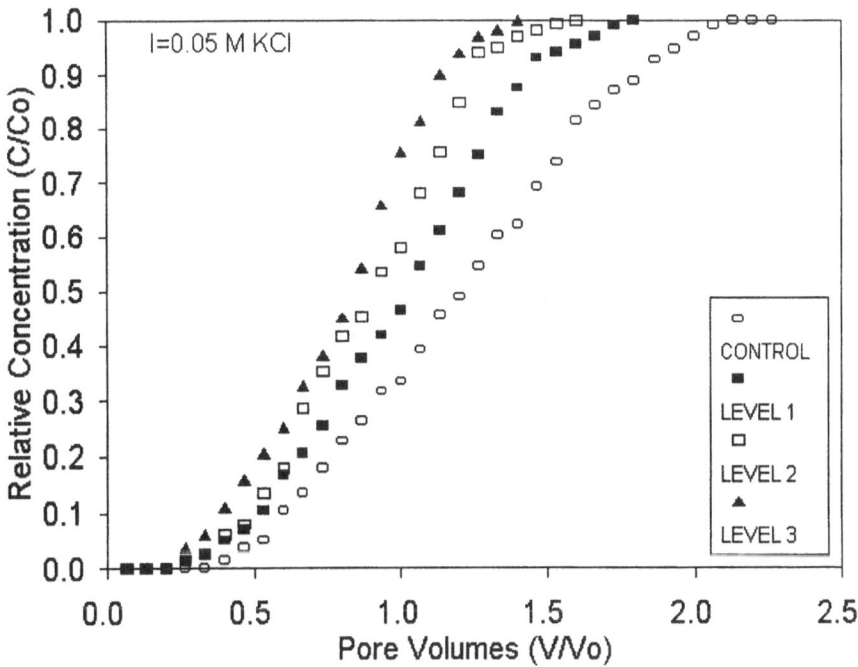

Fig. 8: Transport of Br through Oxisol at four levels of increasing net negative charge

Because anion exclusion and immobile water effects work in the same direction it becomes difficult to separate the extent to which each of them contributes to the movement of anions ahead of the average pore water velocity. Krupp et al. (1972) and Melamed et al. (1994) demonstrated the interaction between electrostatic effects and flow velocity on anion movement. In the latter study, the Br⁻ anion was used as a tracer for NO_3^- due to its similar interaction with the soil surface, similar ionic radii, and because Br is not subjected to biological transformations or oxidation-reduction reactions, as is the nitrate ion (Smith and Davis, 1974; Onken et al., 1977). The increase in negative surface charge of the Oxisol soil used, induced a faster movement of Br⁻ through their columns (Fig. 8), as reflected by the shifts in breakthrough curves to the left. The extent of the effect of increasing the net negative surface charge of the soil on the mobility of the anion, resulted in retardation factors less than 1, implying anion exclusion.

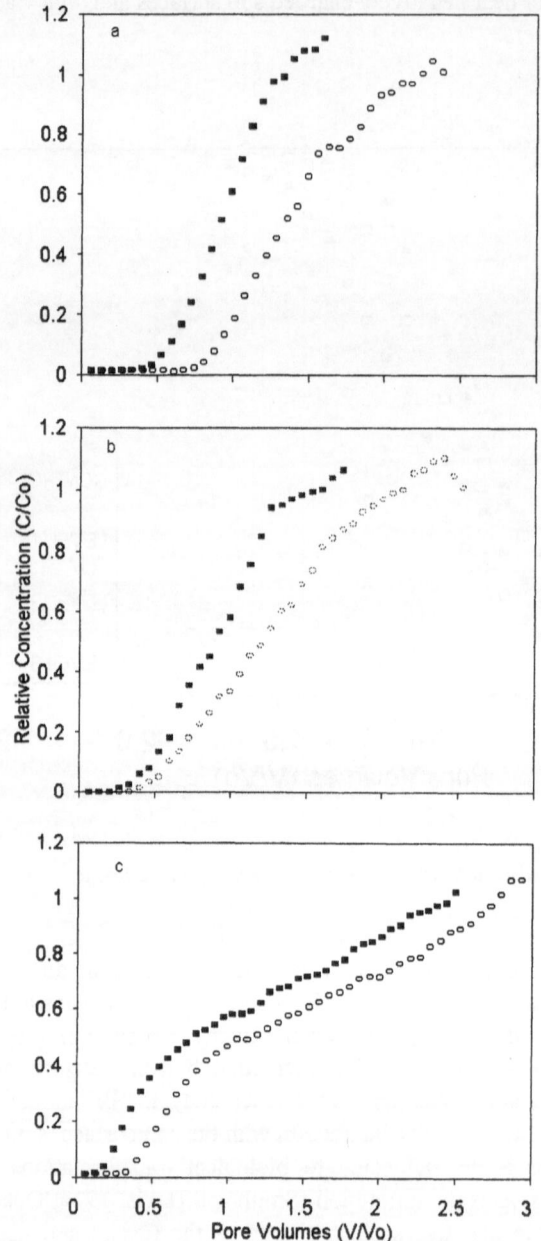

Fig. 9: Transport of Br through Oxisol at two levels of net negative charge (o - low negative charge; ■ - high negative charge) and three pore water velocities: a) 2.9 cm/h; b)8.7 cm/h; c) 48.1 cm/h

When the flow velocity was varied, Br⁻ movement was enhanced under the slowest flow velocity (Fig. 9a), and as the flow velocity progressively increased (Figs. 9b,c), the effect of surface charge on Br⁻ movement diminished, reflecting in progressively reduced separation areas of the BTCs (Fig. 9).

Conclusions

In this chapter, some important reaction mechanisms affecting the transport of arsenic, mercury, and nitrate were highlighted. The effect of the character of the surface, the particular interaction of the compound with the surface, as well as solution speciation on the transport of chemicals through soils were emphasized.

References

Anderson, M.A., and D.T. Malotky. (1979). The adsorption of protolyzable anions on hydrous oxides at the isoelectric pH. *J. Colloid Inter. Sci.* 72:413-427.

Barcellos, C., Lacerda, L. D., Rezende, C. E., Machado, J. (1992). International Seminar Proceedings. As in the Environment its Incidence on Health. p. 59.

Barrow, N. J. (1985). Reaction of anions and cations with variable-charge soils. Adv. Agron. 38:183-230.

Barrow, N.J. (1974). On the displacement of adsorbed anions from soil: 2. Displacement of phosphate by arsenate. *Soil Sci.* 117:28-33.

Barrow, N.J. (1987). *Reactions with variable-charge soils. Developments in Plant and soil sciences*. Martinus Nijhoff Publishers, Dordrecht/Boston/Lancaster.

Barrow, N.J. (1989). Testing a mechanistic model. IX.Competition between anions for sorption by soil. *J. Soil Sci.* 40:415-425.

Bolan, N.S., and N.J. Barrow. (1984). Modelling the effect of adsorption of phosphate and other anions on the surface charge of variable charge oxides. *J. Soil Sci.* 35:273-281.

Bowden, J.W., A.M. Posner, and J.P. Quirk. (1977). Ionic adsorption on variable charge mineral surfaces: Theoretical charge development and titration curves. *Aust. J. Soil Res.* 15:121-136.

Bresler, E. (1973). Anion exclusion and coupling effects in non-steady transport through unsaturated soils: I. Theory. *Soil Sci. Soc. Am. Proc.* 37:663-669.

Camargo, O. A., J. W. Biggar, and D. R. Nielsen. (1979). Transport of inorganic phosphorus in an alfisol. *Soil Sci. Soc. Am. J.* 43:884-890.

Carageorgos, T. (1993). The chemistry of arsenopyrite in some England soils: metals behaviour upon leaching experiment. Ph.D. Thesis, Univ London, Imperial College.

Chapman, D.L. (1913). A contribution to the theory of electrocapillarity. *Phil. Mag.* 25:475-481.

Cleary, D., Thornton, L., Brown, N., Kazantzis, G., Delves, T., Worthington, S., (1994). Mercury in Brazil. *Nature*, 369:613-614.

Craig, P.J. and Moreton, P.A., (1985). The role of speciation in mercury methylation in sediments and water. *Environ. Poll.* series B, 10:141-158.

De Haan, F. A. M., and G. H. Bolt. (1963). Determination of anion adsorption by clays. *Soil Sci. Soc. Am. Proc.* 27:636-640.

Driscoll, C.T., Cheng, Y., Schofield, C.L., Blette, V., Munson, R. and Holsapple, J.G., (1994). The chemistry and bioavailability of mercury in remote Adirondack lakes. *International Conference on Mercury as a Global Pollutant*. July 10-14. Whistler, British Columbia.

Farid, L.H., Machado, J.E.B. and Silva, O.A., (1991). Emission control and mercury recovery from garimpo tailing. In: M.M. Veiga and F.R.C. Fernandes (Editors), *Poconé: Um campo de estudos do impacto ambiental do garimpo*. CETEM/CNPq, Rio de Janeiro, Brazil, pp. 27-44.

Fletcher, P. and Beckett, P.H.T. (1987). The chemistry of heavy metals in digested sewage sludge-II. Heavy metal complexation with soluble organic matter. *Water. Res*, 21:1163-1172.

Fordham, A. W., and K. Norrish. (1979). Arsenate-73 uptake by components of several acidic soils and its implications for phosphate retention. *Aust. J. Soil Res*. 17:307-316.

Frost, R. R., and R. A. Griffin. (1977). Effect of pH on adsorption of arsenic and selenium from landfill leachate by clay minerals. *Soil Sci. Soc. Am. J*. 41:53-57.

Gaudet, J.P., H. Jégat, G. Vachaud, and P.J. Wierenga. (1977). Solute transfer, with exchange between mobile and stagnant water, through unsaturated sand. *Soil Sci. Soc. Am. J*. 41:665-671.

Goldberg, S. (1986). Chemical modeling of arsenate adsorption on aluminum and iron oxide minerals. *Soil Sci. Soc. Am. J*. 50:1154-1157.

Gouy, G. (1910). Sur la constituition de la charge electrique a la surface d'un electrolyte. *Ann. Phys*. (Paris) 9:457-468.

Grahame, D.C. (1947). The electrical double layer and the theory of electrocapillarity. *Chem. Rev*. 41:441-501.

Hem, J.D., (1970). Chemical behaviour of mercury in aqueous media. In: *Mercury in the Environment*. U. S. Geological Survey, Professional paper, no. 713, Washington D.C., pp. 19-24.

Hingston, F.J. (1981). A review of anion adsorption. p. 51-90. *In* M.A. Anderson and A.J. Rubin (ed.) *Adsorption of inorganics at solid-liquid interfaces*. Ann Arbor Science Publishers, Ann Arbor, MI.

Hingston, F.J., A.M. Posner, and J. P. Quirk. (1972). Anion adsorption by goethite and gibbsite: I. The role of the proton in determining adsorption envelopes. *J. Soil Sci*., 23:177-192.

Huckabee, J.W., Elwood, J.W., Hildebrand, S.G., (1979). Accumulation of mercury in freshwater biota. In: J.O. Nriagu. (Editor), *The Biogeochemistry of Mercury in the Environment*, pp. 277-302. Elsevier/North Holland Biomedical Press, Amsterdam.

Iverfeldt, A. and Lindqvist, O., (1986). Atmospheric oxidation of elemental mercury by ozone in the aqueous phase. *Atm. Environ*., vol. 20, no.8, pp. 1567-1573.

Jacobs, L. W., J. K. Syers, and D. R. Keeney. (1970). Arsenic sorption by soils. *Soil Sci. Soc. Am. Proc*., 34:750-754.

James, R.V. and J. Rubin. (1986). Transport of chloride ion in a water-unsaturated soil exhibiting anion exclusion. *Soil Sci. Soc. Amer. J*., 50:1142-1149.

Kissel, D. E., J. T. Ritchie, and E. Burnett. (1973). Chloride movement in undisturbed swelling clay soil. *Soil Sci. Soc. Am. Proc*., 37:21-24.

Krupp, H. K., J.W. Biggar, and D. R. Nielsen. (1972). Relative flow rates of salt and water in soil. *Soil Sci. Soc. Am. Proc*. 36:412-417.

Laguitton, D. (1976). Arsenic removal from gold-mine waste waters: basic chemistry of the lime addition method. *CIM Bull*., p105-109

Langford, C.H. and Cook, R.L., (1995). Kinetic versus equilibrium studies for the speciation of metal complexes with ligands from soil and water. *Analyst*, 120:591-596.

Logan, T. J., and E. O. McLean. (1973). Effects of phosphorus application rate, soil properties, and leaching mode on P movement in soil columns. *Soil Sci. Soc. Am. Proc*. 37:371-374.

Luoma, S.N., (1983). Bioavailability of trace metals to aquatic organisms - a review. *The Sci. Total Environ*., 28:1-22.

MacNaughton, M.G. and James, R.O., (1974). Adsorption of aqueous mercury (II) complexes at the oxide/water interface. *J. Colloid Interf. Sci*., 47:431-440.

Melamed, R, Jurinak, J.J., and Dudley, L.M. (1994). Anion exclusion-pore water velocity interaction affecting transport of bromine through an Oxisol. *Soil Sci. Soc. Am. J.*, 58:1405-1410.

Melamed, R, Jurinak, J.J., and Dudley, L.M. (1995c). Site disposal simulation for arsenopyrite processing waste: arsenic mobility and retention mechanisms. *Proceedings of X Conference on Heavy Metals and The Environment*. Hamburg.

Melamed, R., Jurinak, J.J., and Dudley, L.M. (1995a). Effect of adsorbed phosphate on transport of arsenate through an Oxisol. *Soil Sci. Soc. Am. J.*, 59:1289-1294.

Melamed, R., Villas Bôas, R. C. (1996b). Phosphate-background electrolyte interaction affecting the transport of Hg(II) through a Brazilian Oxisol. *The Science of Total Environment*. In press.

Melamed, R., Villas Bôas, R. C., Gonçalves, G. O., Paiva, E.C. (1995b). Mechanisms of physico-chemical interaction of mercury with river sediments from a gold mining region in Brazil. *Proceedings of X Conference on Heavy Metals and The Environment*. Hamburg.

Melamed, R., Villas Bôas, R. C., Gonçalves, G. O., Paiva, E.C. (1996a). Mechanisms of physico-chemical interaction of mercury with river sediments from a gold mining region in Brazil: Relative mobility of mercury species. *Journal of Geochemical Exploration*. In press.

Miller, D. M., M. E. Sumner, and W. P. Miller. (1989). A comparison of batch- and flow-generated anion adsorption isotherms. *Soil Sci. Soc. Am. J.*, 53:373-380.

Munthe, J. and McElroy, W.J., (1992). Some aqueous reactions of potential importance in the atmospheric chemistry of mercury. *Atmospheric Environ.*, 26A:553-557.

Nishimura, T, Tozawa, K. 1981. Bulletin of Research Institute of Mineral Dressing and Metallurgy, Tohuky University, Sendai, Japan 34, p19

Nkedi-Kizza, P., J.W. Biggar, H.M. Selim, M.T. Van Genuchten, P.J. Wierenga, J.M. Davidson, and D.R. Nielsen (1984). On the equivalence of two conceptual models for describing ion exchange during transport through an aggregated oxisol. Water Resour. Res. 20:1123-1130.

Nkedi-Kizza, P., P. S. C. Rao, R. E. Jessup, and J.M. Davidson. (1982). Ion exchange and diffusive mass transfer during miscible displacement through an aggregated oxisol. *Soil Sci. Soc. Am. J.*, 46:471-476.

Onken, A.B., C.W. Wendt, R.S. Hargrove, and O.C. Wilke (1977). Relative movement of bromide and nitrate in soils under three irrigation systems. *Soil Sci. Soc. Am. J.*, 41:50-52.

Padberg, S. and Stoeppler, M. (1991). Studies of transport and turnover of mercury and methylmercury. In: E. Merian (Editor), *Proceedings of Workshop on Toxic Metal Compounds, Interrelation Between Chemistry and Biology*, Les Diablerets, March 4-8.

Parks, G. A., and P. L. de Bruyn. 1962. The zero point of charge of oxides. J. Phys. Chem. vol. 66:967-973.

Rajan, S.S.S. (1976). Changes in net surface charge of hydrous alumina with phosphate adsorption. *Nature*, 262:45-46.

Robins, R. G., Tozawa, K. (1982). Arsenic removal from gold processing waste waters: The potential ineffectiveness of lime. *CIM Bull.*, 75, p171.

Roy, W.R., J.J. Hasset, and R.A. Griffin. (1986). Competitive coefficients for the adsorption of arsenate, molybdate and phosphate mixtures by soils. *Soil Sci. Soc. Am. J.*, 50:1176-1182.

Smith, S. J. (1972). Relative rate of chloride movement in leaching of surface soils. *Soil Sci.*, 114:259-263.

Smith, S. J., and R. J. Davis. (1974). Relative movement of bromide and nitrate through soils. *J. Environ. Quality.*, 3:152-155.

Sposito, G. (1984). *The surface chemistry of soils*. Oxford Univ. Press, New York.

Stern, O. (1924). Zur Theorie der Elektrolytischen Doppelschicht. *Z. Electrochem.*, 30:508-516.

Thayer, J. S. (1989). Methylation: its role in the environmental mobility of heavy elements. *Applied Organometallic Chemistry*, 3:123-128.

Thomas, G. W., and A. R. Swoboda. (1970). Anion exclusion effects on chloride movement in soils. *Soil Sci.*, 110:163-166.

Van Genuchten, Mh.T., and P.J. Wierenga (1976). Mass transfer studies in sorbing porous media: I. Analytical solutions. *Soil Sci. Soc. Am. J.*, 40:473-480.

Van Riemsdijk, W. H., and J. Lyklema (1980). Reaction of phosphate with gibbsite ($Al(OH)_3$) beyond the adsorption maximum. *J. Colloid and Interface Sci.*, 76:55-66.

Van Riemsdijk, W. H., L. J. M. Boumans, and F. A. M. de Haan (1984). Phosphate sorption by soils: I. A model for phosphate reaction with metal oxides in soil. *Soil Sci. Soc. Am. J.*, 48:537-541.

Veiga, M. M. (1994). *A Heuristic System for Environmental Risk Assessment of Mercury from Gold Mining Operations*. Ph. D. thesis. The University of British Columbia. Canada, pp. 196.

Verta, M., Rekolainen, S. and Kinnunen, K. (1986). Causes of increased fish mercury levels in Finnish reservoirs. In: *Publications of Water Research Institute*, Vesihallitus-National Board of Waters, no.65, Helsinki, Finland, pp. 44-71.

Wagenet, R. J. (1983). Principles of salt movement in soils. p.123-140. *In* D.W. Nelson et al. (eds.) *Chemical Mobility and Reactivity in Soil Systems*. Soil Sci. Soc. Am., Special Publication no. 11, Madison, WI.

Willet, I. R., C. J. Chartres, and T. T. Nguyen (1988). Migration of phosphate into aggregated particles of ferrihydrite. *J. Soil Sci.*, 39:275-282.

Wood, J. M., Cheh, A., Dizikes, L. J., Ridley, W. P., Rakow, S., and Lakowicz, J. R. (1978). Mechanisms for the biomethylation of metals and metalloids. *Federation Proc.*, 37:16-21.

5 Key Role of Flow Velocity and pH in the Lixiviation of Mineral-Forming Elements

José Luis Mogollón
Reservoir Department, INTEVEP S. A., Apdo. 76343, Caracas 1070-A,
Venezuela

Abstract

Two basic aspects recently accepted in the understanding of water composition
and soil evolution are dissolution/precipitation kinetics and how open are natural
systems are to the passage of fluid. In the light of these two aspects a new
generation of geochemical models of coupled chemical reactions and solute
transport has been developed. An equilibrium condition is not assumed a priori in
these models. Despite its importance, validation of these models is still weak, and
in only a few cases have they been applied to the solution of soil lixiviation
problems.

In this paper, different relations between aqueous Al concentrations and flow
velocity are explained in terms of the basic concepts of a coupled
reaction/transport model. These relations were obtained in lixiviation experiments
on a bauxite, rich in gibbsite, by the effect of different initial pH solutions passing
at variable velocities.

It is shown that a numerical model based on these concepts and developed by
Soler et al. (1994) is capable of reproducing, with differences below 20%, Al and
H^+ concentrations obtained in the lixiviation experiments without any data
adjustment. The implications all this could have for natural systems, emphasizing
the effect of the acid rains over tropical soils, are also analyzed.

Introduction

Physico-chemical interactions between aqueous solutions and soils/sediments/
rocks is a very interesting subject for the geochemist, since they determine the
movement of natural and polluting chemical species, in small catchments as at a
planetary level. That is why these interactions are of fundamental importance in
understanding geochemical cycles. In addition, water/rock interaction plays an
important role in mineral reservoir formation and in an efficient oil extraction.

In the tropics, there are big areas covered by acid weathered soils, with low
absorption capacity (Mogollón and Querales, 1995). Just like the presence of acid
rains, this condition could promote lixiviation of Al, which is a toxic element once
the interchangeable cation pool is exhausted.

The composition of water present in soils depends not only on mineral solubility and aqueous species complexing but also on mineral dissolution/precipitation kinetics and the flow velocity of water (Lasaga et al., 1994). Therefore, to obtain a clear vision of the transfer of chemical elements between minerals and solutions, the influence of flow velocity must be investigated as a function of the solution's pH. This is the main aspect studied in this review. First of all, a description of the basic concepts of a coupled reaction-flow model is presented. Then, these basic concepts are applied to interpreting the effect of solution flow through porous media, in column reactors filled with gibbsite. Finally, the model calibration is included.

Dissolution Rate of Alumino Silicates

In the 1960s Garrels and his partners promoted the idea that the chemical composition of ground-waters was controlled by chemical reactions which occur between minerals and water. Some examples of these reactions are shown below:

$$KAlSi_3O_8 + 4H^+ + 4H_2O \Leftrightarrow K^+ + Al^{3+} + 3H_4SiO_4 \tag{1}$$

$$Al(OH)_3 + 3H^+ \Leftrightarrow 3Al^{3+} + 3H_2O \tag{2}$$

Based on a thermodynamic approach, different mineral/solution phase diagrams were used to explain the composition of a great variety of waters. Nevertheless, in some cases, this approach failed. *In fact, in open systems (with input and output of material) there is at least one part of the system which never reaches equilibrium*, (Lasaga and Rye, 1993). All of this conduced, in the last 30 years, to the measurement of great numbers of dissolution velocities of several minerals and to study the pH effect over them, an aspect that will be taken up again later.

Now, let us center our attention in the dissolution rate values (R_{dis}) of some alumino-silicates. Table 1 shows the great variations in R_{dis} depending on the chemical composition and mineral structure. Lifetime, defined as the necessary time to completely dissolve a spherical crystal of ratio r_o, shows the consequences that R_{dis} variations have over permanency or durability of minerals in natural environments. This lifetime was calculated through the following equation (Lasaga, 1995):

$$t_{lifetime} = \frac{r_o}{RV_{molar}} \tag{3}$$

An interesting feature is that the time magnitude and relative stability of minerals (Table 1) support, in a quantitative form, Goldich's weathering series, which was proposed in 1938 (Lasaga, 1984). Nevertheless, what is more important, as shown in Table 1, is that crystalline structure of minerals constitutes a long-term pool which can release, over thousands of years, elements such as Al and Si. It is therefore possible to affirm that mineral dissolution is the process responsible for feeding the soils with interchangeable cations. These cations constitute the short-term pool with which soils quickly respond to acid discharges. Hence the importance of the process of minerals dissolution in the acidification of environmental systems.

In addition, Table 1, which is a brief summary, shows that there are enough kinetic data to establish basic concepts related to alumino-silicate dissolution. The situation is similar for other minerals like oxides (Wieland et al., 1988) and carbonates (Sjöberg and Rickard, 1984).

In general, the dissolution values shown in Table 1 are obtained through batch type experiments. In soils and porous rocks the solid part is not suspended in the solution, but the solution flows through the porous spaces reacting with the solid.

Molecular diffusions and hydrodynamic dispersion phenomena are superposed on chemical interactions. As a consequence, quantitative application to the nature of dissolution velocities is not direct. That is why it is necessary to use models which take as much account of chemical reactions as of fluid movement.

Table 1: Alumino-silicate mineral dissolution rates (R_{dis}) (standardized by BET surface area) and mean lifetime ($t_{lifetime}$) of spherical crystals, 1 mm diameter, at 25° C and pH= 5.0. Modified from Lasaga (1995)

Mineral	Formula	Log R_{dis} (mol/m^2/seg)	V_{molar} (cm^3/mol)	$t_{lifetime}$ (years)
Quartz	SiO_2	-13.39	22.688	34,000,000
Kaolinite	$Al_2Si_2O_5(OH)_4$	-13.28	99.52	6,000,000
Muscovite	$KAl_2Si_3AlO_{10}(OH)_2$	-13.07	140.71	2,600,000
Microcline	$KAlSi_3O_8$	-12.50	108.741	921,000
Gibbsite	$Al(OH)_3$	-11.45	31.956	276,000
Albite	$NaAlSi_3O_8$	-12.26	100.07	575,000
Anorthite	$CaAl_2Si_2O_8$	-8.55	100.79	112

Fluid-Rock Interaction Model

The model described is one which a priori does not consider equilibrium, and is the core over which advanced numerical codes like DGIBBSSPH (Soler et al., 1994), 1DREACT (Steefel, 1993) and OS3D (Steefel and Yabusaki, 1995) are described qualitatively. These codes are simplifications of the complex natural systems; nevertheless, they show a new reference guide in water-mineral interaction studies of hydrodynamic phenomena.

Let us imagine an aqueous fluid that starts moving at time t_{zero} from d_{zero} at a velocity v in a pseudo one dimension, flowing through an isotropic medium, as shown in Figure 1A, and constituted of alumino-silicate reactive and spherical-form minerals. At a time t, the front part of the fluid should be at a distance $d = v \times t$; and if the reaction is an acid hydrolysis [like reactions (1) and (2) above], the H_3O^+ concentration should have decreased an amount depending on the reaction rate. Simultaneously, alumina and silica concentrations in solution should increase, according to the stoichiometry and kinetics of reaction [see reactions (1) and (2) as examples]. The result of this process is that while the top of the reaction is far from the d_{zero} point (and more fluid-mineral contact time has passed) Al and Si concentrations will be bigger and H_3O^+ concentrations will be lower. Once the time t_{eq} necessary to reach equilibrium has passed, no more changes in concentrations should be produced, and the latter should be constant from the distance d_{eq} (Fig. 1B). The change in concentration with distance resulting from this process, and theoretically predicted by Lasaga (1984), Steefel, (1992) and Lasaga and Rye (1993), among others, was experimentally confirmed by Soler et al. (1994) and Mogollón et al. (1995), who in addition were able to predict, through numerical calculus, concentrations in solutions, with differences below 20 %. Later on, in this chapter this point will be further discussed.

Now, let us suppose that, at a moment t, an electrolyte pulse, which is inert to the porous media, is injected to the system. Randomly, some ions will take the longest way and others the shortest. Those ions passing closer to the alumino-silicate surface will be decelerated by friction. This process, known as hydrodynamic dispersion D_{hid} , results in enlargement of the zone occupied by the electrolyte as time goes on or, which is equivalent, while the pulse goes through the porous media, as shown in Fig. 1C. Additionally, in order to equalize chemical potentials, the ions will move from zones of higher concentration to zones of lower concentration; this is known as molecular diffusion D_{mol}, which also contributes to the enlargement of the zone occupied by the electrolyte.

Now, let us go back to the case of a solution capable of reacting with the alumino-silicates. Al^{3+} and H_3O^+ ions as well as H_4SiO_4 will be under the effect of the hydrodynamic dispersion and the molecular diffusion while they are moving through the media. This coupling is described in one dimension by the following equation, which in steady state is equal to zero:

$$\frac{\partial(\phi c)}{\partial t} = -\frac{\partial(\phi v c)}{\partial x} + \frac{\partial}{\partial x}(\phi D \frac{\partial c}{\partial x}) + \phi R_{dis} \qquad (4)$$

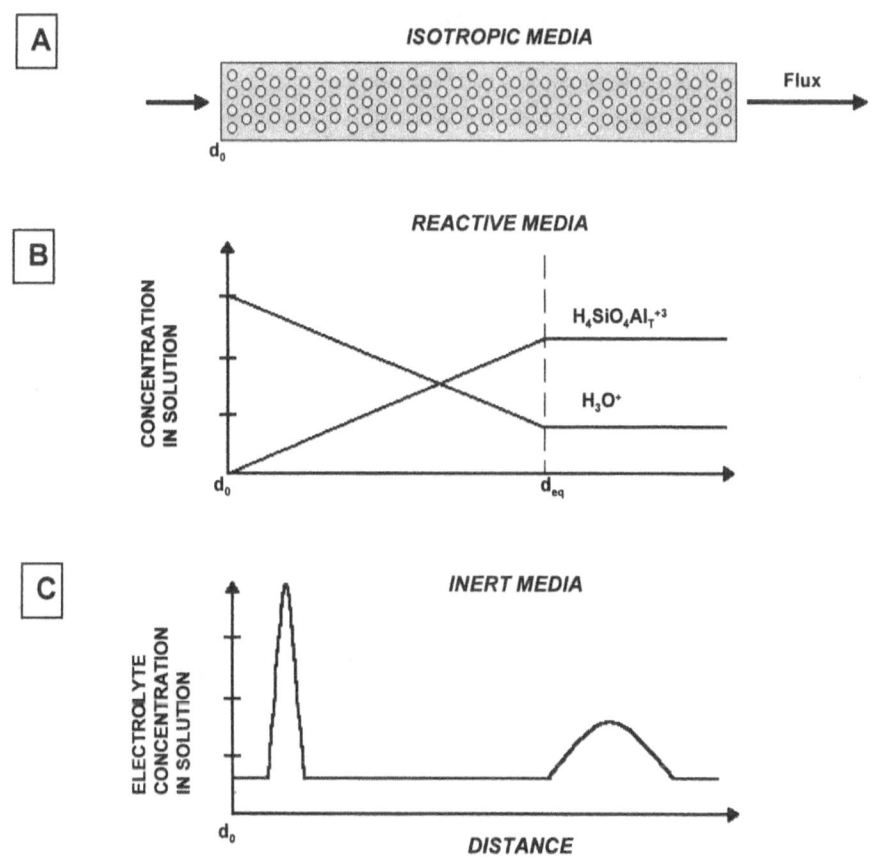

Fig. 1: (A) Isotropic media, pseudo-one-dimensional, composed of spherical alumino-silicate minerals. **(B)** Resulting concentrations from the continuous injection of a reactive acid solution, d_{eq} is the distance at which equilibrium is reached. **(C)** Concentration profile of an electrolyte solution "pulse" as a function of distance

where c is the concentration in the solution; ϕ is porosity; t is the time; v is the advection velocity of the fluid; x is the distance, D is is the sum of D_{hid} and D_{mol}; R_{dis} is the reaction rate. The parameters affecting R_{dis} form the focus of the following section. The other terms in equation (4) have been studied by physicists and engineers, and there is a large literature about it.

General Law of Dissolution Rate

From the study of the different variables affecting the dissolution/precipitation rate a general form of the kinetic law was derived (Lasaga et al., 1994; Mogollón et al., 1994), which is given by :

$$Rate = k_o \, A_{min} \, e^{\frac{E_a}{RT}} a_{H^+}^{n_{H^+}} \prod_H a_i^{n_i} f(\Delta G_r) \qquad (5)$$

where k_o is the specific rate constant; A_{min} is the superficial area of the mineral; E_a is activation energy; T is temperature; a_i refers to the activity of cations or anions species in solution; n_i is reacting order and f (ΔG_r) is a function of Gibbs' ϕ free energy and represents the degree of saturation of the solution.

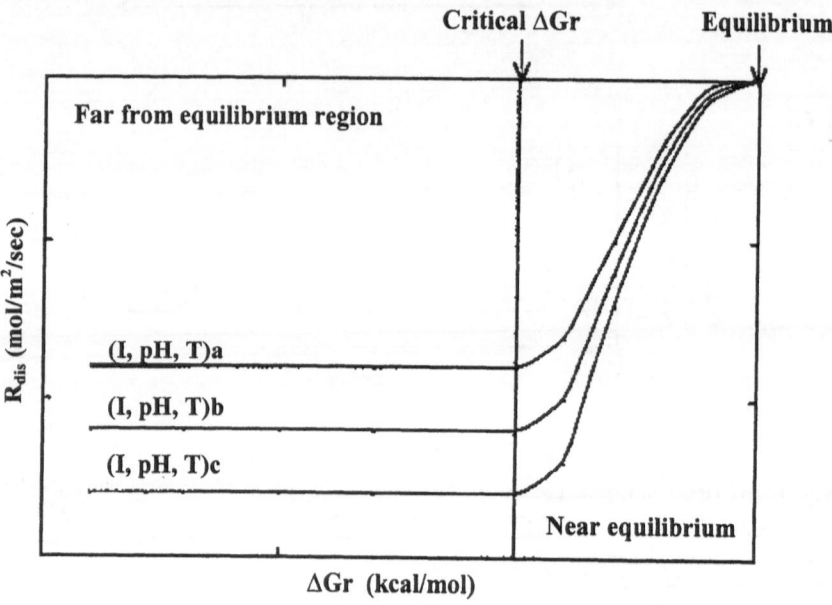

Fig. 2: Sketch of the the dependence of the dissolution rate (R_{dis}) on the Gibbs' free energy of reaction (ΔG_r). The critical ΔG is the value at which R_{dis} becomes independent of ΔG_r. Under this condition, different R_{dis} values are obtained as a function of solution characteristics such as temperature, ionic strength and pH

Of the different variables of equation (5), $f(\Delta G_r)$ and $(a_{H^+})^{n_{H^+}}$ are very important in understanding acidification processes. Of all the effects shown in equation (5) the ΔG_r effect is the most recently studied. Its value is calculated through the relations between ions in solution concentration, from the resulting dissolution reaction, and the equilibrium constant.

For the reaction of gibbsite dissolution [Equation (2)]:

$$\Delta G_r = RT \ln \Omega \qquad (6)$$

where

$$\Omega = \frac{a_{Al^{3+}} / (a_{H^+})^3}{K_{eq}} \qquad (7)$$

K_{eq} being the equilibrium constant of reaction (2).

The measurements of reaction rate, as a function of ΔG_r (Nagy et al., 1991; Nagy and Lasaga, 1992), have shown a strong non-linear dependence in the zone called "close to equilibrium". From a certain value of ΔG_r, called critical ΔG , there is a *plateau* of dissolution rate in which ΔG_r is independent of the saturation degree of the solution and the condition of the system is said to be "far from equilibrium". Specifically, for gibbsite at 80° C and pH 3, Nagy and Lasaga (1992) determined that $f(\Delta G_r)$ is given by the equation:

$$f(\Delta G_r) = 1 - \exp(-8,12 \left[\frac{\Delta G'}{RT} \right]^{3,01}) \qquad (8)$$

Later studies conducted by Mogollón et al. (1994; 1996) suggest that this equation form is independent of the chemical characteristics of the solution in contact with mineral. That is why, in a schematic way, the dependence between the dissolution rate R_{dis} and ΔG_r shown in Figure 2 is proposed. This dependence between R_{dis} and ΔG_r has important connotations in natural system behavior. This aspect will be addressed later.

The a_{H^+} Effect

The catalytic or inhibitor effect of solutes on mineral/water reaction velocities has been the focus of some recent articles. In particular, the effect of H^+ activity, a_{H^+} activity, expressed as pH, is one of the most studied. A succinct summary of this work is found in Ganor et al. (1995).

It was found that the dissolution rate for different minerals at far-from-equilibrium conditions and at constant temperature and superficial areas is proportional to a fractional power $(a_{H^+})^{n_{H^+}}$, where n, the reaction order, remains between 0 and 1 in the acid region and between 0 and -1 in the basic region (Table 2). For pHs close to neutrality, a low dependence of R_{dis} with $(a_{H^+})^{n_{H^+}}$ is observed.

Qualitatively, the dependence of R_{dis} on the pH, for the alumino-silicates, is explained by the amphoteric nature of Al: this element is soluble in both acid and basic media. Fractionary reaction orders have been explained, at least partially by taking into account the quantity of H^+ adsorbed over the mineral surface (Furrer and Stumm, 1986; Blum and Lasaga, 1988; Wieland et al., 1988), instead of the a_{H^+} value in solution.

Under far-from-equilibrium conditions, the dissolution rate of minerals follows this law:

$$R_{dis} = k' a_{H^+}^n \qquad (9)$$

or its equivalent in the terms of OH^- activity:

$$R_{dis} = k' a_{OH^-}^n \qquad (10)$$

where k involves all remaining factors of equation (5).

Table 2: Dependence of the dissolution of Alumino-silicate minerals on pH; values at 25°C. Modified from Lasaga (1996)

Mineral	Reaction order with respect to a_{H^+} (n)	pH range
Quartz	0.0	< 7
Kaolinite	0.4	3 - 4
K-feldspar	1.0	< 7
Albite	0.49	< 6
Albite	0.30	> 8

The importance of equation (9) can be hardly emphasized enough. It shows that an increase in rain acidity must accelerate the rate of mineral dissolution. In fact, for the dissolution reaction of K-feldspar, which has reaction order 1 with respect to the pH, a diminution from 5.6 to 3.6 in rainwater pH produces an increase of 100 times in R_{dis}. That is to say, it is equivalent to raising the liberation rate of K 100 times, Al 100 times and Si 300 times, because of the stoichiometry of the mineral ($KAlSi_3O_8$).

Experimental Approach

In the previous sections, the basic concepts of the processes of coupled reaction-flow in the porous media have been discussed and a model described. The verification of this model was conducted by Mogollón et al. (1993; 1996). To do this, a series of experiments was designed firstly with the aim of determining wether the composition of a solution, resulting from its interaction with the porous medium, could be effectively explained by the previously described concepts; and secondly to calibrate one of the existing numerical models (DGIBSSPH) (Soler et al., 1994).

Since the expectation was to obtain a chemical reaction coupled to the solution flow, column reactors 3.65 and 7.5 cm. long were used. Additionally, with the purpose of obtaining information to contribute to knowledge of acidification effects in the tropics, a bauxite material rich in gibbsite-like column packing was used; and the pH of input solutions was between 3.22 and 4.49.

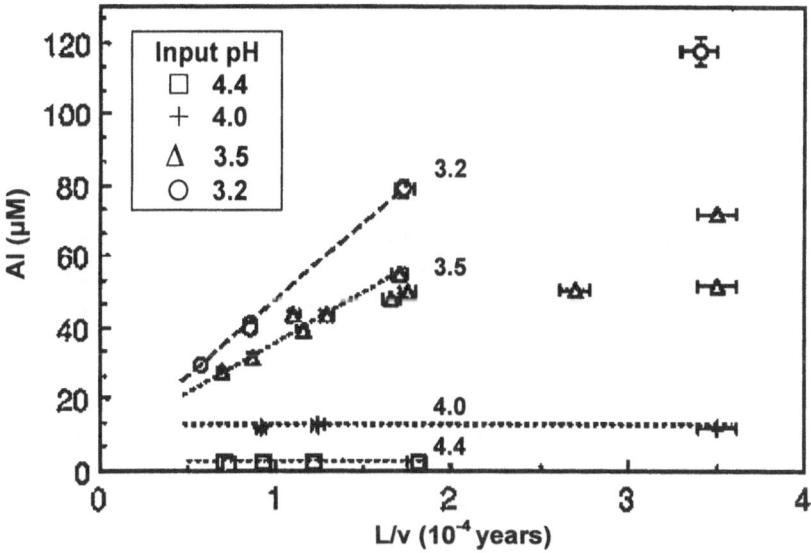

Fig. 3: Gibbsitic-bauxite column experiments: output Al concentrations vs. the fluid-flow velocity ratio for different input pHs at 25 °C. After Mogollón et al. (1993)

Figure 3 shows the effect of fluid flow velocity on the steady state output Al concentration found by Mogollón et al. (1993) for a series of experiments at 25°C. Three different behaviors are observed. For an input pH ranging from 4.0 to 4.5, the output Al concentrations and the output pH are *independent* of the flow velocity. This observation suggests that the dissolution rate is fast enough to enable the solution to reach equilibrium with the solid before the solution reaches the exit point in the column. Considering equilibrium for these runs the log of the equilibrium constant (K_{eq}) for reaction (2), i.e. the gibbsite dissolution, was calculated to be 7.83 ± 0.12. In spite of the iron oxide impurities present in the natural sample used in our study, this value is in remarkable agreement with the pure gibbsite value of 7.74, reported by Palmer and Wesloski (1992).

For input pH in the range 3.2 to 3.5, the output Al concentration varies as an inverse function of the flow rate. This inverse relation reflects a *constant* dissolution rate throughout most of the column under these conditions. A loss of linearity at L/v values higher than 1.75×10^{-4} years for input pH 3.5 experiments indicates close-to-equilibrium dissolution as Al becomes high. The output concentrations found in the experiments of different input pHs at specific flow velocities indicate a pH effect on the rate. In order to explain these output Al concentrations (Fig. 3) the dissolution rate was assumed to be a function of a_{H^+}, as shown in equation (10). Plotting log R_{dis} against pH, the straight line slope predicts a reaction order of 0.29. Therefore, there is a slight dependence of the R_{dis} on pH.

The Gibbs free energy (ΔG_r) of reaction (2) for the output solutions was calculated using equations (6) and (7). The Al values outside the linear trend shown in Figure 3 correspond to experiments in which near-to-equilibrium conditions were reached, i.e. ΔG_r values less than the critical ΔG_r value (-0.7 kcal/mol).

The above results and observations have two important implications. Firstly, in open systems, Al lixiviation can be explained in terms of the input pH and the variations in flow rate. Secondly, considering that in the column experiment far-from-equilibrium conditions required relatively low pHs, high flow rate and short columns, it could be expected that near-equilibrium conditions would prevail in the field.

Another interesting aspect clearly shown is the difference from previous published results of column experiments and batch experiments (without any mass input or output) performed by Mogollón and Querales (1995). In the latter Al lixiviation in acid soils was not found since the final pHs were over 4, a value at which Al solubility is too low. Nevertheless, it is important to notice that according to what was observed in column experiments, in the initial states of batch experiments when the pH was 3.5, a liberation of Al should be produced. Soils are not closed systems of the batch kind, and really Al will always be released when acid solutions are coming into the system. Al liberated from soils can be assimilated by plants and microorganisms living in the zone in which the pH-flow rate relation allows its solubility and that is why it is important to complement batch-type experiments with column ones.

Validation of a Numerical Model of Coupled Flow and Reaction

The results shown in the previous section were used to validate a model of coupled flow and reactions (DGIBSSPH), developed at Yale University (Soler et al., 1994).

The model is based on solving equation (4) in steady state conditions, for Al^{3+} and H^+ at each point of a grid, using a finite differences scheme and Newton-

Raphson approximation. It is assumed that the particles are spheres and the input data are: reaction order of the gibbsite solution (0.29) in respect of a_{H^+}, and equilibrium constant and dissolution rate values just like the characteristics of the solutions (Al^{3+} and H^+ concentrations) before reacting with the solid. Further details can be found in Soler et al. (1994).

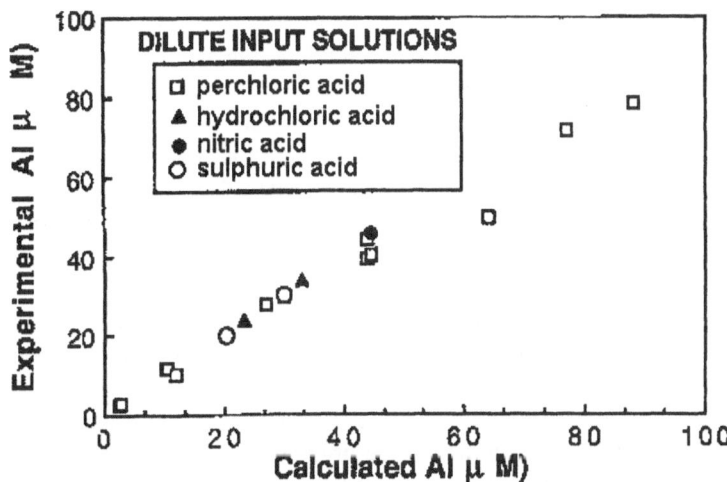

Fig. 4: Coupled reaction-transport model calibration. Experimental data correspond to gibbsitic-bauxite lixiviation in columns at 25°C and cover the whole ΔG_r range. After Soler et al. (1994) and Mogollón et al. (1996)

Al^{3+} and H^+ concentrations calculated by the model are in very good agreement (around 20%) with measurements made at the end of columns, in all the intervals of the experimental conditions, pH_{input} = 3.22 - 4.49; Darcy velocity = 61 - 1050m/year; initial porosity = 0.37 - 0.421 (Fig. 4). The capability of the model in reproducing experimental results, in spite of the different relations of Al vs. L/v shown in Figure 3, is due to the fact that takes into account the dependence of R_{dis} on a_{H^+}. Moreover, inclusion of the non-linear dependence of R_{dis} on ΔG_r allowed reproduction of the experimental results in spite of the non-linearity of the Al vs. L/v relation, to values higher than 1.75×10^{-4} years.

It should be noticed that the model succesfully reproduced experimental results just by using thermodynamic constants and the appropriate kinetic law, without any adjustment of data. This could have important implications at field level. Mogollón et al. (1995) proposed that in nature mineral dissolution can be highly affected by the degree of saturation of the solution, expressed as ΔG_r, which, at least partially, explains the differences found when comparing R_{dis} measurements in the laboratory with field measurements (Brantley, 1992). Additionally, the fact that near-equilibrium R_{dis} is lower, and that non-linear dependence of R_{dis} on the degree of solution saturation could have important implications for modeling water-minerals interactions (Mogollón et al., 1996).

Conclusions

The degree of solution saturation has a strong non-linear effect on the rate of dissolution of gibbsite. Far from equilibrium, the Al concentration is an inverse function of flow velocity. High flow velocities, short distances and acid pHs favor far-from-equilibrium conditions. Under these conditions, the dissolution rate varies according to the pH, with a reaction order of 0.29.

In weathering environments, after short contact periods between the solution and the gibbsite, close-to-equilibrium dissolution kinetics is expected, because of the pH values which normally prevail. In order to understand and predict the effect of acid rain upon Al lixiviation, and upon lixiviation of other mineral-forming elements, the pH effect must be evaluated in a broad range of solution saturations degree and flow velocities.

Al and H^+ concentrations resulting from interactions between acid solutions and gibbsite can be reproduced, with differences below 20%, by computational models which couple solute transport and chemical reactions taking into account the speciation of elements in solution and the kinetic law of mineral dissolution. This gives a powerful tool to investigate how these processes occur and how they are affected by different physical and chemical variables.

Acknowledgments

Thanks are due to Tony Lasaga; many ideas presented in this chapter are the result of discussions with him and his students, and of joint research started some years ago.

Thanks also go to A. Pérez-Diaz and A. Iraldi for providing unpublished data from their thesis work, included in Figure 4.

References

Blum, A.E. and Lasaga, A.C. (1988). Role of surface speciation in the low temperature dissolution of minerals. *Nature* 331, 431-433.

Brantley, S.L. (1992). Kinetics of the dissolution and precipitation. Experimental and field results. *Water-rock interactions*, Y. Kharaka and E. Maest (eds.). Balkema, Rotterdam, pp 3-6.

Furrer, G. and Stumm, W. (1986). The coordination chemistry of weathering: I. Dissolution kinetics of δ-Al_2O_3 and BeO. *Geochim. Cosmochim. Acta* 50, 1847 - 1860.

Ganor, J.; Mogollón, J.L. and Lasaga, A.C. (1995). The effect of pH on kaolinite dissolution rates and on activation energy. *Geochim. Cosmochim. Acta*, 59, 6, 1037 - 1052.

Lasaga, A.C. (1984). Chemical kinetics of water-rock interaction. *J. Geophys. Res.*, 89: 4009-4025.

Lasaga, A.C. and Rye, D.M. (1993). Fluid flow and chemical reaction kinetics in metamorphic systems. *Am. J. Sci.*, 293: 361-404.

Lasaga, A.C., Soler, J.M., Ganor, J., Burch, T., and Nagy, K.L. (1994). Chemical weathering rate laws and global geochemical cycle. *Geochim Cosmochim Acta*, 58: 2361-2386.

Lasaga, A.C. (1996). *Kinetics for Earth Sciences*. Princeton Univ. Press, in press.

Mogollón, J.L., Ganor, J., Soler, J.M. and Lasaga, A.C. (1993). Flow rate and pH effect on Al leaching from a gibbsitic material. *Proceedings of Perspective for Environmental Geochemistry in Tropical Countries Symposium*. Abrão J.J.; Wasserman J.C. and Silva Filho E.V. (eds.). pp.:109-112

Mogollón, J.L.; Pérez, D. A.; Lo Monaco, S.; Ganor, J.; and Lasaga, A.C. (1994). The effect of pH, $HClO_4$, HNO_3 and ΔG_r on the dissolution rate of natural gibbsite using column experiments. *Mineralogical Magazine*, 58A: 619 - 620.

Mogollón, J.L. and Querales, E. (1995) Interactions between acid solutions and Venezuelan tropical soils. *Sci.Total Environ.*, 164: 45-56.

Mogollón, J.L., Ganor, J., Soler, J.M. and Lasaga, A.C. (1996). Column experiments and the complex dissolution rate law of gibbsite. *Am. J. Sci.* (in press).

Nagy, K.L., Blum, A.E. and Lasaga, A.C. (1991). Dissolution and precipitation kinetics of kaolinite at 80° C and pH 3: The dependence on solution saturation state. *Am. J. Sci.* 291: 649-686.

Nagy, K.L. and Lasaga, A.C. (1992). Dissolution and precipitation kinetics of gibbsite at 80° C and pH 3: The dependence on solution saturation state. *Geochim. Cosmochim. Acta*, 56: 3093-3011.

Palmer, D.A. and Wesoloski, D.J. (1992). Aluminum speciation and equilibria in aqueous solutions: II. The solubility of gibbsite in acid sodium chloride solution from 30 to 70° C. *Geochim. Cosmochim. Acta*, 56: 1093-1111.

Sjöberg, E.L. and Rickard, D.T. (1984). Temperature dependence of calcite dissolution kinetics between 1 and 62° C at pH 2.7 and 8.4 in aqueous solutions. *Geochim. Cosmochim. Acta*, 48:485-493.

Soler, J.M., Mogollón, J.L. and Lasaga, A.C. (1994). Coupled fluid flow and chemical reaction: a steady state model of dissolution of gibbsite in a column. *Bol. Soc. Esp. Min.* 17: 17-28.

Steefel, C.I. (1992). *Coupled Fluid Flow and Chemical Reaction: Model Development and Application to Water-Rock Interactions*. Ph.D. thesis, New Haven, Connecticut, Yale University, 234 p.

Steefel, C. I. (1993). *1DREACT, One-Dimensional, Reaction-Transport Model*. User manual and programmer's guide. Battelle Pacific Northwest Lab. 37 p.

Steefel, C.I. and Yabusaki, S. (1995). *Multicomponent Reactive Transport Modelling*. Short Course Manual. Univ. of Minnesota.

Wieland, E., Berhard, W. and Stumm, W. (1988). The coordination chemistry of weathering: III. A generalization on the dissolution rates of minerals. *Geochim. Cosmochim. Acta*, 52: 1969-1981.

6 Spatial Distribution and Seasonal Patterns of Aquatic Emergent Plants in Southeast Brazil: Development of a Tool for Mass Balances.

Aguinaldo Nepomuceno Marques Jr.[1] and Ibra Touré[2]

[1] Universidade Federal Fluminense, Programa de Pós-Graduação em Biologia Marinha, Instituto de Biologia, Morro do Valonguinho s/nº, Niterói, R.J., 24.020-007, Brazil

[2] LGGA/IMG, Université de Nice-Sophia-Antipolis, Parc Valrose, Nice, 06108, France

Abstract

Spatial distribution, biomass, production and phenological data were obtained for two natural populations of the cat-tail *Typha dominguensis* in Lagoa de Maricá, a shallow sub-tropical lagoon. The area covered by *T. dominguensis* was estimated by satellite imagery and aerial photography. The values obtained by the two methods are concordant and close to 13 % of total wetland area. For both sites, plant growth proved to be highly influenced by water depth. Plants grew best at depths over 0.2 m depth. Productivity values (2,022 and 2,227 g m^{-2} y^{-1}) are in the world-wide range for the genus. A classification of the *Typha* swamps in different areas according to flooding exposure is proposed.

Introduction

In the coastal environment, the importance of emergent macrophytes is well established in the literature (Chapman, 1977). However, the mass balances in such environments can be impracticable due to the difficulty in proceeding in the usual methods of mapping as well as biomass measurements. Under these conditions, remote sensing is a remarkable tool for estimation of surfaces, seasonal variation and long term evolution. Wasserman et al. (1991; 1992) have shown that with a simple photograph taken at 9,000 m altitude, a very complete mass balance of a number of elements could be done.

The cat-tail (*Typha* sp.) is a common and dominant macrophyte in many wetlands of both tropical and temperate regions. High natural productivity and also a high ability to absorb nutrients have been frequently reported in studies of these perennial emergent aquatic plants (Hotchkiss and Dozier, 1949; Hill, 1979). This has led to investigations into the utilization of these wetland species for the treatment of industrial and domestic waste and the production of energy from their biomass (Von-Oertzen and Finlayson, 1984).

Studies on the role of these plant communities in temperate wetlands were largely carried out in comparison to tropical areas. In temperate wetlands *Typha* has a well defined growing season and during this period it retains large amounts of nutrients. Several factors, like nutrient availability (Boyd and Hess, 1970; Smart and Barko, 1983), salinity (Smart and Barko, 1983), pH (Hotchkiss and Dozier, 1949), light, competition ability and adequate water supply (Gopal and Sharma, 1982; Grace, 1985), oxygen deficiency (Boyd and Hess, 1970; Drew, 1983) and organic matter loads in bottom sediment (Barko and Smart, 1983) have been demonstrated to influence the distribution and growth of these plants. However, establishment of direct cause-effect relationships in natural populations are frequently difficult to establish, since several factors may act simultaneously in these populations. Particularly water depth and soil moisture have been reported to strongly influence *Typha* temperate species.

In the present chapter, data on natural populations of the emergent macrophyte *Typha dominguensis* Pers. in Lagoa de Maricá, a tropical Brazilian lagoon, are discussed together with estimates of the area of the lagoon margins that is covered by this macrophyte. The behavior of two *Typha* populations, growing on different kinds of sediments as related to flooding, are also compared.

Location and Sampling Sites

Maricá Lagoon (Fig. 1) is located at latitude S 23° and longitude W 42° on the southern coastline of Brazil. The internal lagoon of the Maricá-Guarapina System is composed of four inter connected shallow Holocene lagoons. Each lagoon is separated from the ocean by a beach barrier ("restinga"). The low-lying area adjacent to the lagoon is covered by dense stand of *Typha dominguensis* Pers.

Maricá Lagoon has an area of 17.9 km², a watershed of 240 km², a mean depth of 1 m, and the salinity ranges between 0.5 % - 0.6 % (Oliveira et al., 1955; Azevedo, 1984; Knoppers et al., 1989). Variations of water depth are mainly controlled by the inputs through Rio Mombuca, according to the amount of precipitation in the period. According to the classification of Koeppen, the climate is Awa, with an annual mean rainfall of 1,300 mm (Barbiére, 1981). High and low water phases characterize seasonal fluctuations of the water depth, respectively in the winter (June to August) and summer (December to February) months. Anthropogenic influence is characterized by fishing, agricultural activities in the drainage basin and some domestic wastewater discharges.

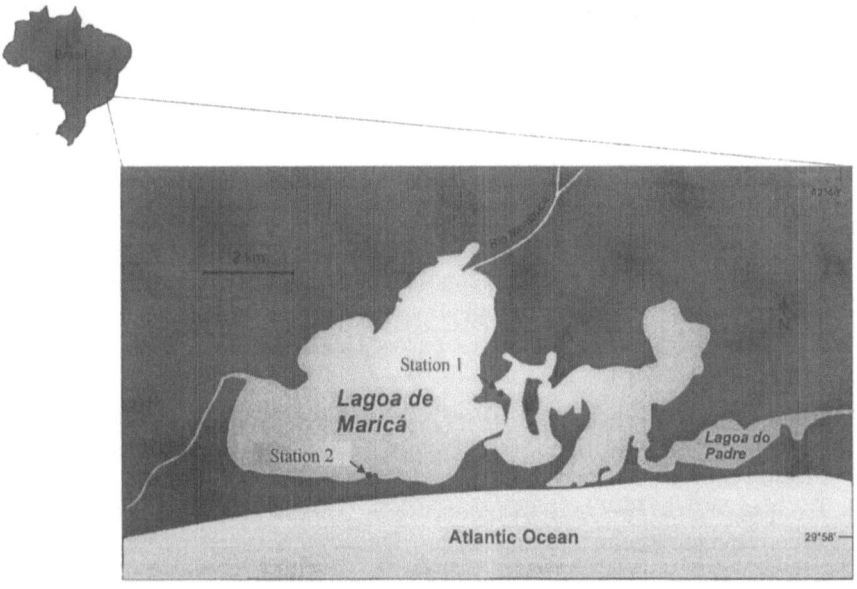

Fig. 1: Study area location and sampling sites

Fig. 2: Normalized Difference Vegetation Index (NDVI). Areas with high chlorophyll activity appear in light gray

The area covered by *Typha* was estimated by interpretation of a Landsat TM (172-73) image (date: July, 10, 1986). Numerical manipulation was performed on TM2, TM3 and TM4 spectral bands. Aerial photograph and ground observation were also checked. Image treatment involved the following steps:

1. Standard manipulations like contrast and trichrome color combination adjustments were carried out in order to characterize leading environmental features;
2. A general classification of the vegetation was made by calculating, the "Normalized Difference Vegetation Index" (NDVI, Tucker, 1979). By this analysis, areas with high chlorophyll activity appear in light gray. *Typha*, no *Typha* vegetation, water, urban area, exposed sand and soil were mapped by this index (Fig. 2);
3. Using a color classification derived from the NDVI as reference, sample portions were used to make a final analysis, "Maximum Likelihood Classification".
4. The area covered by *Typha* was estimated by masking and defining the limits of its corresponding color.

Ecological and general field data were obtained between July 1988 and July 1989. Plant material was sampled at bimonthly intervals in two locations within the lagoon margin: *Zacarias (S1)*, located in a confined section of the lagoon at the eastern margin; and *Restinga (S2)*, located in the southern margin of the lagoon, close to the beach barrier (Fig. 1).

The soil is silty and rich in organic matter at S1 compared to the S2 site. At this last site, sediments are sandy in the first top 5 cm. Sedimentation rates estimated for S1 ranges from 1.2 mm y^{-1} to 1.4 mm y^{-1} (Fernex et al., 1992). The chemical features of the cores to 30 cm depth in S1 give the average values presented in Table 1.

Comparing both sites, site S2 presents concentrations of Ca and Mg that are some 30 times less and Na and K, 6 times less than S1 (Marques, 1991).

Typha distribution in the lagoon and the area were estimated by aerial photograph and field observations.

Table 1: Chemical features of cores from Maricá Lagoon (Marques, 1991)

	Concentration
Ca (mg g^{-1})	1.7
Mg (mg g^{-1})	3.4
Na (mg g^{-1})	2.9
K (mg g^{-1})	0.7
Organic matter (%)	28.8
Silt-clay (%)	51.8

Biomass and phenological data of plants were collected from 1 m^2 quadrats along a lagoon-continental transect. Four quadrats (Q1, Q2, Q3 and Q4) were sampled at the S1 and three (Q1, Q2, Q3) at S2.

During the study, Q3 was sampled 13 times (monthly) and the other quadrats were sampled 4 times. At this quadrat (Q3), above-ground biomass was harvested, by cutting emerging leaves at the sediment-water interface. Samples of plant material were oven-dried (90°C), then weighed for dry weight estimations. For the others quadrats, above-ground biomass was estimated by a phenometric technique. In this case, 99 ramets were cut, measured and separated into fresh and dead leaves. The material was oven-dried, weighed and a phenometric relation *dry weight x height of the leaf* observed. Productivity in the two Q3 sampling points were estimated according to Smalley's technique (Hopkinson et al., 1980).

Results and Discussion

The overall quality of the satellite image analysis was considered satisfactory. The interpretation procedure used ("Maximum Likelihood Classification") included the classification of 93.4% of the pixels (Table 2) and the *Typha* area was well defined. However, some pixels with the same signature as *Typha* (light-gray) are also observed in continental areas. These "artifact pixels" (3.6 ha in the urban area and 0.09 ha in the forest area) were substracted from the total *Typha* area.

Table 2: Matrix of Maximum Likelihood Classification

Class name	Class number	Area (ha)	Class Performance				
			Water 1	*Typha* 2	Urban 3	Forest 4	SP/SN 5
Water	1	13.59	13.41	0.00	0.18	0.00	0.00
Typha	2	3.87	0.00	3.87	0.00	0.00	0.00
Urban	3	19.62	0.00	3.60	15.66	0.00	0.36
Forest	4	33.57	0.00	0.09	0.63	32.85	0.00
SP/SN	5	3.42	0.00	0.00	0.00	0.00	3.42
	Total	74.07	13.41	7.56	16.47	32.85	3.78

Overall performance (769/823) = 93.4 %   SP/SN – unidentified pixels

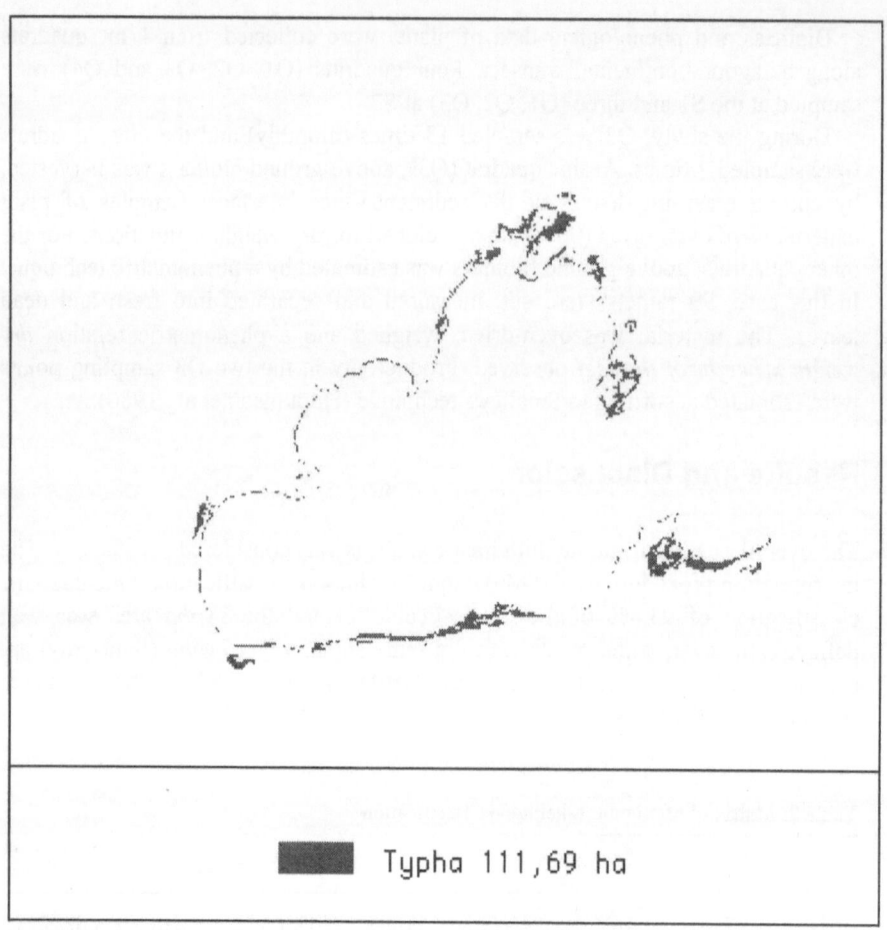

Fig. 3: Cartography of *Typha* community at the Maricá margin lagoon (source: Landsat TM images (172-73 ; July 10, 1986)

From the mapping treatment it could be calculated that *Typha* occupies 111.69 ha of the lagoon area and its growth is more intense in two sectors: at the mouth of the Rio Mombuca and at the S1 site (Fig. 3). Nevertheless, these data represent only the high water phase of the lagoon since the image was taken in July, a "high water level month". Since the phenological state of plants changes during the year, data analysis concerning the low water phase would be required. This would allow the determination of the annual expansion-retraction rates of *Typha* populations in the lagoon and also establish a general pattern of spatial distribution. This supplementary information is necessary for enhanced planning and management of the lagoon.

Table 3: Average ramet density, flowering incidence and above-ground biomass against water depth gradient in *S1* and *S2* sites

Station	Quadrat	Average water depth (cm)	Number of flowering ramets	Leaf ramet density (ram. m^{-2})	Dead ramet density (ram. m^{-2})	Average biomass (g m^{-2})
	Q1	20 (15-30)	0	19 ± 7	0	686 ± 440
S1	Q2	16 (0-30)	9	20 ± 8	4 ± 5	1061 ± 481
(Zacarias)	Q3	5 (1-10)	0	19 ± 14	43 ± 20	498 ± 388
	Q4	0	0	13 ± 6	4 ± 4	69 ± 35
	Q1	24 (0-49	3	11 ± 6	0	104 ± 124
S2	Q2	8 (0-22)	10	15 ± 6	6 ± 4	1114 ± 489
(Restinga)	Q3	0	0	6 ± 1	3 ± 1	40 ± 24

Average density, biomass and flowering for the measured water-depths are presented in Table 3. Higher densities and lower biomass values were frequently observed in areas of low water depth. Such behavior was also observed by other authors in this same lagoon system (Couto, 1989). In scarcely flooded and unflooded stands (S1 - Q4; S2 - Q3), the growth of the plants and consequently the biomass and the plant density were very low. The incidence of flowering in *Typha* was generally low in Maricá and restricted to the zones with constant flooding. This may suggest that the mobilization of the resources to maintain the populations is vegetative reproduction (Dykyjová and Kvet, 1978). Height of plants has also been shown to be influenced by water depth.

Flooding state normally affects the population phenotype of wetland plants (Gopal and Sharma, 1982). Results of a study on the competition abilities of *T. latifolia* and *T. dominguensis* (Grace, 1985), reveals that although the second plant can be observed in stands with up to 0.8 m - 0.9 m depth, in deeper areas, the former dominate. Our results show that the sub-tropical macrophyte *T. dominguensis* grows well in shallow waters (0.2 m), which could be an adaptation to the absence of other competing species. The low flowering observed suggests that *Typha* could be exposed to a physical stress like temperature (Grace, 1985).

Even though soils are different in both locations, water depth seems also to affect the height and density of plants. Plants showed a general trend increasing height and decreasing density with increasing water depth (Fig. 4). This pattern was also observed for the same species in Guarapina Lagoon (Couto, 1989), the external lagoon of the Maricá-Guarapina system which is connected to the sea.

The temporal variation of biomass obtained by phenometric estimation and by weight is plotted in Fig. 5 and shows significantly similar trends (P < 0.01, t test).

The equation of the phenometric curve, adjusted for the minimum quadrat method, was

$$y = 2.86 \times 10^{-1} + 5.25 \times 10^{-1} \times \ln x$$

where x = height (m) and
y = biomass (g).

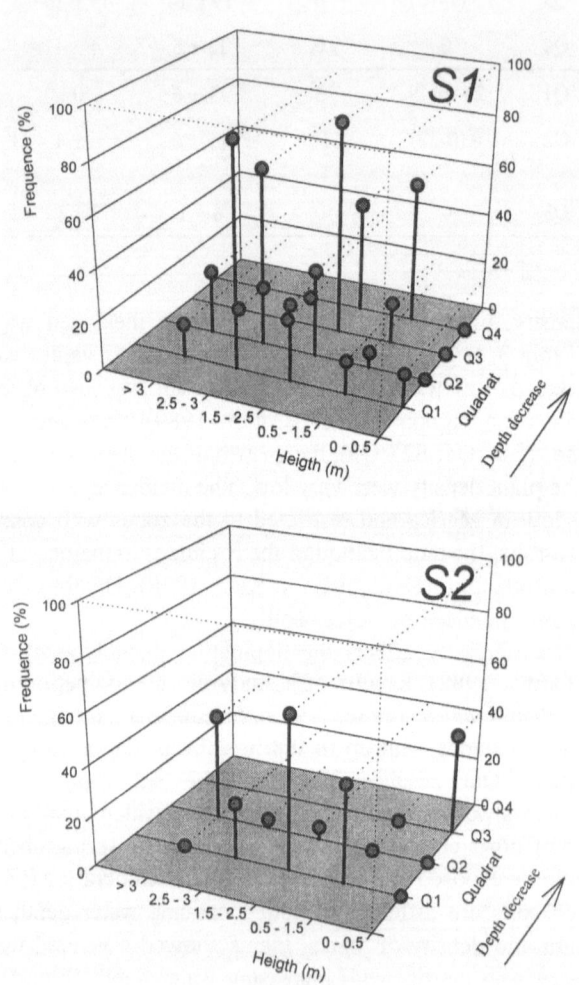

Fig. 4: Frequency distribution of plant height against water depth in sites S1 and S2

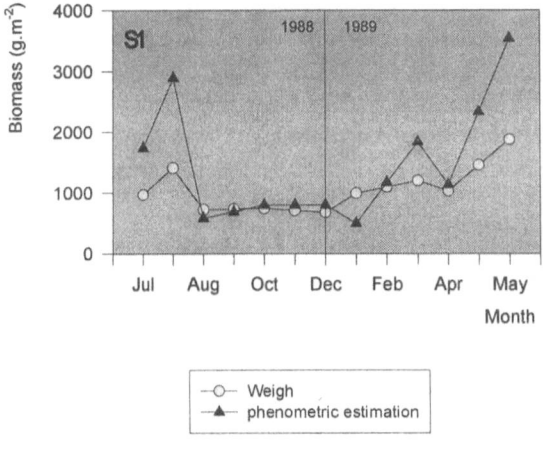

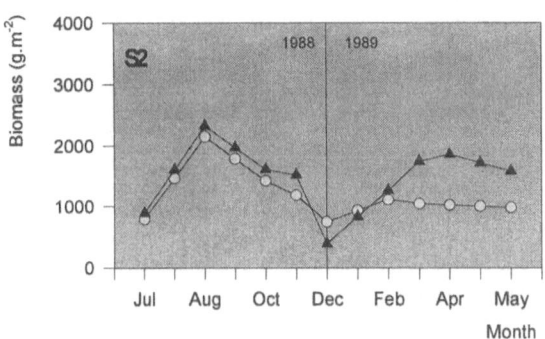

Fig. 5: Temporal variation of above ground biomass of Q3 stands in the sites S1 and S2

Our results show that the phenometric technique can be used for further studies on the population dynamics and chemistry of *T. dominguensis* stands. The temporal variations of water flooding seem to strongly affect above-ground biomass. The highest values were obtained in the months of maximum water depths (July to September) in both sites.

Production values were similar in both stations: station S1 yielded 2,022 g m^{-2} y^{-1} and S2 yielded 2,227 g m^{-2} y^{-1}. These values represent normal production rates when compared to other aquatic macrophytes (200 to 10,000 g m^{-2} y^{-1}; Westlake, 1965) and to the same species in Lagoa de Guarapina (Couto, 1989). The observed reduced differences on the production values in both sites (S1 and S2) suggest that organic matter is efficiently recycled within these swamps. Data on nutrient release during leaf decomposition, concentration factors (Marques, 1991) and ^{13}C data on the lagoon sediments (Knoppers et al., 1991) support this interpretation.

Our results indicate that the phenotype and spatial distribution of *T. dominguensis* in Maricá Lagoon may be strongly affected by water depth. Variation of the phenological parameters allows Maricá swamps to be divided into sub-areas according to flood exposure of the populations. In *extremely and moderately flooded areas,* biomass and height are high and density is low. In *weakly flooded areas*, although plant density is high, the biomass and height are very low. Finally, in *virtually dry areas*, biomass, height and density values are the very low.

Acknowledgments

We thank Dr. Bastian Adrian Knoppers for dedicated technical assistance and Drs. François Fernex for manuscript review and Robert Campredon for providing satellite images. Thanks are also due to Mr. Ricardo Pollery and Mrs. Lucia Beatriz F. Oliveira for assistance in sample collection and illustrations, respectively. Financial support for this work was provided by CAPES (Coordenação de Aperfeiçoamento de Pessoal de Ensino Superior) and FAPERJ (Fundação Amparo e Pesquisa do Estado do Rio de Janeiro).

References

Azevedo, L.S.P., (1984) Considerações geoquímicas das lagunas do litoral leste do Estado do Rio de Janeiro. In: Lacerda, L.D.; Cerqueira, A.R. and Turcq, B. (eds.). *Restingas: Origem, estrutura e processos*; Niterói, CEUFF. p.123-135.

Barbiére, E.B. (1981). O fator climático nos sistemas territoriais de recreação. *Rev. Bras. Geogr.* 43(2):145-265.

Barko, J.W. and Smart, R.M. (1983). Effects of organic matter additions to sediment on the growth of aquatic plants. *J. Ecology.* 71:161-175.

Boyd, C.E. and Hess, L. (1970). Factors influencing shoot production and mineral nutrient levels in *Typha latifolia. Ecology*, 51:296-300.

Chapman, V.J. (1977). *Wet Coastal Ecosystems*. Elsevier Sci. Pub. Co., Amsterdam. 428 p.

Couto, E.G.C. (1989). *Produção, decomposição e composição química de Typha dominguensis Pers (THYPHACEAE) no Sistema Lagunar de Guarapina, RJ.* Master's dissertation, Dept. Geoquímica, Universidade Federal Fluminense, 150p.

Drew, M.C. (1983). Plant injuries and adaptation to oxygen deficiency in root environment: a review. *Plant and Soil.* 75:179-199.

Dykyjová, D. and Kvet, J. (1978). *Pond littoral ecosystems*. Springer-Verlag, Heidelberg.

Fernex, F.F., Bernat, M., Ballestra, S., Fernandes, L.V. and Marques Jr., A.N. (1992). Ammonification rates and 210Pb in sediments from a lagoon under a wet tropical climate: Maricá, Rio de Janeiro state, Brazil. *Hydrobiologia* .242:69-76.

Gopal, J.B. and Sharma, K.P. (1982). Studies of wetlands in India with emphasis on structure, primary production and management. *Aquatic Botany*, 12:81-91.

Grace, J.B. (1985). Juvenile vs. adult competitive abilities in plants: size-dependence in cat-tails (*Typha*). *Ecology*, 66(5):1630-1638.

Hill,B.H. (1979). Uptake and release of nutrients by aquatic macrophytes. *Aquatic Botany*, 7:187-193.

Hopkinson, C.S., Gosselink, J.G. and Parroundo, P.J. (1980). Production of coastal Louisiana marsh plants calculated from phenometric techniques. *Ecology*, 61(5):1091-1098.

Hotchkiss N. and Dozier, H.L. (1949). Taxonomy and distribution of N. American cat-tails. *Am. Mdl. Nat.* 41:237-254.

Knoppers B.A., Machado, E.C., Moreira, P.F. and Turcq, B. (1989). A physical and biogeochemical description of Lagoa de Guarapina, a subtropical Brazilian lagoon. *Internat. Symp. Global Changes in S. America during the Quatern.*; São Paulo:275-281

Knoppers, B. A., Erlenkeuser, H., Pollehne, F. (1991). A origem da matéria orgânica depositada no sedimento da lagoa de Guarapina, RJ. *In: Congresso Brasileiro de Geoquímica*, São Paulo, vol.1:393-397.

Marques Jr., A.N. (1991). *Ecologia e dinâmica de elementos nutrientes em dois brejos de Typha dominguensis Pers (THYPHACEAE) na Laguna de Maricá, RJ.* Master's dissertation, Dept. Geoquímica, Universidade Federal Fluminense, 114p.

Oliveira, L., Nascimento, R. and Krau, L. (1955). Observações biogeográficas sobre a Lagoa de Maricá. *Mem. Inst. Oswaldo Cruz.* 53(2/4):171-225.

Smart, R.M. and Barko, J.W. (1983). Influence of sediment salinity and nutrients on the physiological ecology of selected salt marsh plants. *Est. Coast. Mar. Sci.*, 7:487-495.

Tucker, C.J. (1979). Red and photographic infrared linear combinations for monitoring vegetation. *Rem. Sens. Environm.*, 8:127-150.

Von Oertzen I. and Finlayson, C.M. (1984). Wastewater treatment with aquatic plants: ecotype differentiation of *Typha dominguensis*, seedlings. *Environ. Pollut.*, Ser. A, 35:259-269.

Wasserman, J.C., J.C. Dumon et C. Latouche (1991). Importance des zostères (*Zostera noltii* Hornemann) dans le bilan des métaux lourds du Bassin d'Arcachon. *Vie et Milieu*, 41(2/3):81-86.

Wasserman, J.C., J.C. Dumon et C. Latouche (1992). Le bilan de 18 éléments traces et 7 éléments majeurs dans un environnement peuplé par des Zostères (*Zostera noltii* Hornemann). *Vie et Milieu*, 42(1):15-20.

Westlake, D.F. (1965). Some basic data for investigation of productivity of aquatic macrophytes. *Mem. Inst. Ital. Ichrobiol.*, 18(suppl.):229-248.

7 Instrumental Multi-Element Analysis in Plant Materials: A Modern Method in Environmental Chemistry and Tropical Systems Research.[1]

Bernd Markert

Zittau International Hoschschul, Zittau, Germany

Abstract

This chapter describes possibilities of using instrumental multi-element methods to answer various ecological questions. Emphasis is placed on the ecosystem-related approach and thus on comparison of various ecosystems and their compartments. The basis for this is a project by the International Union of Biological Sciences (IUBS) aimed at establishing "element concentration catalogues in ecosystems" (ECCEs). In principle, the intention is to collect data on the world-wide distribution of individual elements and element species in various ecosystems. Synthetic reference systems (reference plant, reference freshwater, etc.) in the form of chemical fingerprints can provide important aids for the chemical characterization of many different environmental specimens. From the point of view of both toxicology and nutrient physiology such a characterization of ecosystems is expected to provide important information on the concentrations, effects and reactions of individual elements in different systems. Interest has so far been focused mainly on data from the northern hemisphere, but from the biogeochemical viewpoint the scope urgently needs to be extended to cover tropical and subtropical systems.

Introduction

In the early 1980s the application of instrumental multi-element techniques to geological, ecological and medical questions opened up fields of scientific work which, as they continued to develop, provided a multitude of new insights in the sphere of modern research (Bowen, 1979; Sansoni, 1985; Markert, 1987; Zeisler et al., 1988; Lieth and Markert, 1988, 1990; Kovacs et al., 1990, 1993; Markert, 1993a). The field of "biogeochemistry", for example, could not have been opened up without efficient methods of instrumental analysis (Adriano, 1992). At the centre of interest are questions of nutrient physiology and ecotoxicology as well as problems of pollutant minimization and disposal (Förstner, 1990; Frankenberger and Karlson, 1992).

---

[1] Also published in *Tecnologia Ambiental*, CETEM No. 8, 1995, 33 pp.

Analysis of the inorganic composition of our environment has special priority, for at the ecotoxicological level elements such as lead, cadmium, arsenic and mercury have played an important role for years (Nriagu and Pacyna, 1988). Moreover, in respect to physiology of metabolism, many chemical elements are known to be of essential significance especially at the trace and ultra-trace level (Baker, 1981; Salomons and Förstner, 1984; Thornton, 1990; Davies, 1992; Lepp, 1992; Brooks, 1993; Ernst, 1993; Verrleis, 1993). This has already been well investigated for elements such as cobalt, nickel, chromium, molybdenum and selenium. The continuing development of instrumental multi-element techniques, resulting especially in lower detection levels, better reproducibility and greater accuracy of the measured signal, now makes it possible to determine the concentration of almost all the 88 naturally occurring elements in any matrix.

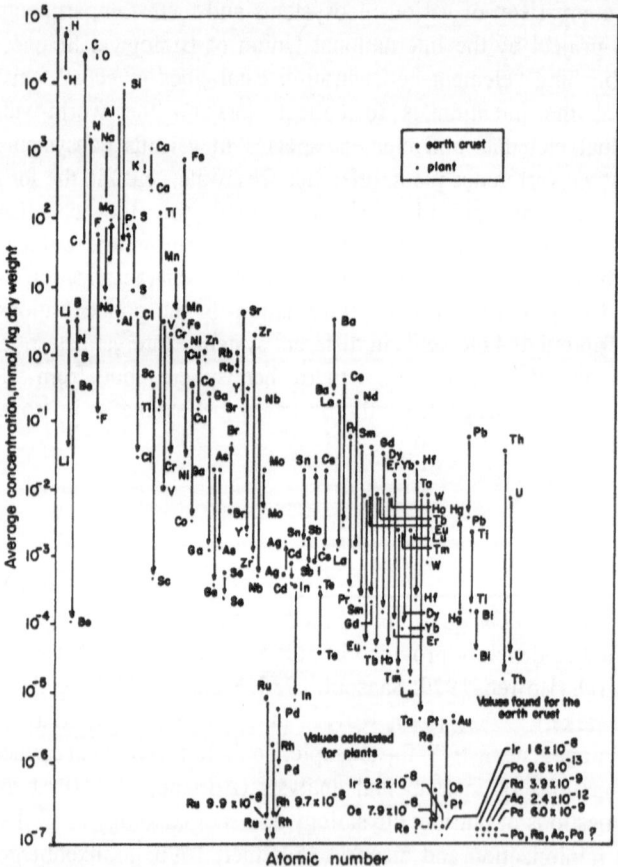

Fig. 1: Average concentrations (in mmol kg⁻¹ dry weight) of all elements (except the noble gases) occurring naturally in the earth's crust and in plants, plotted according to increasing atomic numbers

The earth's crust may be regarded as a natural reservoir of all the chemical elements of the biosphere (Kabata Pendias and Pendias, 1992). But a remarkable feature is that 99% of the total mass of the earth's crust consists of only 8 of the 88 naturally occurring elements. It is made up of 46.4% oxygen, 28.15% silicon, 8.32% aluminium, 5.63% iron, 4.14% calcium, 2.36% sodium, 2.33% magnesium and 2.09% potassium. Of the 8 most common elements of the earth's crust, oxygen is the only non-metal. The remaining 80 elements in the periodic table account for less than 1% of the structure (Markert, 1992 and 1993a).

The greater part of the fresh weight of living plant organs, i.e. those with an active metabolism, consists on average of 85-90% water. The solid matter of the plant body consists mainly of the following elements: carbon (44.5%); oxygen (42.5%); hydrogen (6.5%); nitrogen (2.5%); phosphorus (0.2%); sulphur (0.3%) and the alkaline and alkaline-earth metals potassium (1.9%), calcium (1.0%) and magnesium (0.2%).

If the average element concentrations in the earth's crust and the dry solids of plants are plotted on a graph against increasing atomic mass as in Fig. 1, the initial result is a confused picture; but this does show once more that the bulk of organic life consists chiefly of non-metals, in contrast to the earth's crust. A more interesting picture is presented in Fig. 2, where the molar masses of individual element concentrations in the plant body have been plotted against the molar masses of average single element concentrations in the earth's crust. Most of the element correlations are to be found along the bisector of the angle between the two concentration axes. This shows on the one hand that the extraterrestrial origins of the elements are reflected both in the earth's crust and in the living biomass, and on the other hand that some elements have acquired a special significance in the process of evolution and especially in the biological life of today. These are the elements shown in group I of Fig. 2, namely C, H, O, K, Ca, Si, Na, Fe, P, S, N, Mn, B, Zn, Cu, Ni, Cr, Co, Cl, V, F, Rb, Sr, Ba, Ti and Al. All but the last five have an essential function at least in certain groups of organisms; that is, they are necessary for the life of these organisms. The last five elements - Rb, Sr, Ba, Ti and Al - are also assumed to have an essential function, but we are unable to define it precisely at present. Some of the elements in group II are known to have essential functions (I, Mo, Se and Sn), but they are characterized by a high degree of toxicity, in most cases even at low concentrations. They include most importantly the heavy metals Pb, Cd, As, Ti, Hg etc. Group III consist of elements that have not been able, in the course of evolution, to break away from their passive role in the earth's crust and integrate themselves as active components of organic life. The most important members of this group are the lanthanides and the platinum metals.

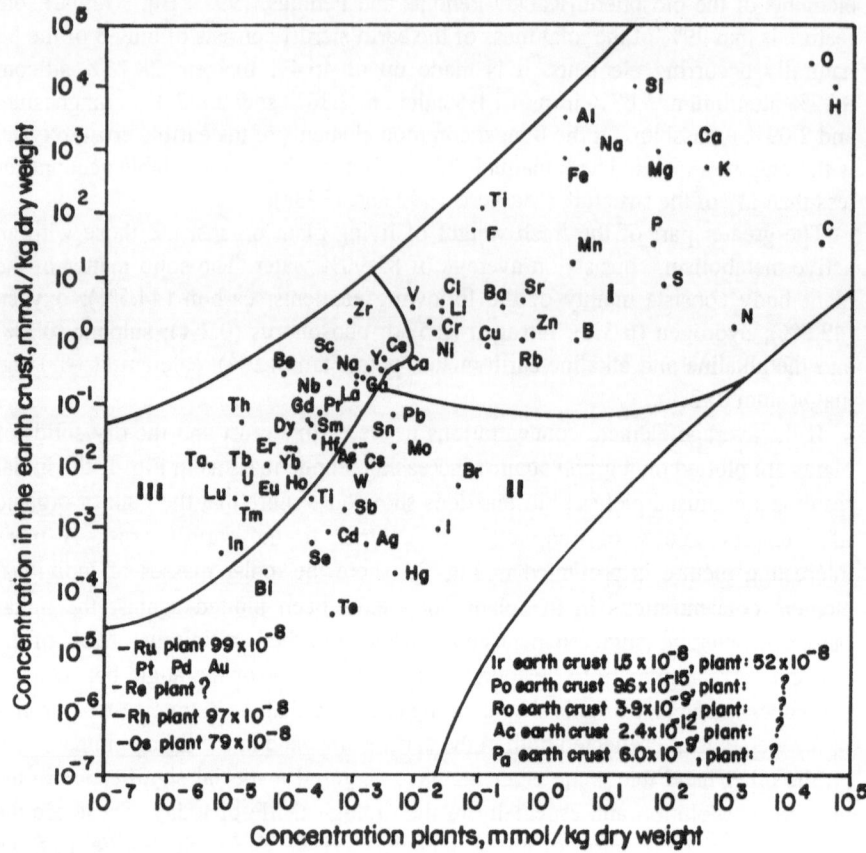

Fig. 2: Average concentrations of individual chemical elements in plants, plotted against the average element concentration in the earth's crust. All figures in mmol kg[-1] dry weight

This concept of a selection of chemical elements enabling organic life, based on the evolutive development, led to the establishment of a first Biological System of Elements (BSE) which is shown in Fig. 3. This Biological System of Elements, originally made out only for glycophytic terrestrial plant, draws on individual parameters relating to the essentiallity and mode of uptake of individual elements and their ability to correlate with each other (Markert, 1994c). It is assumed with the BSE that the elements hydrogen and sodium played a very special role in the development of terrestrial life. This can be seem from their "extraordinary" position outside the periodicity of the BSE. As a structural element of carbohydrates, fats and proteins, an electron donor in redox and thus respiration processes and as a pH regulator, hydrogen may possibly be the chemical element that played the decisive part in the development of aerobic life. Similar element constellations have now been found for halophytes, so that the establishment of a comparable BSE may be expected for these too.

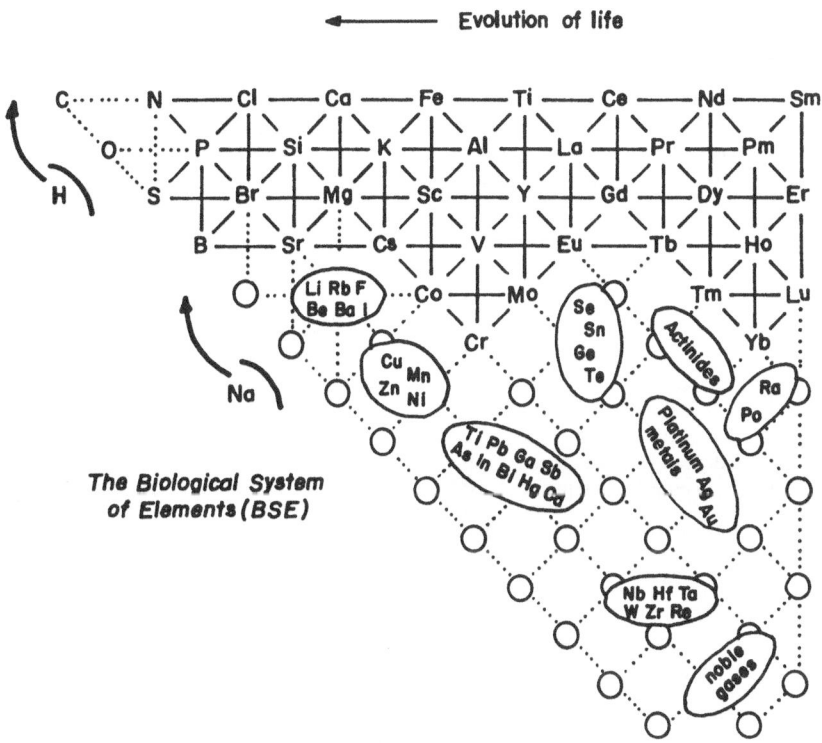

Fig. 3: The Biological System of the Elements (BSE) compiled from data on correlation analysis, physiological function of the individual elements in the living organism, evolutionary development out of the inorganic environment and with respect to the form of their uptake by the plant organism as a neutral molecule or a charged ion. The elements H and Na exert various functions in the biological system so that they are definitively fixed. The ringed elements can at present only be summarized as groups of elements with a similar physiological function, since there is a lack of correlation data or these data are too imprecise (from Markert, 1994c)

The work described above was only made possible by the advent of a method of instrumental analysis that yielded accurate and reproducible results at all the stages of analysis (Fig. 4). That this progress towards accurate results in environmental analysis was not easy - nor is it still! - is illustrated by the words of an outstanding analytical chemist, the late Professor Dr. H.W. Nurnberg, in 1984:

It is evident that competent and efficient analytical chemistry is a key factor in accurate and reliable assessment of the environmental load, especially as a prerequisite for sound ecotoxicology. It is the analyst's task to conduct it with skill and inspiration. But in this connection the enthusiasm for environmental issues that is pervading many branches of research is also fraught with special risks. These are the ignorance and naiveté so often observed in the face of the difficulties to be overcome when carrying out the necessary highly skilled analysis of inorganic environmental chemicals. Many institutions without analytical expertise are starting to buy expensive instruments and indulge in environmental analysis. So we have to expect that in the next few years a glut of false results will be produced that will swell the amount of inaccurate data that already exists. Since the accurate results will then be in the minority there is a danger that for the time being they will simply be drowned in this flood of incorrect data. Severe setbacks of an ecological, ecochemical, social and economic nature may well be the grave long-term results of such an abortive development.

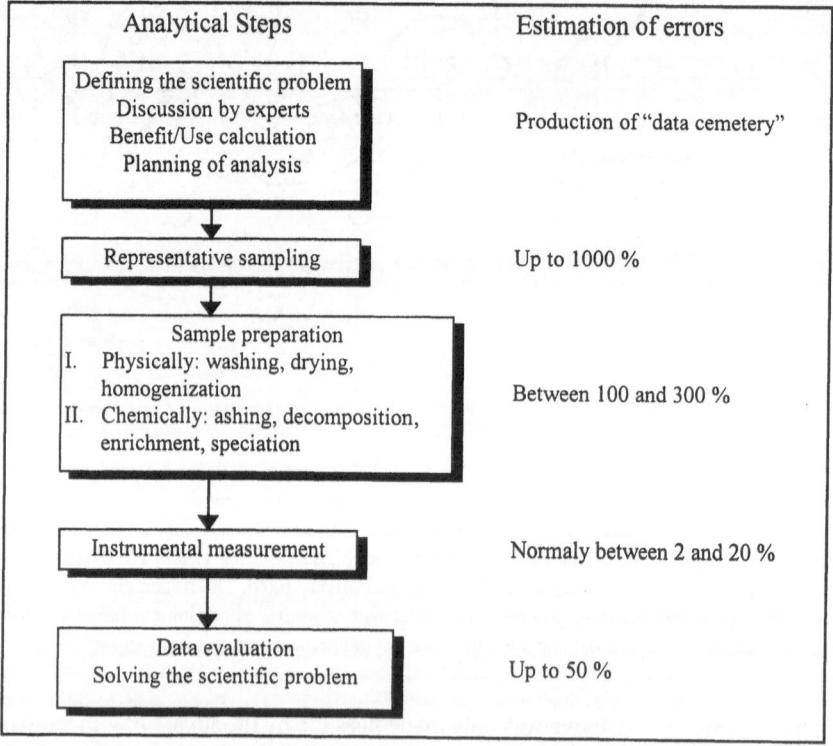

Fig. 4: Simplified analytical flow chart for chemical multi-element analysis (from Markert, 1993b). The full flow chart can be found in Markert (1993a)

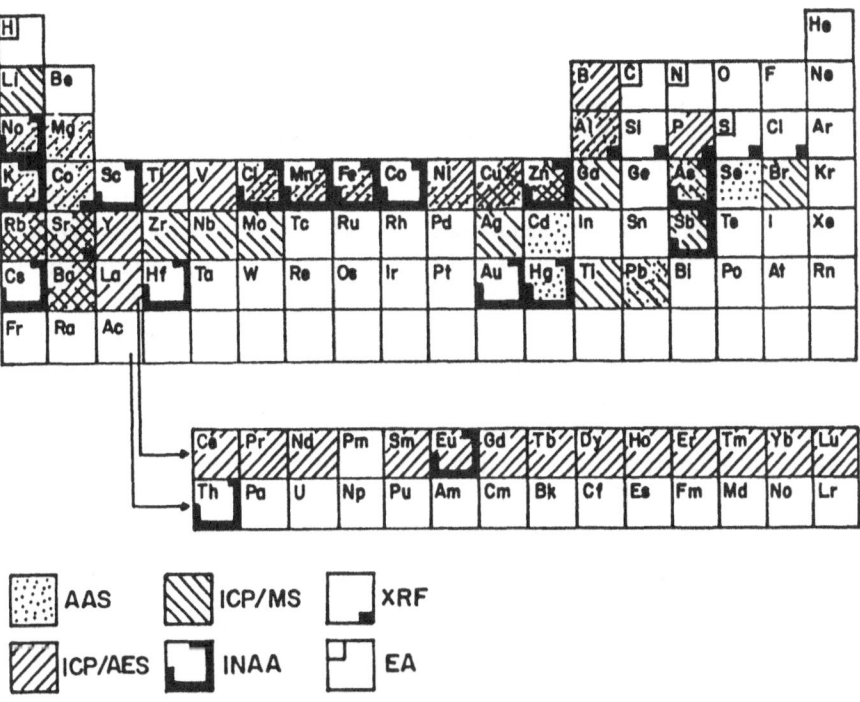

AAS   ICP/MS   XRF

ICP/AES   INAA   EA

Fig. 5: Matrix-dependent detection sensitivity of different instrumental measuring procedures. Uncontaminated plant material served as the matrix (from Markert, 1993b). Total reflection fluorescence analysis (TRFA) was not included in this overview. However, for quite a number of elements it already achieves detection limits such as those of ICP/MS

It is true that there has been a meteoric increase in the amount of equipment acquired in recent years. Nowadays routine analytical methods are available for quite a number of elements, as Fig. 5 shows. In many cases the estimation of errors from sample preparation to instrumental measurement (Fig. 4) can now be carried out satisfactorily by using certified standard reference materials or independent methods, although there is still a great need for new reference materials for quite a number of individual elements and matrices. A much more difficult task is to develop satisfactory methods of sampling and integrate these in an overall concept (Pendias et al., 1991; Kabata- Keune et al., 1991; Roth, 1992; Wilken, 1992; Djinguva and Ruleff, 1993; Jayasekera, 1993; Wagner, 1993; Steinnes, 1993; Ernst, 1994; Fränzel, 1994; Markert, 1994).

The IUBS / ECCE Project

In many cases investigations into element patterns and distributions only make sense at ecosystem level, as Fig. 6 aims to show. In an ecosystem the paths and whereabouts of the elements may be influenced in a specific manner by organism activity, for example by selective uptake and accumulation. Elements that occur together may have a positive and/or negative effect on their transportation or accumulation in the organism (Markert, 1992; 1993a). It may be possible to explain antagonistic element behaviour by competition for the same binding site in the organism. Interactions may be assumed to exist between organism activity and flow rates and flow patterns of the elements through the various components and compartments of an ecosystem which can only be interpreted accurately if all the elements are included in the analysis (Lieth and Markert, 1988; 1990). In the case of most elements and ecosystems the nature and extent of these mutual influences doubtless depends on abiotic factors such as the weather. Flow rates and flow patterns therefore vary with these factors. This must always be taken into account when interpreting ecochemical data. It means that the results are only characteristic of the constellation of factors under which sampling was carried out (Fig. 4).

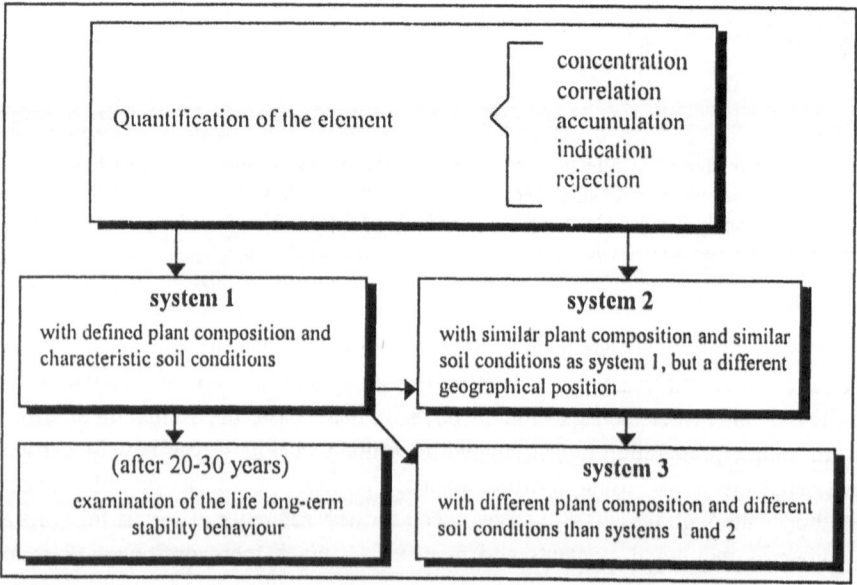

Fig. 6: Comparable aspects of instrumental multi-element analysis in ecosystems. Comparability of single system data will be achieved by harmonizing sampling procedure, sample preparation, instrumental measurement and data evaluation (from Markert and Wtorowa, 1992)

The following are possible approaches to interpreting the data, depending on which parts of an ecosystem are investigated, how often sampling takes place and what opportunities there are of comparison with other ecosystems of the same kind (Fig. 6).

1. If the system components and compartments analysed are directly successive stations in the flow of substances, conclusions on the following can be drawn from the results of a single multi-element analysis:

   - The concentrations at which individual elements occur
   - Whether and to what extent individual elements occur in correlation in the samples analysed
   - Whether the samples show accumulative, indicative or rejective behaviour for certain elements

2. If a multi-element analysis is carried out for a certain constellation of factors in each of several ecosystems of the same kind (characterized, for example, by similar soil conditions and vegetation) which are subject to different inputs of elements, a comparison of the data gives a first indication of how various similar ecosystems may react to different inputs of substances. Not only do mere changes in concentration have to be taken into account; just as important are modifications of the accumulative behaviour of certain plant species, shifts in element correlations etc. Traditionally, special attention is given to elements known to be of ecotoxicological significance.

3. When compared with systems that have different vegetation and different soil conditions, the results of multi-element analyses for a constellation of factors recognized to be characteristic of a particular ecosystem form a reliable basis (possibly with different data) for an analysis of cause and effect (Markert, 1992; 1993a). This provides information on:

   - Whether, to what extent and under what conditions the element correlations in the individual plant samples change
   - Whether, to what extent and under what conditions changes in accumulative behaviour can be traced back directly to different soil conditions or plant-specific element patterns
   - Under what constellations of factors and to what extent changes in concentrations are observed

   In particular such a comparison may include systems in different climatic zones. Comparison between tropical and European virgin forests, or between South American secondary eucalyptus forests and Central European spruce monocultures, are likely to produce interesting results directly relevant to the market economy.

4. If we compare the results of multi-element analyses conducted over a lengthy period on similar ecosystems differing only in respect of their input of substances, we may be able to draw conclusions concerning the long-term stability of the systems. Investigations of this kind are being carried out in the

context of environmental specimen bank projects, especially in Germany and the USA (Rossbach et al., 1992).

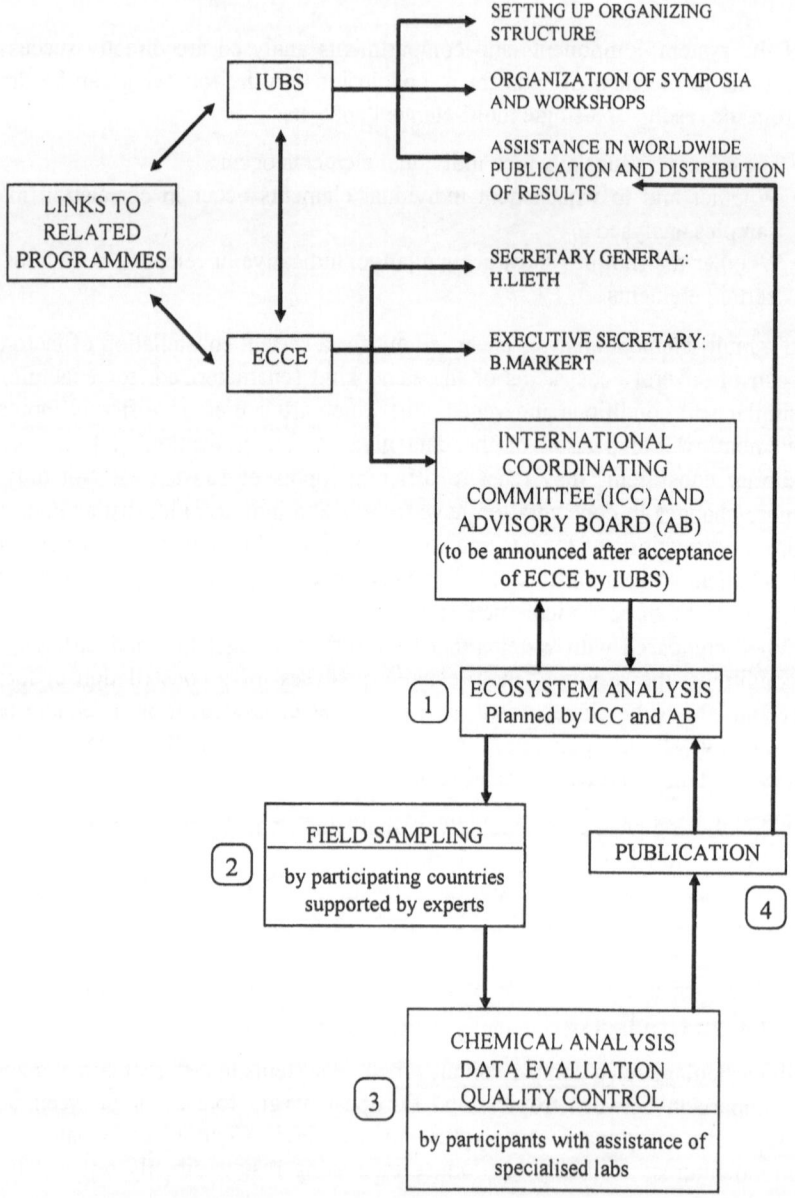

Fig. 7: Structure of the administrative and scientific part of the Element Concentration Catalogue in Ecosystems project (ECCE) supported by the International Union of Biological Sciences (IUBS). From Markert (1994a)

The test plan shown in Fig. 6 shows that multi-element analyses of ecosystems must not be regarded as synonymous with mere measuring of element concentrations. The validity and interpretability of concentration data within a series of measurements should always be seen in relation to the specific question to be answered and the particular constellation of parameters. Since August 1991 the International Union of Biological Sciences (IUBS) has been supporting a world-wide multi-element programme with the title "Element Concentration Catalogue in Ecosystems" (ECCE). The organizational structure of IUBS/ECCE is shown in Fig. 7. The objectives of the long-term programme may be formulated as follows:

1. Biological objectives:

- To establish "element concentration catalogues" of the most common plant species in a forest ecosystem in the form of chemical fingerprints (concentration aspect)
- To establish basic data for a natural ecosystem and determine "natural" baseline concentrations, preferably of elements that have rarely been investigated in the past (concentration aspect)
- To discover inter-element relationships and other ways in which plant organisms conform to definite laws (correlation aspect)
- To determine the ability of different plant species to select single elements or groups of elements in the periodic table and to discover possible accumulation properties (accumulation or rejection aspect)
- To use plant systems for monitoring pollutants in ecosystems over lengthy periods of time (indication aspect)

2. Analytical objectives:

- To improve representative sampling and developed individual environmental sampling programmes
- To optimize sample preparation techniques (especially washing, homogenizing, drying, decomposition and ashing)
- To use and compare different multi-element methods suitable for chemical characterization of environmental samples in respect of the reproducibility, accuracy and concentration dependence of the analytical data

Presentation and Evaluation of the Data

Synthetic Reference Systems

One of the problems of inorganic chemical analysis that receives far too little attention is the manner in which the results are presented. The difficulties lie in the large amount of individual data and the distribution of these data over concentration ranges that are usually between 10^6 and 10^{-9} mg kg^{-1}. In the majority of cases the data are presented in the form of complicated tables, and these generally only permit "element to element" comparisons. An example is Table 1, that shows a multi-element comparison of seven plant species in two forest ecosystems in Germany and Russia.

A first approach to establishing a reference system for plants to which all the multi-element data produced world-wide could in future be related was the development of a "reference plant" on the lines of the "reference man" created back in the 1960s. The average chemical composition of this plant is shown in Fig. 8. In establishing this reference plant it did not very much matter whether it was a moss, a fern, a flowering plant or a tree. The aim was rather to specify its inorganic chemical composition in order to enable a comparison of multi-element data acquired by chemical analysis from all manner of investigations conducted on plants. Like the data for the element composition of reference man, the data for the reference plant shown in Fig. 8 will often have to be modified. One reason is that it will increase enormously in the future as a result of further multi-element analysis; another are that the data shown here is not adequately backed up by statistics.

Table 1: Element content of plants investigated in the Kalinin area in comparison with similar plant species taken in a forest ecosystem in Germany (Grasmoor near Osnabrück). Data are given, unless otherwise specified, in mg kg^{-1} dry weight. (From Markert and Wtorowa, 1992). VacVit: *Vaccinium vitis-idaea*; VacMyr: *Vaccinium myrtillus*; PolCom: *Polytrichum commune*; PolFor: *Polytrichum formosum*; SphSpe: different *Sphagnum* species; HypPhy: *Hypogymnum physodes*; PicAbi: *Picea abies*; PinSyl: *Pinus sylvestris*

	VacVit		VacMyr		PolCom	PolFor	SphSpe		HypPhy	PicAbi	PinSyl
	USSR	FRG	USSR	FRG	USSR	FRG	USSR	FRG	USSR	USSR	FRG
Ac	-	-				-		-	-	-	-
Ag	-	-	-	0.14	-	0.302	-	0.3	-	-	0.174
Al	130.4	87	273.6	158	305	306	214.6	482	638	99.5	142
As	0.5	0.108	0.090	0.22	0.095	0.198	0.965	0.23	1.31	0.090	0.214
Au										0.0026	0.004
B	-	14	-	21	-	0.2	-	0.2	-	-	20
Ba	34.8	31	60	29	7.8	4.5	10	5.4	23	15	1.1
Be	-	-	-	-	-	-	-	-	-	-	-
Bi	-	-	-	-	-	-	-	-	-	-	-
Br	-	0.91	1.54	3.61	2.5	3.16	2.27	6.09	11.5	1.54	2.91
C	51.7%	50.2%	45.9%	47.8%	44.7%	44.8%	43.6%	42.6%	46%	50.1%	51.4%
Ca	3240	4900	9900	6500	2710	1000	194	550	810	6100	1200
Cd	0.085	-	0.185	-	0.205	-	0.31	-	0.245	0.32	
Ce	0.11	0.39	0.217	0.29	0.26	0.51	0.442	141	1.865	0.208	0.31
Cl	-	100	-	1530	-	670	-	670	-	-	431
Co	0.185	0.037	0.083	0.06	0.289	0.300	0.241	0.8	0.355	0.083	0.124
Cr	2.0	0.51	-	0.95	1	3.67	0.72	6.4	2.23	0.73	1.72
Cs	0.147	0.371	0.215	0.317	0.36	0.33	0.218	0.74	0.185	0.099	0.098
Cu	9.07	5.5	6.14	4.9	16.48	12.7	21	6.65	10.9	4.0	4.28
Dy	0.008	0.030	0.014	0.026	0.017	0.041	0.042	0.099	0.10	0.011	0.024
Er	0.005	0.020	0.008	0.017	0.010	0.028	0.025	0.066	0.062	0.007	0.016
Eu	0.002	0.008	0.006	0.007	0.0085	0.008	0.010	0.013	0.030	0.007	0.006
F	-	-	-	-	-	-	-	-	-	-	-
Fe	91.8	82	111.4	130	165	184	207	364	664	79.32	118
Ga	0.31	0.054	0.11	0.14	-	0.14	0.42	0.12	0.405	0.42	0.117

Table 1: Continued

	VacVit		VacMyr		PolCom	PolFor	SphSpe		HypPhy	PicAbi	PinSyl
	USSR	FRG	USSR	FRG	USSR	FRG	USSR	FRG	USSR	USSR	FRG
Gd	0.010	0.04	0.016	0.034	0.020	0.054	0.050	0.13	0.12	0.014	0.032
Ge	-	-	-	-	-	-	-	-	-	-	-
H	6.6	6.5%	5.84%	6.02%	5.99%	6.05%	5.95%	5.91%	6.27%	6.12%	6.77%
Hf	-	0.276	-	0.415	-	0.079	-	0.631	0.165	-	0.058
Hg	-	0.072	-	0.067	-	0.28	-	0.417	0.260	-	0.0995
Ho	0.002	0.008	0.002	0.007	0.003	0.011	0.008	0.025	0.020	0.002	0.006
I	-	-	-	-	-	-	-	-	-	-	-
In	-	-	-	-	-	-	-	-	-	-	-
K	7966	5850	8280	7000	15390	9000	11.470	9100	3655	5600	4400
La	0.057	0.2	0.16	0.15	0.24	0.26	0.265	0.72	0.91	0.194	0.16
Li	-	0.003	0.2	0.2	-	0.37	0.28	0.4	0.885	0.29	0.475
Lu	0.008	1100	0.001	0.003	0.001	0.004	0.004	0.01	0.01	0.001	0.003
Mg	1590	674	4047	1600	240	439	696	452	293	650	733
Mn	1200	-	3981	604	179	30	272	22	149	718.5	208
Mo	0.425		0.4	0.4	0.2	1.6	0.5	2.34	0.4	0.24	0.62
N	1.48%	1.15%	2.29%	2.15%	1.84%	1.65%	1.39%	1.72%	1.13%	1.0%	2.24%
Na	272	29	440	50	60.4	180	427	430	129	425	101
Nb	0.085	-	0.065	0.021	0.07	0.037	0.07	0.031	0.08	0.08	0.0275
Nd	0.044	0.17	0.088	0.12	0.095	0.22	0.02	0.6	0.58	0.074	0.13
Ni	4.0	1.1	2.2	-	1.34	0.6148	3.48	1.7	3.405	3.95	1.4
O	-	40%	-	41.4%	-	45.5%	-	46.9%	-	-	37.4%
P	-	746	-	1090	-	1200	-	916	-	-	950
Pa	-	-	-	-	-	-	-	-	-	-	-
Pb	5.75	0.8	3.005	2.25	2.675	0.35	459	7.25	8.59	4.075	4.75
Po	-	-	-	-	-	-	-	-	-	-	-
Pr	0.014	0.044	0.026	.033	.029	.058	0.061	0.1	0.18	0.023	0.035
Ra	-	-	-	-	-	-	-	-	-	-	-
Rb	17.0	22.5	26.3	30	37	23	29.25	45	7.37	18	8.1
Re	-	-	-	-	-	-	-	-	-	-	-
S	1000	2000	2300	3140	800	2100	300	1940	1300	-	2200
Sb	-	-	-	0.071	-	0.16	0.053	0.11	0.265	-	0.084
Sc	0.0046	0.009	0.0125	0.015	02279	0.025	0.0377	0.036	0.196	0.0173	0.015
Se	-	-	-	-	-	-	-	-	0.39	-	-
Si	-	700	-	670	-	1390	-	1120	-	-	966
Sm	0.009	0.030	0.018	0.020	0.018	0.039	0.040	0.11	0.11	0.015	0.021
Sn	-	-	-	0.002	0.061	0.037	0.066	0.016	0.193	0.028	0.004
Sr	-	-	-	-	-	-	0.01	-	-	0.01	-
Ta	-	5.1	-	4.7	-	6.4	-	9.5	-	-	5.1
Tb	0.001	0.003	0.002	0.003	0.002	0.005	0.005	0.011	0.012	0.002	0.006
Te	0.045	-	-	-	-	-	0.11	-	-	0.11	-
Th	-	-	-	0.002	0.061	0.037	0.066	0.016	0.193	0.028	0.004
Ti	-	-	-	-	-	-	0.01	-	-	0.01	-
Tl	-	5.1	-	4.7	-	6.4	-	9.5	-	-	5.1
Tm	0.001	0.003	0.002	0.003	0.002	0.005	0.005	0.011	0.012	0.002	0.006
U	0.045	-	-	-	-	-	0.11	-	-	0.11	-
V	0.08	0.33	0.13	-	0.37	0.574	0.16	1.2	1.5	0.16	0.65
W	2.6	-	0.44	-	0.22	-	0.065	-	-	0.2	-
Y	0.046	0.18	0.076	0.16	0.091	0.25	0.23	0.59	0.58	0.064	0.15
Yb	0.005	0.02	0.008	0.017	0.010	0.027	0.024	0.065	0.062	0.007	0.016
Zn	28.6	42	13.87	20	46.45	58	36.25	36	72	29.65	53
Zr	0.045	-	0.07	0.13	0.125	0.21	0.2	-	0.525	0.07	0.13

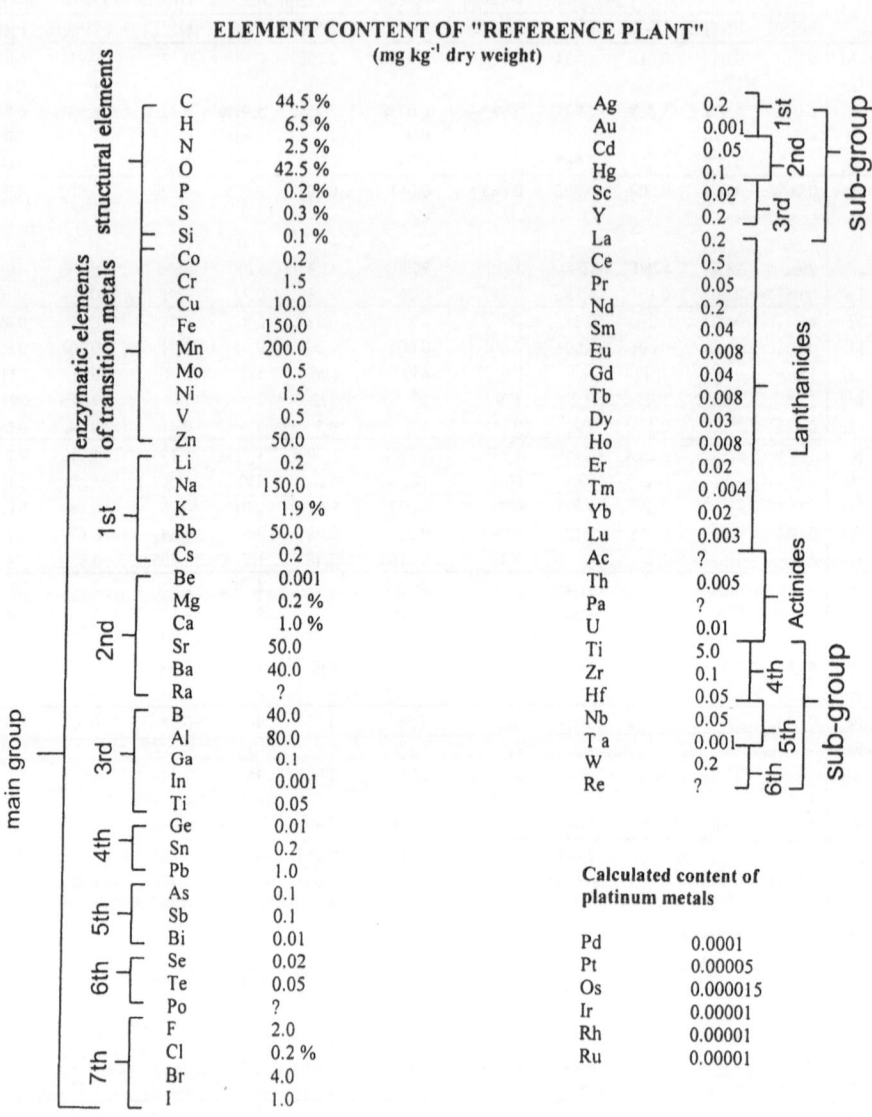

ELEMENT CONTENT OF "REFERENCE PLANT"
(mg kg⁻¹ dry weight)

Fig. 8: Data for "reference plant". No data for typical accumulator plants were used. Data were mainly extracted from the analytical work of Markert (1993a). If data for single elements were not available they were collected from Bowen (1979) or Kabata Pendias and Pendias (1992). The sequence of the elements is based on their position in the chemical periodic table. Exceptions are the biological structural elements C, H, N, O, P and Si and the transition metals Co, Cr, Cu, Fe, Mn, Mo, Ni, V and Zn which have an enzymatic effect (from Markert, 1993a)

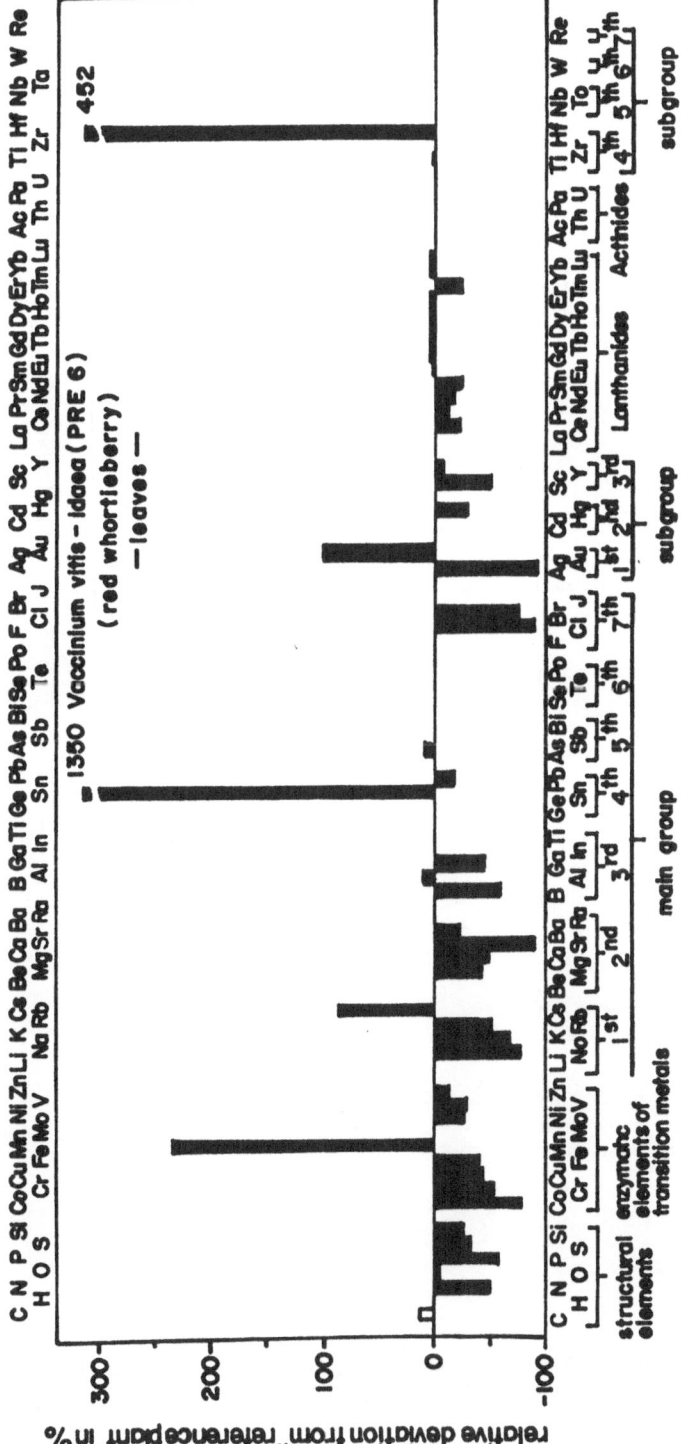

Fig. 9: Chemical fingerprint of *Vaccinium vitis-idaea* (red whortleberry, leaves) after normalization against "reference plant". The samples were collected in the Grasmoor near Osnabrück, Northwest Germany (from Markert, 1993a)

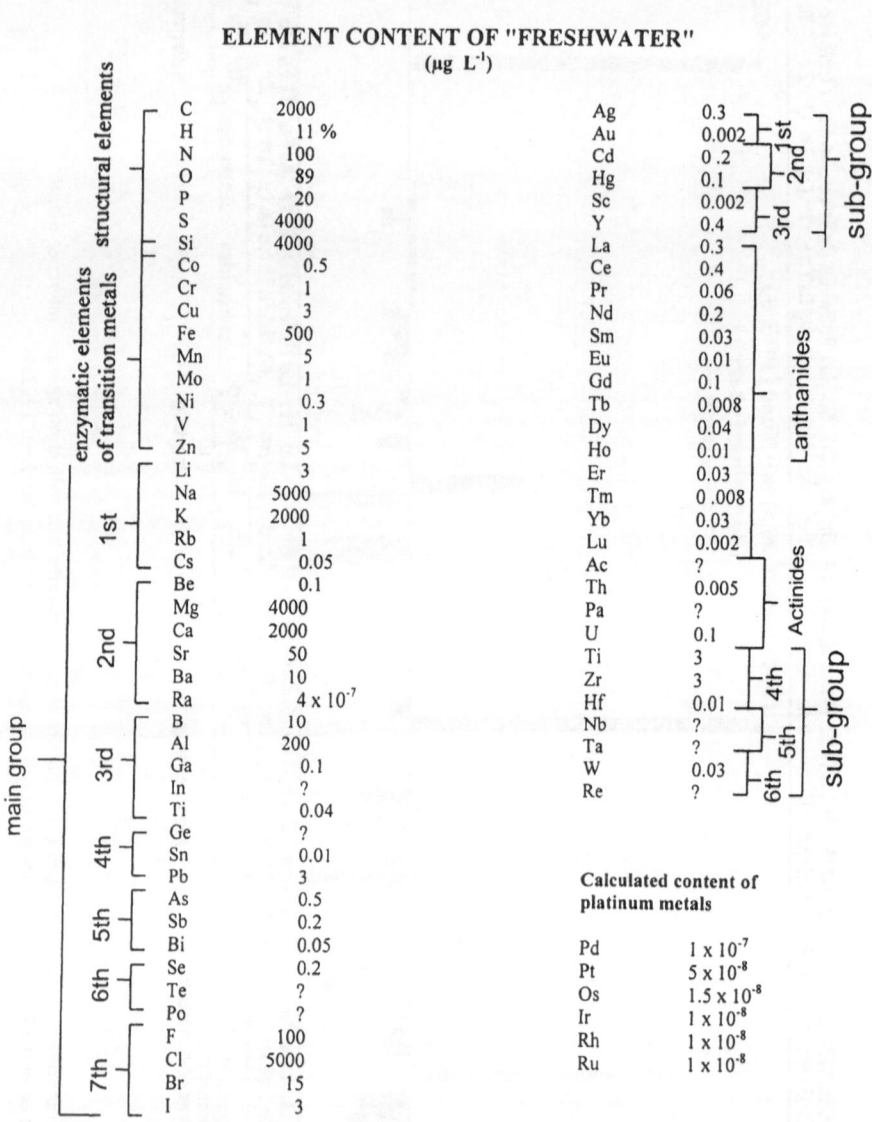

Fig. 10: Data for "reference freshwater". Note: No data for highly polluted water were used. Further explanations given in Fig. 8 (from Markert and Geller, 1994)

By normalizing against the values of the reference plant and depicting the positive and negative deviations of the individual plant species from the standard values of the reference plant as percentages on a graph it was possible to produce so-called "chemical fingerprints". Figure 9 is such a fingerprint for *Vaccinium vitis-idaea* (red whortleberry). In this chapter it is not possible to go into details of the possibilities offered by chemical fingerprinting. These are adequately described in the literature (e.g. Zeisler et al., 1988; Rossbach, 1992; Markert, 1993a;). Let us only say this much here by way of a summary: through normalization against the reference plant, inorganic fingerprints permit direct comparison of individual plant species. A great advantage is that they enable comparisons over any number of orders of magnitude of individual concentration ranges; such comparisons plainly indicate not only an accumulation of single elements but a reduction in these as well. If the detection limit of analysis is reached or some element concentrations are missing because they have not been determined, this does not affect the comparability of individual fingerprint graphs. Fingerprint graphs make it possible, in general, to demonstrate any marked characteristics of individual plant species, thus enabling these plants to be identified by their typical element composition. Comparisons of individual species make it possible to quantify certain chemical distribution patterns and recognize kinships. The success of this chemical system depends very much on "proper" representative sampling of the individual plant species and "proper" overall conduct of the analysis.

Similar reference systems have since been devised for other environmental compartments as well. Figure 10 shows such a system for fresh water. Where sampling is sufficiently representative and the constellation of factors is known, this should give rise to totally new ways of comparing both stagnant and flowing waters. Initial investigations have recently started in South American systems (Markert and Geller, 1994).

The data described above will now permit global calculations leading to model statements on the occurrence and available resources of individual elements in the different compartments and in the ecosystems themselves. In this way the total content of chemical elements in the plant biomass of the earth was recently calculated (Markert, 1992) (Table 2). This calculation revealed, for example, that a total of 90 tonnes of platinum is bound in the overall plant biomass of the earth. This figures are subject to an error factor of anything up to two or three orders of magnitude because of uncertainty in the analysis of elements, inadequate statistical backup and insufficient quantities of data on which to base them, especially in tropical and subtropical regions (Golley et al., 1978; Payne, 1986; Zauke et al., 1992; Abrão et al., 1993; Duursma, 1993; Jayasekera, 1993; Lacerda and Barcellos, 1993; Müller, 1993; Pedruzo et al. 1993). Nevertheless, this material offers an initial fundamental basis on which to calculate global ecosystem models.

Table 2: Estimation of the total element content in the world plant biomass in tonnes (from Markert, 1992)

Element	Total content		Element	Total content		Element	Total content	
Ac	?		Hf	9.2	$\times 10^4$	Ra	?	
Ag	3.682	$\times 10^5$	Hg	1.841	$\times 10^5$	Rb	9.2	$\times 10^7$
Al	1.47	$\times 10^8$	Ho	1.472	$\times 10^4$	Re	?	
As	1.841	$\times 10^5$	I	5.523	$\times 10^6$	Rh	1.84	$\times 10^1$
Au	1.841	$\times 10^3$	In	1.841	$\times 10^3$	Ru	1.84	$\times 10^1$
B	7.364	$\times 10^7$	Ir	1.841	$\times 10^2$	S	5.523	$\times 10^{10}$
Ba	7.364	$\times 10^7$	K	3.497	$\times 10^{10}$	Sb	1.841	$\times 10^5$
Be	1.841	$\times 10^3$	La	3.682	$\times 10^5$	Sc	3.682	$\times 10^4$
Bi	1.841	$\times 10^4$	Li	3.682	$\times 10^5$	Se	3.682	$\times 10^4$
Br	7.364	$\times 10^6$	Lu	5.523	$\times 10^3$	Si	1.841	$\times 10^9$
C	8.19	$\times 10^{11}$	Mg	3.682	$\times 10^5$	Sm	7.364	$\times 10^4$
Ca	1.841	$\times 10^{10}$	Mn	3.682	$\times 10^8$	Sn	3.682	$\times 10^5$
Cd	9.2	$\times 10^4$	Mo	9.2	$\times 10^5$	Sr	9.2	$\times 10^7$
Ce	9.2	$\times 10^5$	N	4.602	$\times 10^{10}$	Ta	1.841	$\times 10^3$
Cl	3.682	$\times 10^9$	Na	2.76	$\times 10^8$	Tb	1.472	$\times 10^4$
Co	3.682	$\times 10^5$	Nb	9.2	$\times 10^4$	Te	9.2	$\times 10^4$
Cr	2.7615	$\times 10^6$	Nd	3.682	$\times 10^5$	Th	9.2	$\times 10^3$
Cs	3.682	$\times 10^5$	Ni	2.76	$\times 10^6$	Tl	9.2	$\times 10^4$
Cu	1,841	$\times 10^7$	O	7.824	$\times 10^{11}$	Ti	9.2	$\times 10^6$
Dy	5.523	$\times 10^4$	Os	2.7615	$\times 10^1$	Tm	7.364	$\times 10^3$
Er	3.682	$\times 10^4$	P	3.682	$\times 10^{10}$	U	1.841	$\times 10^4$
Eu	1.472	$\times 10^4$	Pa	?		V	9.2	$\times 10^3$
F	3.682	$\times 10^6$	Pb	1.841	$\times 10^6$	W	3.682	$\times 10^5$
Fe	2.76	$\times 10^8$	Pd	1.841	$\times 10^2$	Y	3.682	$\times 10^5$
Ga	1.841	$\times 10^5$	Po	?	?	Yb	3.682	$\times 10^4$
Gd	7.364	$\times 10^4$	Pr	9.2	$\times 10^4$	Zn	9.2	$\times 10^7$
Ge	1.841	$\times 10^4$	Pt	9.2	$\times 10^1$	Zr	1.841	$\times 10^5$

References

Abrão, J.J.; Wasserman, J.C. and Silva Filho, E.V. (1993). *Proceedings of the International Symposium on Perspectives for Environmental Geochemistry in Tropical Countries*, held in Niterói, Rio de Janeiro, Brazil, 29/11/93 - 3/12/93, Universidade Federal Fluminense, Geochemistry Department. 485p.

Adriano, D.C. (1992). *Biogeochemistry of the trace metals*. Lewis Publishers, Boca Raton. 513p.

Baker, A.J.M. (1981). Accumulators and excluders - strategies in the response of plants to heavy metals. *J. Plant Nutrit.*, 3:643-654.

Bowen, H.J.M. (1979). *Environmental chemistry of the elements*. Academic Press, London. 333p.

Brooks, R.R. (1993). Geobotanical and biogeochemical methods for detecting mineralization and pollution from heavy metals in Oceania, Asia, and The Americas. In: *Plants as biomonitors - Indicators for heavy metals in the terrestrial environment*. Markert, B. (ed.). VCH-Publisher, Weinheim, New York, Tokio. pp.:127-154.

Davies, B.E. (1992). Trace metals in the environment Retrospect and Prospect. In: *Biogeochemistry of trace metals*. Adriano, D.C. (ed.). Lewis Publishers, Boca Raton. pp.:1-17.

Djingova, R. and Kuleff, I. (1993). Monitoring of heavy metal pollution by Taraxacum officinale. In: *Plants as biomonitors - Indicators for heavy metals in the terrestrial environment*. Markert, B. (ed.). VCH, Weinheim, New York, Tokio. pp.:435-460.

Duursma, E.K. (1993). Are tropical estuaries environmetal sinks, sources, or neither of them. *In*: *Proceedings of the International Symposium on Perspectives for Environmental Geochemistry in Tropical Countries*. Abrão, J.J.; Wasserman, J.C. and Silva Filho, E.V. (eds.). Universidade Federal Fluminense, Geochemistry Department, Niterói, Rio de Janeiro, Brazil. pp.:319-322.

Ernst, W.H.O. (1990). Element allocation and (re)translocation in plants and its impact on representative sampling. *In*: *Element concentration cadaster in ecosystems*. H. Lieth and B. Markert (eds). VCH, Weinheim, New York, Tokio. pp.:17-40.

Ernst, W.H.O. (1993). Geobotanical and biogeochemical prospecting for heavy metal deposit in Europe and Africa. *In: Plants as biomonitors - Indicators for heavy metals in the terrestrial environment*. Markert, B. (ed.). VCH, Weinheim, New York, Tokio. pp.:107-126.

Ernst, W.H.O. (1994). Sampling of plants for environmental trace analysis in terrestrial, semi-terrestrial and aquatic ecosystems. In.: *Sampling of Environmental Materials for Trace Analysis*. Markert, B. (ed.). VCH, Weinheim, New York, Tokio.

Förstner, U. (1990). *Umwertschutztechnik*. Springer-Verlag, Berlin, Heidelberg, New York. 462 p.

Frnzle, O (1994). Representative soil sampling. In: *Sampling of Environmental Materials for Trace Analysis*. Markert, B. (ed.). VCH, Weinheim, New York, Tokio.

Frankenberger, W.T. Jr. and Karlson, U. (1992). Dissipation of soil selenium by microbiological volatilization. *In: Biogeochemistry of Trace Metals*. Adriano, D.C. (ed.). Lewis, Boca Raton. pp.:365-381.

Golley, F.B., Richardson, T. and Clements, R.G. (1978). Element concentrations in tropical forests and soils in northwestern Columbia. *Biotropica*, 10:144-151.

Jayasekera, A. (1993). Concentrations of selected heavy metals in different compartments of a mountain rain forest ecosystem in Sri Lanka. *In: Plants as biomonitors - Indicators for heavy metals in the terrestrial environment*. Markert, B. (ed.). VCH, Weinheim, New York, Tokio. pp.:613-622.

Jayasekera, A. (1994). Sampling of tropical terrestrial plants with particular reference to the determination of trace elements. *In*: *Sampling of Environmental Materials for Trace Analysis*. Markert, B. (ed.). VCH, Weinheim, New York, Tokio.

Kabata-Pendias and Pendias, H. (1992). Trace Elements in Soils and Plants. 2nd edition, CRC Press, Boca Raton. 365 p.

Kabata-Pendias, A., Pietrowska, M. and Dudka, S. (1993). Trace metals in legumes and monocotyledons and their suitability for the assessment of soil contamination. *In*: *Plants as biomonitors - Indicators for heavy metals in the terrestrial environment*. Markert, B. (ed.). VCH, Weinheim, New York, Tokio. pp.:485-494.

Keune, H., Murray, A.B. and Benking, H. (1991). Harmonization of Environmental Measurement. *Geojournal*, 23:249-255.

Kovacs, M., Turcsanyl, G., Nagy, L., Koltay, A., Kaslab, L. and Szke, P. (1990). Element concentration cadaster in *Quercetum petreaeae-cerris* forest. *In*: *Element concentration cadaster in ecosystems*. H. Lieth and B. Markert (eds). VCH, Weinheim, New York, Tokio. pp.:255-264.

Kovacs, M., Penkaska, K., Turcsanyl, G., Kaszab, L. and Szke, P. (1993). Multielements-Analyse der Arten eines Waldsteppen-Waldes in Ungarn, *Phytoecologia*, 23:257-267.

Lacerda, L.D. and Salomons, W. (1991). *Mercury in the Amazon: A Chemical Time Bomb?* Dutch Mininistry of Housing/Physical Planning and Environments, Haren. 46 p.

Lacerda, L.D. and Barcellos, C. (1993). Cadmium and zinc pathway differentiation in coastal environments. *In*: *Proceedings of the International Symposium on Perspectives for Environmental Geochemistry in Tropical Countries*. Abrão, J.J.; Wasserman, J.C. and Silva Filho, E.V. (eds.). Universidade Federal Fluminense, Geochemistry Department, Niterói, Rio de Janeiro, Brazil. pp.:137-142.

Lepp, N.W. (1992). Uptake and accumulation of metals in bacteria and fungi. *In*: *Biogechemistry of Trace Metals*. Adriano, D.C. (ed.). Lewis, Boca Raton. pp.:227-298.

Lieth, H. and Markert, B. (1988). *Aufstellung und Auswertung Kosystemarer Element-Konzentrationskataster*. Springer-Verlag, Berlin, Heidelberg, New York. 193 p.

Lieth, H. and Markert, B. (1990). *Element Concentration Cadaster in Ecosystems, Methods of Assessment and Evaluation*. VCH, Weinheim, New York, Tokio. 448 p.

Markert, B. (1992). Presence and significance of naturally occurring chemical elements of the periodic system in the plant organism and consequences for future investigations on the inorganic environmental chemistry in ecosystems. *Vegetatio*, 103:1-30.

Markert, B. (1993a). *Instrumentelle Multielementanalyse von Pflanzenproben*. VCH, Weinheim, New York, Tokio. 269 p.

Markert, B. (1993b). *Plants as biomonitors - Indicators for heavy metals in the terrestrial environment*. VCH, Weinheim, New York, Tokio. 645 p.

Markert, B. (1994a). Occurrence and Distribution of Chemical Elements in Plants - Outlook and Further Research Plans. *Toxicol. Environm. Chem.*

Markert, B. (1994b). *Sampling of Environmental Materials for Trace Analysis*. VCH, Weinheim, New York, Tokio.

Markert, B. (1994c). The biological system of the elements (BSE) for terrestrial plants (glycophytes). *Sci. Tot. Environm.*

Markert, B. and Geller, W (1996). Multielement analysis in tropical lakes In: Giani, A., Von Sperling, E. and Pinto-Coelho, R. *Ecology and Human Impact on Lakes and Reservoirs in the State of Minas Gerais*. Cooperation Project Brazil/Germany.

Markert, B. and Wtorova, W. (1992). Inorganic chemical investigations in the forest biosphere reserve near Kalinin, USSR, Part III: Comparison of the multi-element budget with a forest ecosystem in Germany - aspects of rejection, indication and accumulation of chemical elements. *Vegetatio*, 98:43-58.

Müller, G (1993). Long distance mercury transport in Rio Tapajós, Pará, Brazil. *In: Proceedings of the International Symposium on Perspectives for Environmental Geochemistry in Tropical Countries.* Abrão, J.J.; Wasserman, J.C. and Silva Filho, E.V. (eds.). Universidade Federal Fluminense, Geochemistry Department, Niterói, Rio de Janeiro, Brasil. pp.:363-364.

Nriagu, J.O. and Pacina, J.M. (1988). Quantitative assessment of worldwide contamination of air, water and soils by trace metals. *Nature*, 333:134-139.

Nurnberg, H.W. (1984). Editorial on "Inorganic analysis in environmental research and protection". *Fresenius Zeitschrift für Analytische Chemie*, 317:197-199.

Payne, A.I. (1986). *The Ecology of Tropical Lakes and Rivers*. John Wiley, Chichester, UK.

Pedroso, F., Chillrud, S., Temporetti, P. and Diaz, M. (1993). Chemical composition and nutrient limitation in rivers and lakes of northern Patagonian Andes (39.5°S - 42°S; 71°W, Republic Argentina). *Verh. Internat. Verein. Limnol.*, 25:207-214.

Rossbach, M., Schladot, J.D. and Ostapzuk, P. (1992). *Specimen Banking, Environmental Monitoring and Modern Analytical Approaches*. Springer-Verlag, Berlin, Heidelberg, New York. 242 p.

Roth, M. (1992). Metals in invertebrate animals of a forest ecosystem. *In: Biogeochemistry of Trace Metals*. Adriano, D.C. (ed.). Lewis, Boca Raton. pp.:299-328.

Salomons, W. and Förstner, U. (1984). *Metals in the Hydrocycle*. Springer-Verlag, Berlin, Heidelberg, New York.

Sansoni, B. (1987). Multi-element analysis for environmental characterization. *Pure and Appl. Chem.*, 59(4):579-610.

Steinnes, E. (1993). Some aspects of biomonitoring of air pollutants using mosses as illustrated by 1976 Norwegian surveys. *In: Plants as biomonitors - Indicators for heavy metals in the terrestrial environment*. Markert, B. (ed.). VCH, Weinheim, New York, Tokio. pp.:381-395.

Thornton, I. (1990). A survey of lead in the British urban environment: An example of research in urban geochemistry. *In: Element concentration cadaster in ecosystems*. H. Lieth and B. Markert (eds). VCH, Weinheim, New York, Tokio. pp.:221-233.

Verkleij, J.A.C. (1993). The effects of heavy metal stress on higher plants and their use as biomonitors. *In: Plants as biomonitors - Indicators for heavy metals in the terrestrial environment*. Markert, B. (ed.). VCH, Weinheim, New York, Tokio. pp.:415-424.

Wagner, G. (1993). Large-scale screening of heavy metal burdens in higher plants. *In: Plants as biomonitors - Indicators for heavy metals in the terrestrial environment*. Markert, B. (ed.). VCH, Weinheim, New York, Tokio. pp.:425-434.

Wilken, R.D. (1992). Mercury analysis: A special example of species analysis. *Fresenius J. Anal. Chem.*, 342:795-802.

Zauke, G.P., Niemeyer, R.G. and Gilles, K.P. (1992). *Limnologie der Tropen und Subtropen, Grundlagen und Prognoseverfahren der limnologischen Entwicklung von Stauseen*. Ecomed-Verlag, Landsberg.

Zeisler, R., Stone, S.F. and Sanders, R.W. (1988). Sequential determination of biological and pollutant elements in marine bivalves. *Anal. Chem.*, 60:2760-2765.

8 Chemistry and Distribution of Trace Elements in the Patos Lagoon, South Brazil

Paulo Roberto Baisch[1] and Julio Cesar Wasserman[2]

[1] Departamento de Geociências, Fundação Universidade do Rio Grande, Caixa Postal 474, Rio Grande, RS, Brazil, 96200-000

[2] Departamento de Geoquímica - UFF. Outeiro de São João Batista s/n°, Niterói, RJ, Brazil, 24020-150

Abstract

This chapter reports a thorough study of the geochemistry of some metals in the sediments of the Patos lagoon, one of the biggest coastal lagoons in the world (and the biggest in Brazil). After a screening study of the sediments, which considered the granulometry, organic carbon, organic nitrogen and sulphur contents, Pb, Cu, Zn and Cr concentrations of more than 100 samples collected all over the system, the relationships between these variables were considered in order to explain their distribution. The distribution of metals within different granulometric fractions was also studied. Finally a sequential extraction procedure was performed in selected samples to provide information on the geochemical partitioning of the metals in the sediments. The absence of direct anthropogenic sources of metals to the lagoon was established. The main source of metals is the Guaiba system (northern portion), whose material is spread all over the lagoon and can be concentrated in the areas where physico-chemical parameters are favourable. The Camaquã river also constitutes a significant source of Cu and Pb for the southern portion of the Patos Lagoon, originating from the mining of metallic sulphides.

Introduction

Most of the urban centres and industries of the Rio Grande do Sul State are situated within the region of the Patos lagoonal-riverine system (South Brazil). Coal and metal mining are also important activities in the region, significantly contributing to the contamination of the lagoon. The continuous disposal of metallic elements into the aquatic environment may enrich the sediments, constituting an important stock that would represent a risk to the ecosystem and to the neighbouring populations.

Despite the efforts of the government sanitary control organs that have monitored the water quality of the Guaiba river system and its tributaries, few studies have been carried out in the lagoonal sector and in the drainage complex.

Most of the studies concerning heavy metal concentrations in bottom sediments were done in the estuarine region of the lagoon (Baisch et al., 1982; Baisch, 1985a; Baisch, 1985b, Baisch and Niencheski, 1985; Baisch and Niencheski, 1986; Baisch et al., 1988).

Some other works have also focused on specific areas of the lagoon (Baisch, 1987; Baisch, 1989; Baisch et al., 1989). These works have established the very low impact of anthropogenic contaminants, but they could not be compared to background values. Although Baisch (1987) and Baisch et al. (1988) carried out sequential extractions in bottom sediments of the estuarine region, the lack of background values renders the interpretation of data rather difficult.

Few studies concerning fluxes of metallic trace elements transported by the rivers to the Patos lagoon were also done. Most of these studies exclusively concern the dissolved phase, aiming to establish levels of water contamination (CEEIG, 1981; DMAE, 1983; Baumgarten et al., 1990). However, Vilas-Boas (1990), Baisch (1987; 1994) and Baisch et al. (1989) observed in the northernmost part of the Patos lagoon, a significant inflow of anthropogenic metals originated in the Guaiba system.

In the Camaquã river the existing data are still very scarce and are mainly related to the impact of mining on the upstream drainage basin (Laybauer, 1995; Pestana et al., 1995; Laybauer and Bidone, this volume). A very recent work by Baisch et al. (1996) is a comprehensive contribution concerning the transfer of trace metals in particulate and dissolved phases from this river to the Patos lagoon.

This chapter presents a large scale study of the concentrations of metals in the sediments of the Patos lagoon, considering that the main sources of anthropogenic enrichment are provided by the Guaiba system and to a lesser extent by the Camaquã river. The total concentration of metals will be presented in a north/south axis, indicating a pollution gradient.

The modifications in the chemical dynamics of these contaminants as they evolve in the system are also tested, using their statistical interrelations with the geochemical carriers and performing sequential extractions in selected samples.

Study Area

Geographic Outline

The Patos system is located in south Brazil (30° - 32° S, 51° - 52° W), with its main axis parallel to the Atlantic Ocean. Most of its drainage basin lies in the territory of Rio Grande do Sul State, with a small area in the northern part of Uruguay (figs. 1 and 2).

The whole drainage basin covers an area of 197,770 km^2 (Herz, 1977), circa 30,000 km^2 are in the Uruguayan territory. It is limited in the south and west by the drainage basin of the La Plata river and in the north by the drainage basin of the Uruguay river.

The Patos lagoon is a rather narrow lagoon, with a north-south length of about 300 km and a surface area of 10,360 km². It is the largest lagoon in South America and one of the largest in the world (Kjerfve, 1986).

In the southernmost portion, the lagoon communicates with the Atlantic Ocean through a narrow channel (channel of Rio Grande, 700 m wide), forming an estuarine environment of circa 900 km². In this same area, the Patos lagoon is connected to the Mirim Lagoon, through a 70 km long channel, forming the Patos-Mirim system (Fig. 1).

The Patos lagoon is a sedimentary environment that receives fresh water and suspended material from eight sub-basins, but the hydrographic system is organised in such a manner that almost all of the freshwater inputs are provided by two main pathways: the Guaiba system and the Camaquã river.

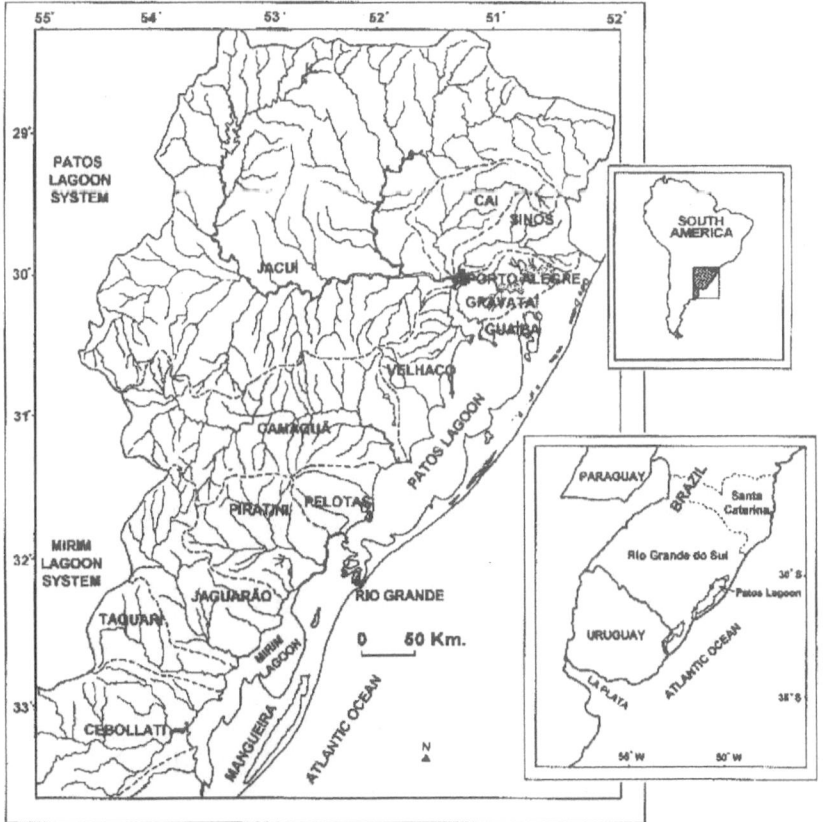

Fig. 1: Map of the Patos lagoon drainage basin

• *The Guaiba system*: This system flows into the far north end of the lagoon and is considered the most important freshwater supply. The Guaiba river is the pathway for water and solid matter originating in the rivers Jacui/Taquari, Cai, Sinos and Gravatai (fig. 1), constituting all together a drainage basin of 90,000 km^2. These rivers drain Cretaceous basalt-rhyolitic terrains of the high plateau as well as Triassic and Permian sedimentary rocks of the peripheral sedimentary depression. The main urban centres and almost all of the industrial plants (tanning, metallurgy, siderurgy, general chemicals, petro-chemicals, papers, coal mining) are present in this area. The anthropic activities are concentrated in the urban area of Porto Alegre, an important urban and industrial centre.

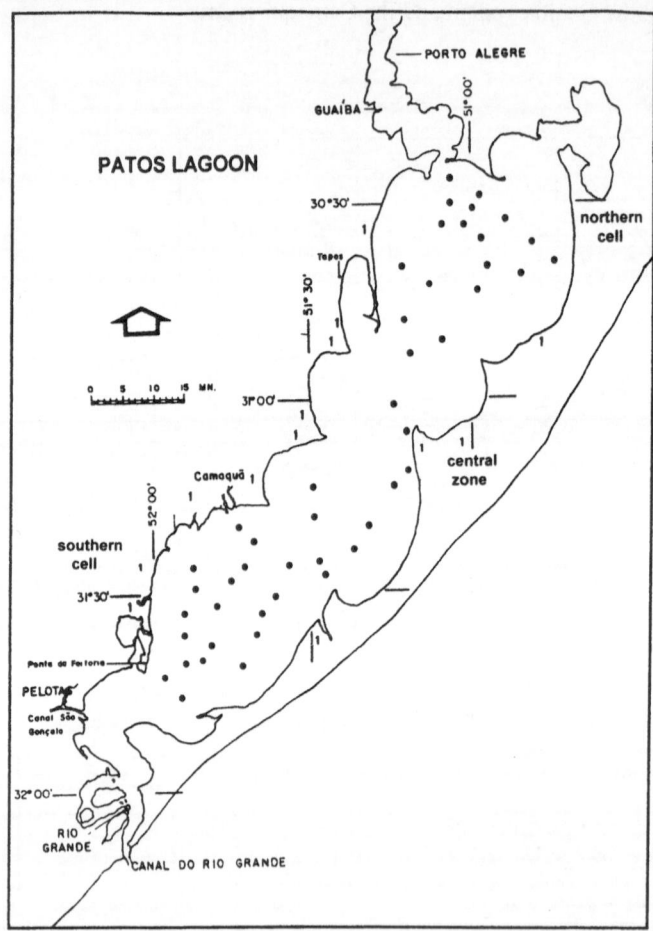

Fig. 2: Map of the Patos lagoon with the surface sediments and Holocene lagoonal core sampling sites

• *The Camaquã river*: This river discharges its waters in the southern sector of the Patos lagoon (fig. 1). Its drainage basin (19,000 km^2) covers the lowlands of the Pre-Cambrian shield, with a varied lithology, where gneissic-granitic rocks, co-paleozoic sedimentary rocks and low grade metamorphic rocks are present. Oppositely to the Guaiba, the drainage basin of the Camaquã presents a very low urban density and the only significant industrial activity is the exploitation of metallic sulphide minerals.

These two systems constitute the only significant sources of metals for the Patos lagoon. The margins of the lagoon are almost natural environments. Agricultural activities are developed in the coastal plain between the lagoon and the ocean, but do not constitute an important source of metals for the lagoon.

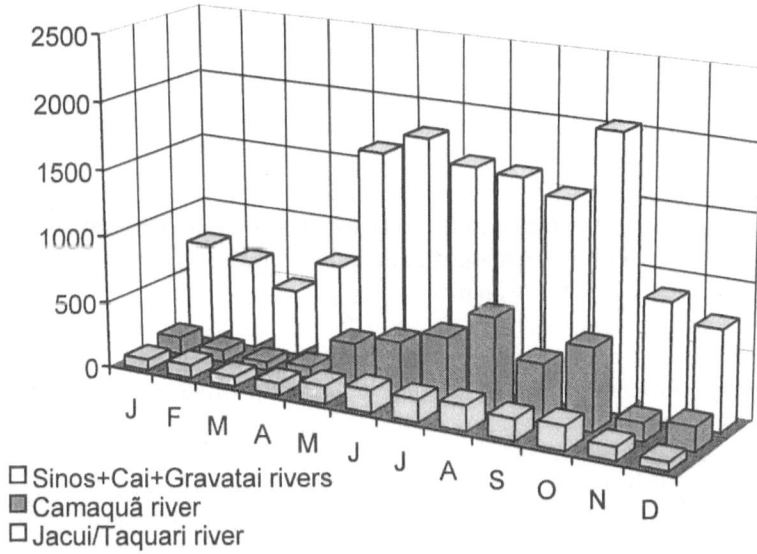

Fig. 3: Monthly mean discharges of the Guaiba system (sum of the discharges of the rivers Jacui/Taquari, Cai, Sinos and Gravatai) and of the Camaquã river. The data are means of 14 years

Hydrologic Regime

• *The Guaiba system*: The rivers included in this system have similar hydrologic behaviour (Herz, 1977). Winter (from late June through September) until spring (October through December) is the period of highest monthly mean discharges, corresponding to the period of more intense rains (fig. 3). Summer (January through March) until autumn (April through June) is a period of lower monthly mean discharges. The river Jacui/Taquari has the highest discharge ($1,030 m^3 s^{-1}$) and is responsible for 86.3% of the discharge of the Guaiba system. The mean discharge of the Guaiba is $1,200 m^3 s^{-1}$, with a maximum of $14,000 m^3 s^{-1}$ and a minimum of $50 m^3 s^{-1}$.

• *The Camaquã river*: This river presents a mean liquid discharge of $316 m^3 s^{-1}$, accounting for about one fourth of that of the Guaiba system. The highest discharges (maximum $5,300 m^3 s^{-1}$) have been observed during winter and spring, while the lowest discharges (minimum $6 m^3 s^{-1}$) have been reported in the summer and the first half of the autumn (Fig. 3).

Material and Methods

The sediment samples were collected in pre-established sites in the Patos lagoon, in order to provide a good overview of the concentration distribution and also a good gradient of contamination, considering that the main sources of metals are localised in the Guaiba system and Camaquã river (Fig. 2). In these areas the sampling site density was improved. Bottom sediment samples were collected at depths greater than 5 m, in order to avoid the influence of marginal material. Holocene sediment cores were also collected in depositional terraces, preferably close to the margin, in order to have good estimates of the background levels.

The surface sediments were collected using a metal free van Veen grab sampler. The chemical analyses were performed in the fine fraction of the sediments (< 63 µm), was separated by sieving with nylon sieves and dried at 80°C.

In the laboratory, samples were prepared for the analysis of organic carbon (Etchebert, 1981), organic nitrogen (micro-kjeldahl, Bremner, 1965), and granulometry (conventional sieving and pipette method). Dried samples were also analysed by x-ray fluorescence for total sulphur and major element contents (Lapaquellerie and Maillet, 1984).

Total metal contents of the fine fraction (< 63 µm) were analysed following the method of Loring (1986) using $HF:HNO_3:HCl$ (6:1:3). A partial extraction with HCl (0.1N), also called bio-available phase extraction, was performed (Fiszman et al., 1984). Samples were selected to perform a four-phases sequential extraction. The sequential extraction procedure used was a simplification of the one developed by Meguelatti (1983). The only difference between Meguelatti's scheme and ours is that we have eliminated the first step. Therefore, the first step extracts the exchangeable and the carbonatic phases all together. This first phase can be called the acid-soluble phase. The scheme of the sequential extraction is described in Fig. 4. All of the extracts were read in conventional flame and graphite furnace atomic absorption spectrophotometry (AAS).

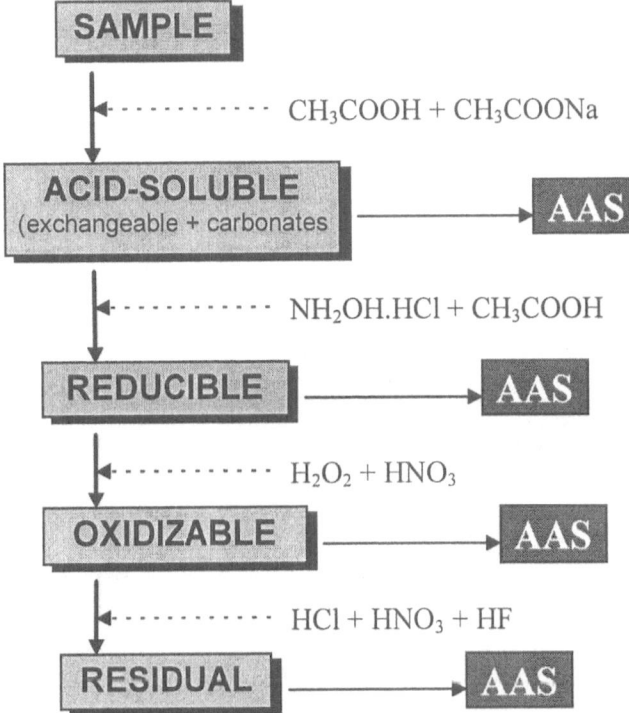

Fig. 4: Scheme of the sequential extraction (after Baisch, 1994)

Results and Discussion

Granulometry

The results presented geographically in the figures are related to distances in km from the Guaiba system mouth (0 km) to the limit of the estuarine region (180 km).

Figure 5 shows a clear diminution of the silt and clay contents from the northern to the southern portion of the lagoon. This behaviour had previously been observed (Herz, 1977; Martins et al., 1987; Baisch et al., 1989). The results also show that in this same direction there is an increase of the very fine grained fraction (fine grained clay - <2µm - and coarse grained clay - 2 to 4µm). On the other hand, the silt grain contents decrease. The very high silt contents (> 90 %) in the sediments within the mouth of the Guaiba (northern lagoonal portion) are the first stage of a sedimentary process at a local level represented by the Patos lagoon. Farther to the south, particularly in the southernmost portion of the lagoon, the remaining fine fraction will be enriched in argillaceous material.

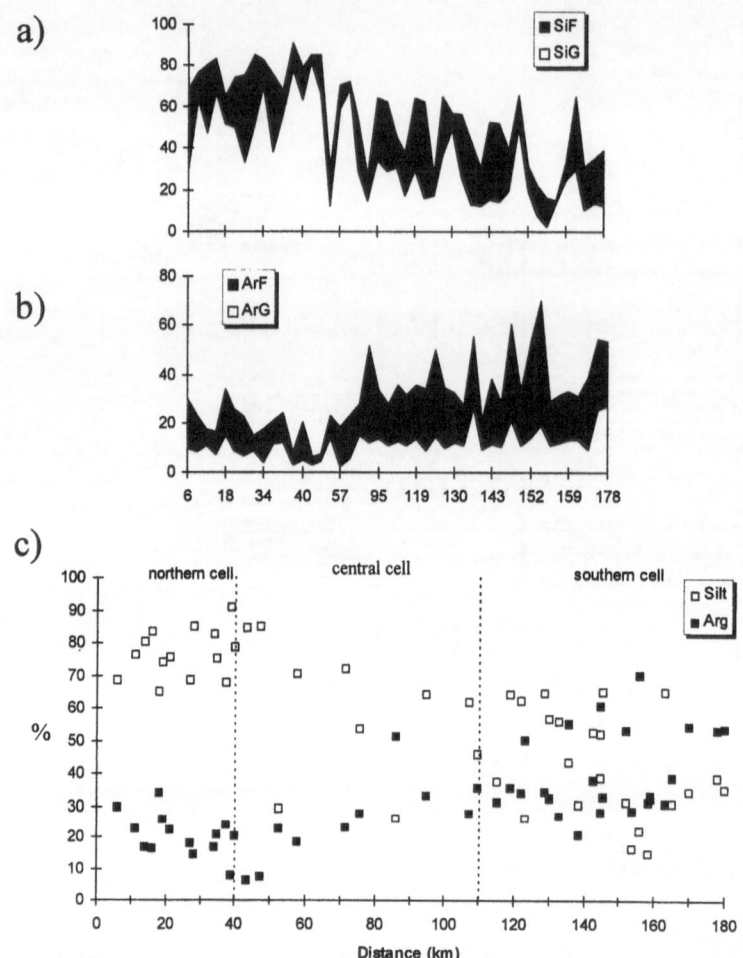

Fig. 5: North/south profiles: a) SiF (particle size 4-16µm) and SiG (particle size 16-64µm); b) ArF (particle size < 2µm), ArG (particle size 2-4µm) and c) silt and clay granulometry

Although the Guaiba system seems to be the main factor controlling the granulometry of the sediments, contributing 82 % of the suspended matter of the lagoon (Baisch, 1994), in the distance of 130 km a considerable increase can be observed in the silt fraction furnished by the Camaquã river. In this southern cell, are observed a mixture of fine grained sediments originating in the Guaiba and a coarse grained fraction originating in the Camaquã. The sedimentation of very fine grained sediments in the southern part of the lagoon is also due to a flocculation process occurring as a consequence of the occasional intrusion of the salt-wedge.

Organic Carbon, Organic Nitrogen and Total Sulphur

Carbon contents were shown to be low for a coastal lagoon, not exceeding 2%. The general trend, as can be seen in Fig. 6, is a gradual reduction in the concentrations in the middle of the system. The highest concentrations were observed close to the Guaiba outlet. However, in the southernmost part, where the lagoon is mostly influenced by the marine environment, the organic carbon concentrations increase again.

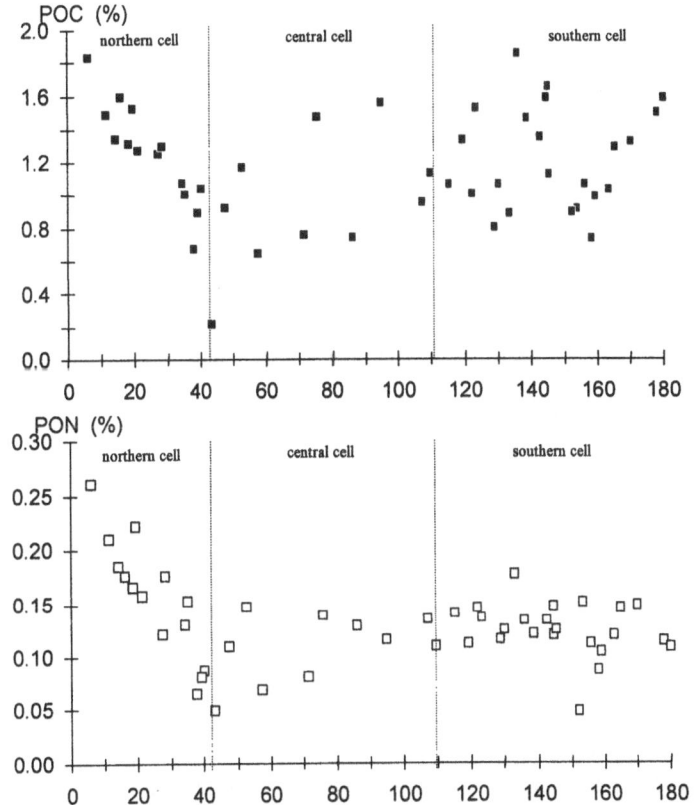

Fig. 6: North/south profile of organic carbon and organic nitrogen in the Patos lagoon

Organic nitrogen shows similar behaviour organic carbon (Fig. 6), but nitrogen seems more enriched in the Guaiba outlet, steeply decreasing to the southern portion.

For both organic carbon and organic nitrogen it can be observed that concentrations get enriched at distance 130 km, just after the outlet of the

Camaquã. This enrichment, which is widespread over the system until the oceanic water is probably not only due to the contribution of the Camaquã but also to a process of coagulation of the organic matter that has been well described in the zones of increasing ionic strength (Sholkovitz et al., 1978).

In order to obtain a better defined outline of the evolution of the organic fraction in the sediment, the Organic Sediment Index (OSI) as established by Ballinger and McKee (1971) was calculated. The classification of Ballinger and McKee (1971) is presented in Fig. 7 and the results of the calculations are plotted in the graph in Fig. 8.

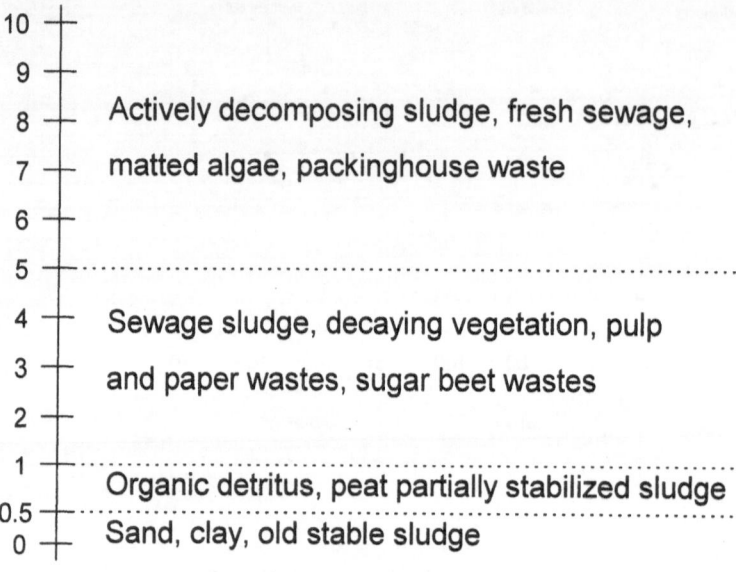

Fig. 7: The Organic Sediment Index (after Ballinger and McKee, 1971)

Fig. 8, showing the North-South profile of the OSI, presents a general trend of reduction of values, confirming that the main source of organic matter is the Guaiba system. On the other hand, the values obtained confirm the non-organic character of the sediments. This behaviour can be considered unusual since the coastal lagoons are systems which are characterised by high rates of accumulation of continental material, especially organic matter, that under the reducing conditions of such environments is quite well preserved. This condition can be explained by dilution, and considering the dimensions of Patos lagoon this hypothesis cannot be excluded. Another point that probably contributes to the dilution is that organic matter entering the lagoon does not precipitate immediately and is transported far from the outlet of the Guaiba system. This hypothesis can only be confirmed with a better understanding of the organic matter cycle in the

Patos lagoon and further studies are necessary. Besides that, the absence of organic matter in the sediments has important implications for the geochemistry of metals and other pollutants in the studied system.

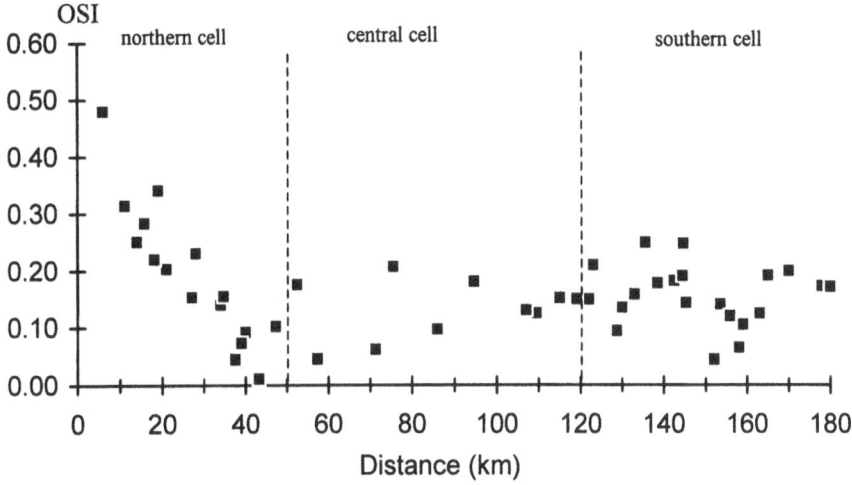

Fig. 8: North/south profile of the calculated Organic Sediment Index in Patos lagoon

Sulphur shows an inverse behaviour to carbon and nitrogen (Fig. 9), being enriched in the southernmost portion of the lagoon and less enriched in the north. Unlike nitrogen and carbon, sulphur shows important variations in concentrations. The highest concentrations (south) were almost 16 times higher than the lowest (north).

The lack of sulphur enrichment in the northern portion of the lagoon is somewhat surprising, since it is a reducing environment and one would expect that sulphate reduction would cause precipitation, increasing sediment concentrations. Apparently, this process does not occur in the north, probably because there is not enough sulphate in the Guaiba system. On the other hand, the enrichment observed in the southern portion can be attributed to the precipitation of marine or Camaquã river sulphates, or both. Furthermore, this river is a possible source of sulphate to the lagoon as a result of the process of oxidation of the metallic sulphides that are exploited in this basin. The barite tailings produced in the mining process also contribute to the enrichment observed in this area.

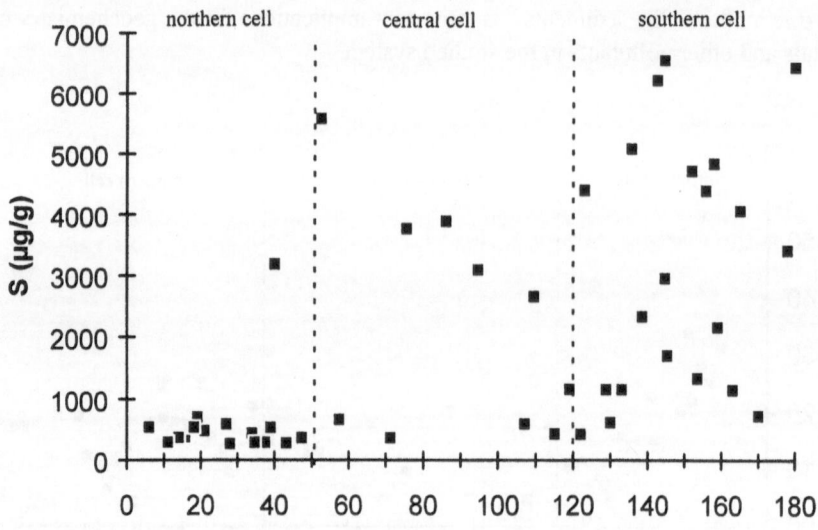

Fig. 9: North/south profile of the sulphur contents in the Patos lagoon

The sulphate enrichment in the interstitial water engendering sulphide formation and further precipitation has been observed in a number of aquatic environments, like lagoons (Tessier et al., 1989), estuaries (El-Ghobary, 1983), coastal-marine (Gaillard et al., 1986; Edenborn, 1987) and marine environments (Filipek and Owen, 1980). The metal-sulphide formation will severely affect the chemistry of these metallic pollutants, and in the Patos lagoon it can be especially important given the low concentrations of organic matter.

Distribution of Fe_2O_3 and MnO

The contents of iron oxides and manganese oxides measured in the sediments are considered elevated for this kind of environment (particularly the iron oxides). The southern and the northern portions both present very high values, reaching 13% of Fe_2O_3 (Fig. 10). These high values show that oxi-hydroxides may play an important role on the chemistry of the metallic pollutants in the lagoon. The natural process of leaching of basalt-rhyolitic rocks present in the drainage basin of the Guaiba system is responsible for an important input of oxi-hydroxides of iron and manganese (in the form of colloids and particles). The weathering of those rocks, mainly the basaltic one yields extremely high levels of iron and manganese oxides in the riverine sediments of the Guaiba (Baisch, 1994). In the southern portion, the lagoon also receives hematites and colloidal oxi-hydroxides from the Camaquã, having been dumped in the mining plants upstream.

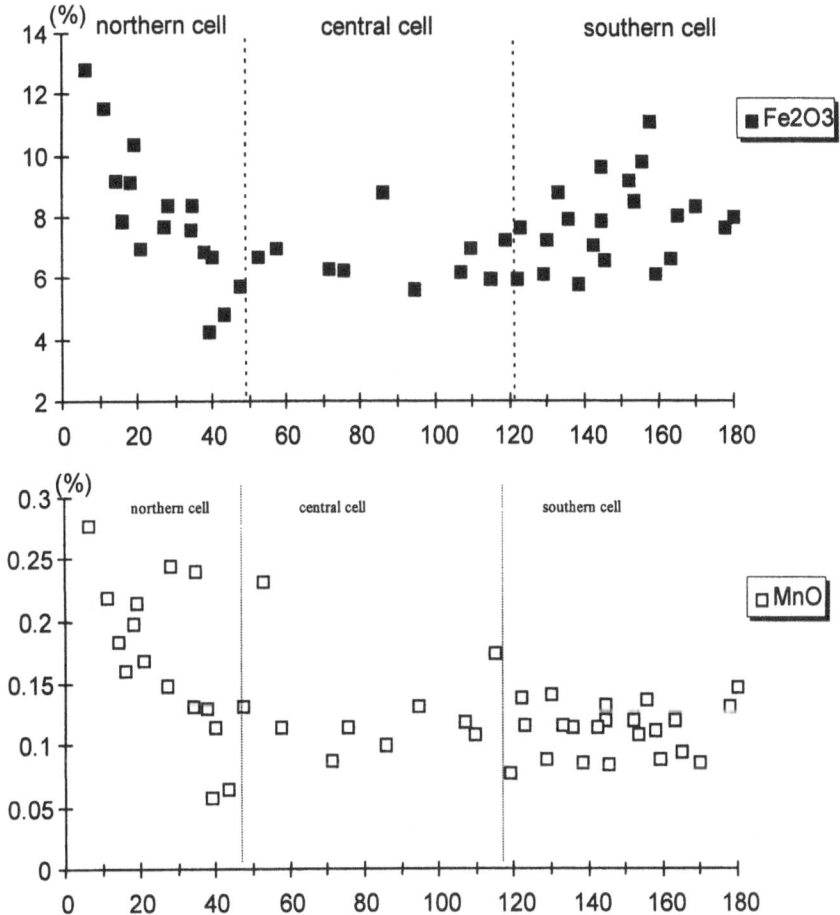

Fig. 10: North/south profile of Fe₂O₃ and MnO contents in sediments from the Patos lagoon

Metal Distribution Throughout the Lagoon

In order to study the distribution of metals within the lagoonal system and considering the quantity of available data, the calculations given below were applied to eliminate noise due to granulometry, local background, etc. The metal concentrations were normalised by Al. This procedure has been successfully applied in several pollution studies (e.g. Bruland et al., 1974; Duce et al, 1976; Jouanneau, 1982; Trefry et al., 1985; Quevauviller et al., 1989). Enrichment factors (Ef) were also calculated, following the procedure described in Baisch (1994). The enrichment factors are based on the reference levels obtained by establishing a mean of the concentrations of 13 sediment core samples of an early Holocene age (Baisch, 1994). This procedure is proposed as an estimation of the background values (Förstner and Wittmann, 1979; Förstner and Salomons, 1980; Salomons et al., 1988) and is frequently used in the study of the impact of metallic elements (e.g. Förstner and Müller, 1974; Förstner, 1978; Middleton and Grand, 1990; Grand and Middleton, 1990; Baisch, 1994). The calculations use the following formulations:

$$RL_{(x)} = \frac{[Me_x]_H}{[Al]_H}$$

Where:

$RL_{(x)}$ = Reference level for the metal x (Pb, Cu, Zn and Cr)

$[Me_x]_H$ = Concentration of the metal x in the early Holocene sediments

$[Al]_H$ = Concentration of Aluminium in the early Holocene sediments

$$Ef = \frac{[Me_x]_S/[Al]_S}{RL_{(x)}}$$

Where:

Ef = Enrichment factor

$[Me_x]_S$ = Concentration of the metal x in surface sediment

$[Al]_S$ = Concentration of aluminium in surface sediment

Lead

Figure 11 presents total and partial concentrations of Pb (Fig. 11a) as well as Pb enrichment factors (Fig. 11b).

Figure 11 shows that Pb is enriched all over the lagoon. Although these Ef are not as elevated as in other works presented in the literature, it is to be considered that our results are real enrichment factors, since they are calculated on the basis of realistic background levels and not, as presented elsewhere, mean shale, mean composition of neighbouring rocks, etc. On the other hand, the lagoon is an environment of considerable dimensions and dilution of metal enrichments is to be expected.

The highest concentrations are observed (as expected) in the northern part of the system, close to the Guaiba system outlet. Farther to the south, the concentrations decrease, showing a considerable dilution process. Around 100-120 km, concentrations start to increase again. Our results indicate recent anthropogenic enrichment, since Ef increase in this area, showing that background levels are maintained constant. This enrichment can be attributed to Pb originating in the Camaquã river mining activities. This point is further confirmed by the increase in the bio-available phase concentration (Fig. 11). Going into the marine environment, levels tend to decrease and Ef tend to approach the unit. Another important trend in the southern part of the lagoon is the enrichment of the very fine grained fraction (< 4μm, see Fig. 4) that would provide good geochemical support for the metals. Figure 12 shows an assay of the distribution of metals among the granulometric fractions for two samples picked at km 119 and at the km 133. As can be observed in this figure, granulometry is a very important factor controlling distribution of Zn and Cr, but does not explain enrichments for Pb and Cu.

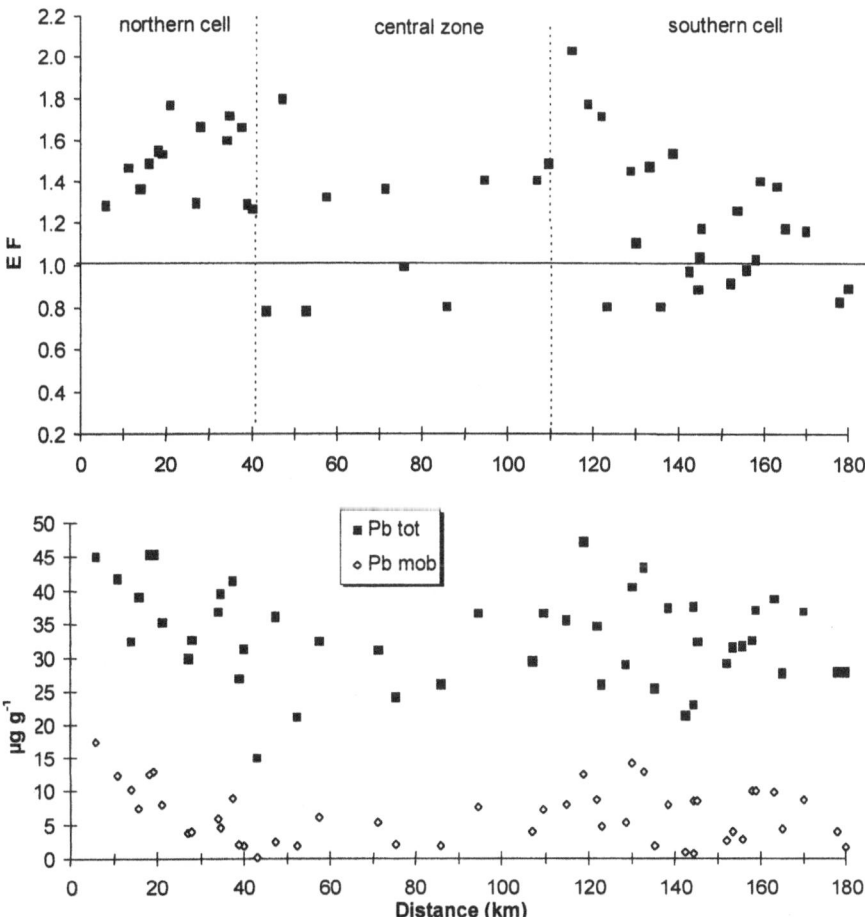

Fig. 11: Top, Pb enrichment factors (Ef) in the Patos lagoon and bottom, north/south profile of the Pb total (tot) and partial (mob) concentrations

The bio-available fraction constitutes circa 20 % of the total concentrations and its behaviour is similar to the total concentrations. Furthermore, the good correlation between the bio-available fraction and enrichment factors indicates that most of the metals present in this fraction are of anthropogenic origin.

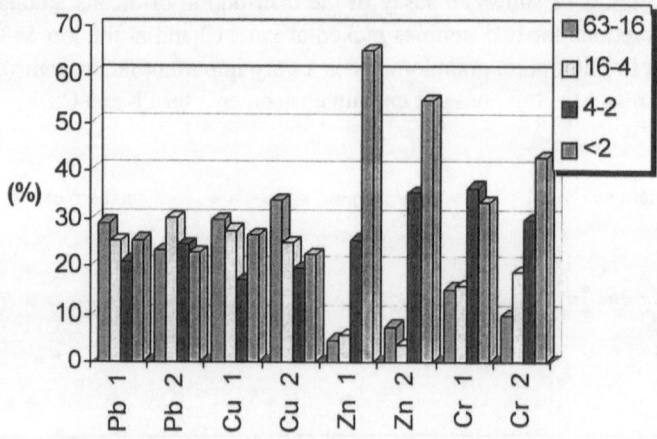

Fig. 12: Distribution of metals among the granulometric fractions for two samples picked up at km 119 (1) and km 133 (2)

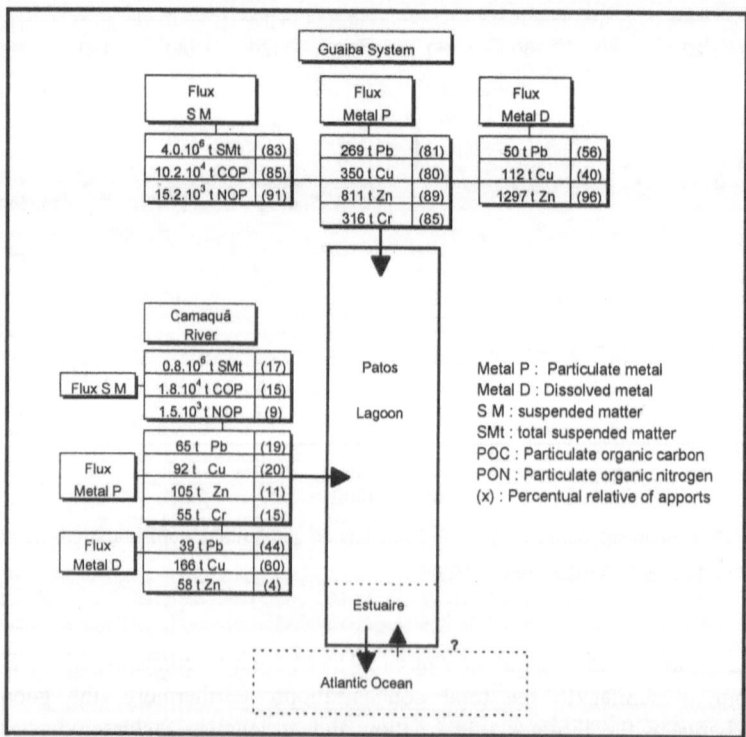

Fig. 13: Particulate and dissolved metals and organic matter fluxes into the Patos lagoon, measured in the Guaiba system and Camaquã river (after Baisch et al., 1996)

The anthropogenic enrichment observed in the southern portion cannot be fully explained by the inputs provided by mining activities in the Camaquã river, because this enrichment can be observed out of the area of influence of the mouth of this river. Furthermore, Baisch et al. (1996) show that the inputs of particulate and dissolved Pb are considerably smaller than those of the Guaiba (Fig. 13). Therefore, it can be suggested that the enrichment observed in this region may be due to the combination of anthropogenic inputs from Camaquã and Guaiba. Further studies are necessary to elucidate this point.

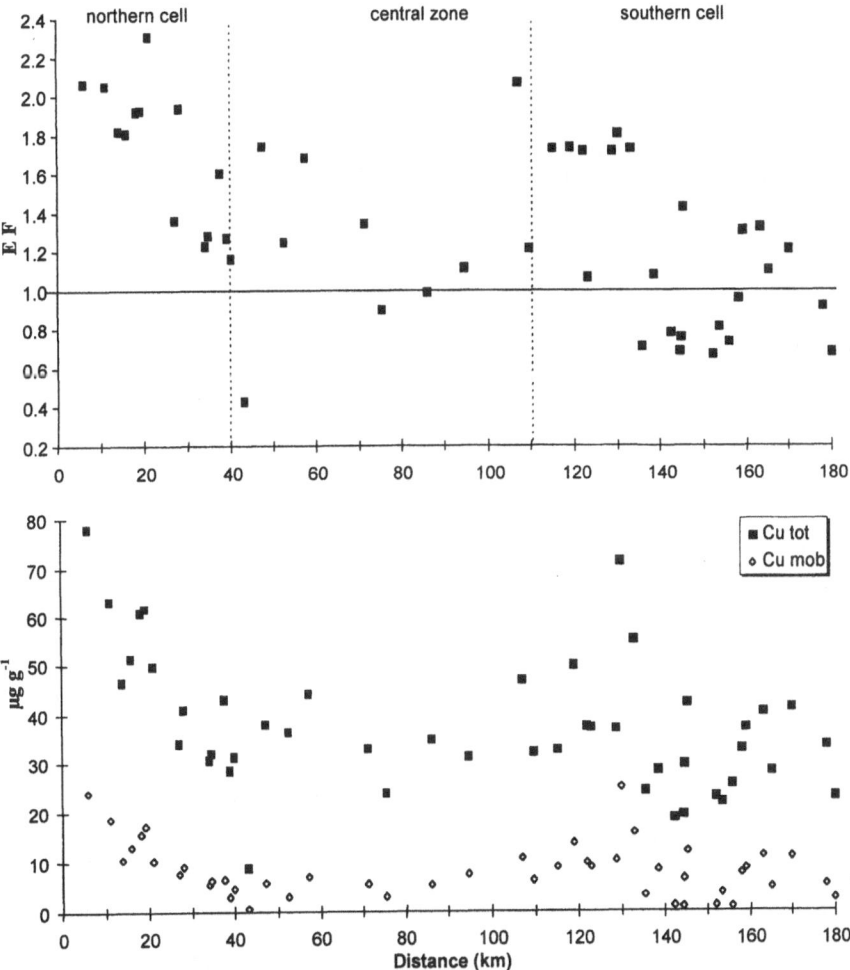

Fig. 14: Top, Cu enrichment factors (Ef) in the Patos lagoon and bottom, north/south profile of the Cu total (tot) and partial (mob) concentrations

Copper

The general trend observed for Pb, is here repeated for Cu (Fig. 14), showing that the behaviours of these two elements are similar ($p > 99\%$, Table 1) in the studied environment. Cu did show high concentration values at the mouth of the Guaiba, that steeply fall toward the central cell of the lagoon, underlining the importance of the anthropogenic inputs of this river and confirming the dilution/mixing processes throughout the lagoon. from the north to the south. Beside this general feature, increase in concentration values is observed around km 100. It is also possible to observe an enrichment in the end of the middle sector that can be attributed to the anthropogenic inputs from the Camaquã river. The influence of this pollutant is underlined by the peak concentrations in the total and bio-available Cu (Fig. 14). These latter concentrations are the highest found in the lagoonal environment.

Table 1: Correlation coefficient matrix (r) of the considered variables. Coefficients (r) $p>95\%$, coefficients (*r) $p>99\%$. ArF (fine clay, < 2 μm); ArG (coarse clay, 2-4 μm); SiF (fine silt, 4-16 μm) and SiG (coarse silt, 16-64 μm)

	Pb	Cu	Ba	Zn	Cr
Cu	* 0.74				
Ba		* 0.45			
Zn					
Cr		* 0.40		* 0.64	
S	*-0.55	*-0.51	*-0.67	* 0.39	
SiO_2			0.31	*-0.66	-0.32
Al_2O_3		0.32		* 0.65	0.29
Fe_2O_3	0.34	* 0.43		* 0.53	* 0.55
MnO	0.35	* 0.52		0.36	* 0.54
MgO			*-0.55	* 0.55	
CaO	*-0.42		-0.32		
TiO_2	* 0.57	* 0.65	* 0.45		* 0.44
K_2O			0.35	*-0.43	-0.35
Na_2O			0.36		
POC				* 0.78	* 0.53
PON	* 0.42	* 0.60		* 0.48	* 0.61
ArG			-0.32	* 0.46	
ArF			*-0.38	* 0.40	
SiG			* 0.44	-0.30	
SiF					

The good correlations between Cu/Ba and Pb/Ba ($p > 99\%$) show the anthropogenic origin of these metals, since Ba is enriched in the mining tailings of the Camaquã river (Baisch, 1994). As for Pb, the increase in concentration in the southern part of the lagoons cannot be explained by the particulate Cu fluxes from the Camaquã river. Although the dissolved metal fluxes can become important in the Camaquã river (mainly Cu, but also Pb), the hypothesis of dissolved metal becoming trapped in the geochemical carriers of the sediments cannot be discarded. The longer residence time of the water masses of the lagoon and the modifications on the hydrogeochemical conditions, compared to the river conditions, are also important factors to be considered.

Zinc

Before starting to discuss Fig. 15, where Ef and total and bio-available Zn concentrations are presented, it is necessary to consider Fig. 12, where metal distribution among granulometric phases are presented. Two groups of elements are represented here, Pb/Cu and Zn/Cr. The first group presents a homogeneous distribution while the second group presents enrichment in the very fine grained phase ($< 4\mu m$). Zn is the element that presents the most heterogeneous distribution, being strongly enriched in the fine clay phase ($< 2\mu m$), showing strong positive correlations with the clay fractions of the sediments. Another important figure is the dissolved Zn load from Guaiba system, that contributes 96% of the lagoonal fluxes (Fig. 13). The sediments did not show enrichment factors accordant with the significant fluxes of dissolved Zn, indicating that this metal has a tendency to be kept in this phase in the lagoonal environment.

These features unravel in a more widespread contamination by Zn in the Patos lagoon (although the enrichment factors are lower than observed for other metals). Figure 15 shows a more scattered distribution of the contaminant, which does not present any well defined trend. In the light of the two above statements, granulometry (see Fig. 5) tends to be enriched in the fraction $< 4\mu m$, which would engender increasing concentrations of Zn. This is counterbalanced by the distance of the main source (Guaiba system), thus maintaining concentrations more or less homogeneous. On the other hand, another factor that certainly contributes to this homogeneity is the large load of dissolved Zn, that do not settle immediately (trapped by the lagoon sediments) but can be transported throughout the lagoon. The dissolved Zn will only sediment near steep ionic gradients, i.e., far to the South of the lagoon, more specifically in the estuarine region that was not considered in this study. As to the bio-available fraction, Zn presents the highest concentrations in the northern portion of the lagoon, steeply reducing southward.

The Ef for Zn are fully consistent with the above, and can also be considered quite homogeneous throughout the lagoon. On the other hand, it would be expected that bio-available Zn should be a more important fraction of the sediments. The very reducing conditions should be responsible for the complexation of Zn with organic matter (as indicated by the good correlations between Zn/POC and Zn/PON - $p > 99\%$) and neo-formation of Zn sulphides ($p > 99\%$). However, this is not corroborated by the results of the sequential extraction, which will be discussed later on in this chapter.

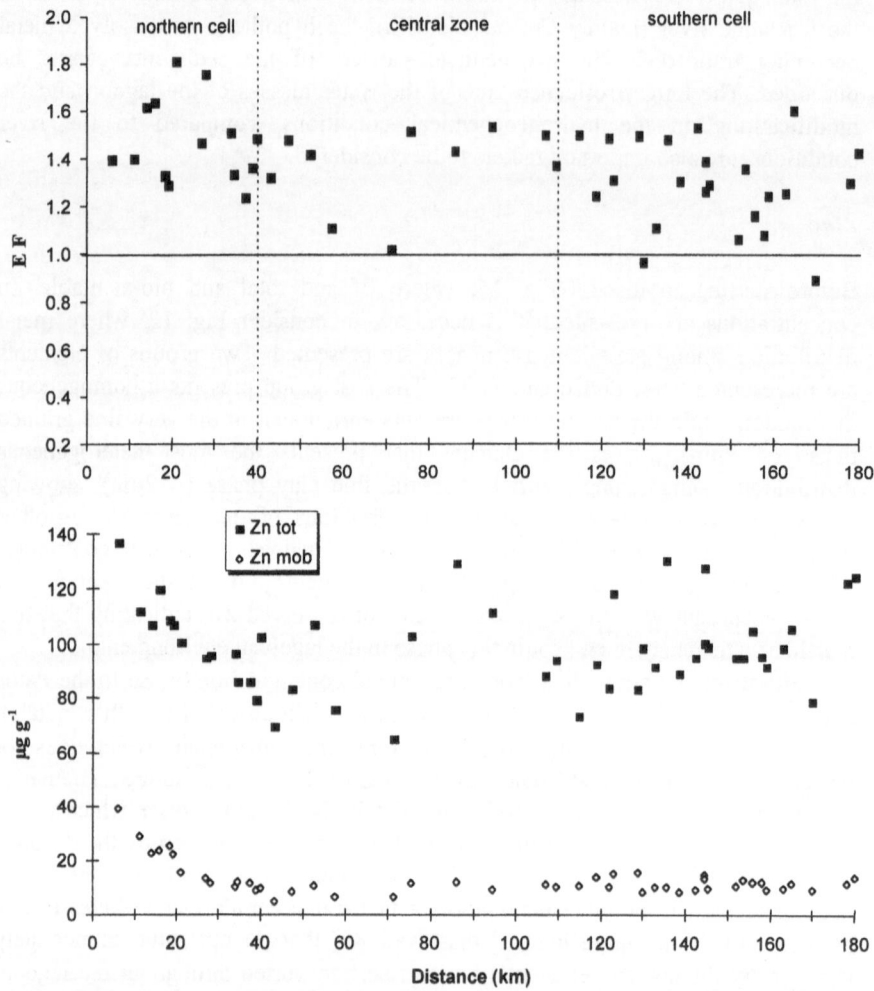

Fig. 15: Top, Zn enrichment factors (Ef) in the Patos lagoon and bottom, north/south profile of the Zn total (tot) and partial (mob) concentrations

Chromium

Although the influence of the very fine grained sediment fraction is important for Cr, this metal does not show quite the same behaviour as Zn. Probably, the geochemical control of the granulometry is not as important for Cr as it is for Zn. Cr did not present significant correlations with the silt and clay fractions (Table 1). Nevertheless, an association can be established between this metal and the fine grain fractions of the sediments (Fig. 12), corroborated by the good correlation between Cr and Al_2O_3 (Table 1).

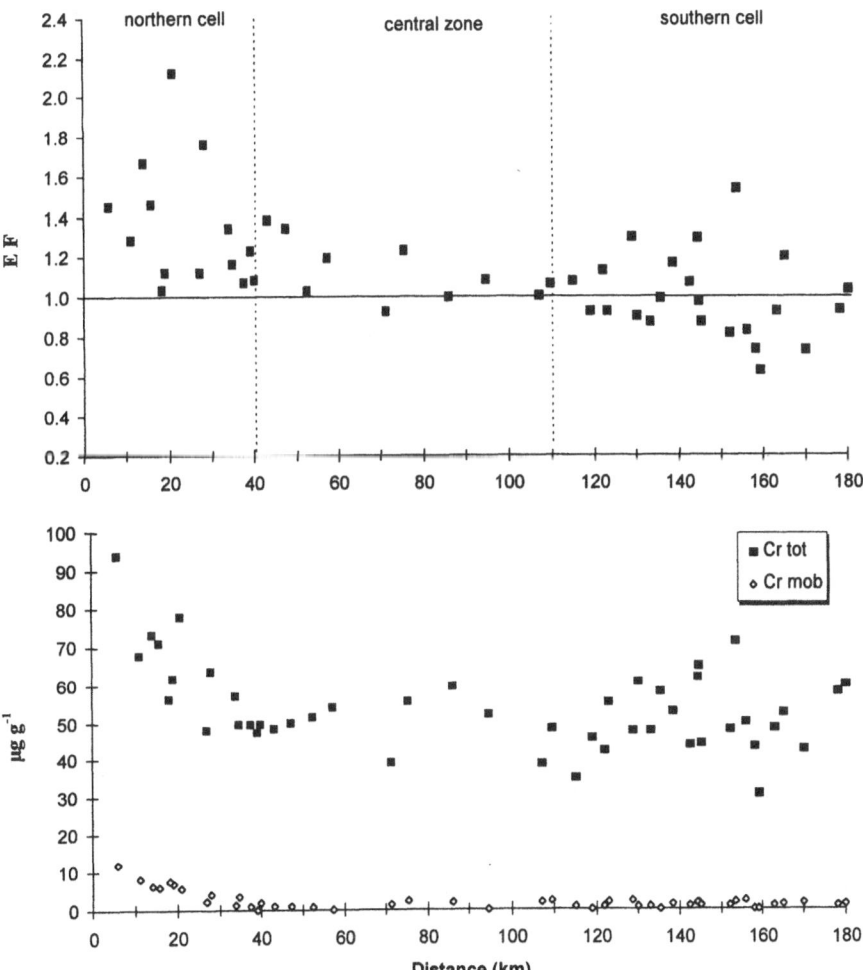

Fig. 16: Top, Cr enrichment factors (Ef) in the Patos Lagoon and bottom, north/south profile of the Cr total (tot) and partial (mob) concentrations

This implies a more simple dilution model for the distribution of Cr concentrations, as in Fig. 16. This behaviour is clearer on the Ef graph, where a tendency for a strong depletion is observed after km 120. Another point to be underlined for Cr is the low concentrations of the bio-available fraction (seldom exceeding 10%), indicating the low mobility of this metal in the lagoon. This is not in agreement with the data of Baisch (1994) which identify important sources of Cr in the drainage basin of the Camaquã, where the inputs of particulate matter are considered elevated (Fig. 13).

Apparently, the good correlation for POC/Cr and for PON/Cr indicates that organic matter is an important geochemical carrier for Cr. The metal would be immobilised with the organic matter in the reducing sediments of the lagoon, as has already been established for a number of other similar environments (e.g. Carvalho and Lacerda, 1992; Barrocas and Wasserman, this volume). This results are not corroborated by sequential extractions, which will be discussed in the next session.

Geochemical Partitioning

Figure 17 shows the sites where the samples were collected for sequential extraction. The selection of these sites was based on the fact that there are two main fluvial sources of metals (Guaiba and Camaquã) and that these are spreading metals that evolve within the lagoonal environment.

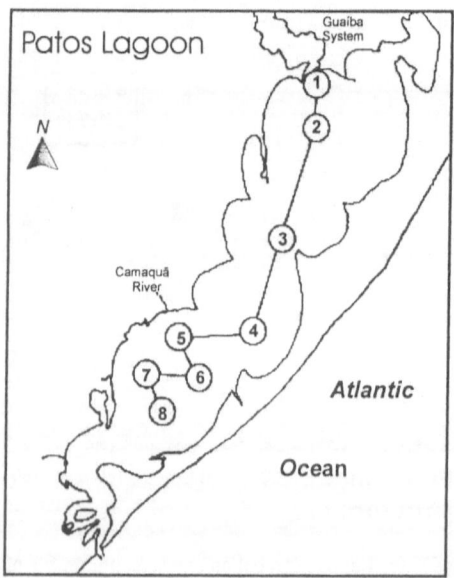

Fig. 17: Sampling sites for the geochemical partitioning

Figure 18 shows the partitioning of the metals studied in the Patos lagoon. In the northern portion (sites 1, 2 and 3), considering all of the studied metals (Cu, Pb, Zn and Cr), there is a clear reduction of the mobile phases (acid-soluble + oxidizable + reducible), concomitant with the increase of the residual fraction. These data confirm that the Guaiba system is the main anthropogenic source of metals and that the sediments close to the mouth of this system are the most contaminated in the whole lagoon. On the other hand, this behaviour also confirms that there is a steep dilution of the anthropogenic metallic load within the lagoonal environment. It is interesting to note that the northern portion of the lagoon is the only region where it is possible to find metals associated with the oxidisable or to the acid-soluble phases.

The increasing importance of the residual phase, together with the low Ef, corroborates the uncontaminated character of the central portion of the lagoon, which shows total concentrations very close to background.

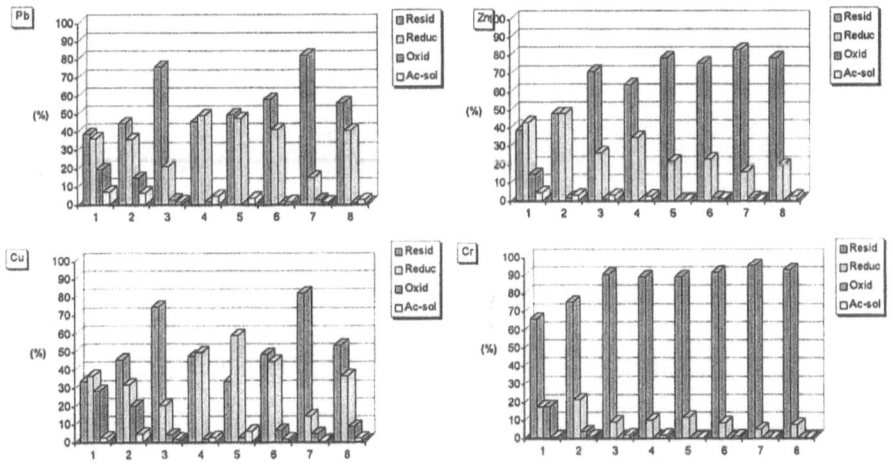

Fig. 18: Sequential extraction results for Pb, Cu, Zn and Cr in the Patos lagoon

In the southern portion, near the mouth of the Camaquã river (sites 4, 5 and 6), an important change can be observed in the geochemistry of Cu and Pb, owing to the increasing importance of the reducible phase, which can retain more than half of those metals. At site 5, the nearest to the mouth of the Camaquã river, 59 % of the Cu is associated with the oxi-hydroxides (reducible phase). Furthermore, 5.5 % of this metal is present in the acid-soluble phase (the highest percent contents of Cu associated with this phase in the whole lagoon), confirming the anthropic inputs of metals via the Camaquã river. As established elsewhere, metals tend to accumulate in the oxi-hydroxides of the sediments in the surroundings of the mining plants of the Camaquã river (Baisch, 1994; 1996).

There are significant geochemical differences between the north and the south of the lagoon. This can be confirmed by a closer analysis of the correlation coefficients for variables of the southern portion and of the northern portion (Table 2). Within the northern portion, all of the metals are strongly correlated with POC, PON, Al_2O_3 and with the clay fraction (Pb, Cu and Zn) or the silt fraction (Cr).

In the southern portion, Zn and Cr show good correlations with most of the parameters that represent the fine fraction of the sediment (Zn/clay, Zn/Al_2O_3 and Cr/Al_2O_3). On the other hand, Cu and Pb are associated with Ba (originating from the mining activities) and to the coarse fraction (silt/Cu, K_2O/Cu, Na_2O/Cu, K_2O/Pb and NaO/Pb). These data are somewhat conflicting since, as Pb and Cu are of an anthropogenic origin, they should accumulate preferentially in the fine fractions.

Table 2: Correlation coefficient matrix (r) of the considered variables for the northern portion and for the southern portion. Coefficients (r) $p>95\%$, coefficients ($*r$) $p>99\%$. ArF (fine clay, < 2 μm); ArG (coarse clay, 2-4 μm); SiF (fine silt, 4-16 μm) and SiG (coarse silt, 16-64 μm)

Northern part	Pb	Cu	Zn	Cr	Ba
Ba					-
SiO_2	*-0.80	*-0.90	*-0.90	*-0.67	
Al_2O_3	* 0.77	* 0.83	* 0.80		
K_2O		*-0.64	*-0.86	*-0.75	
Na_2O				*-0.57	
POC	* 0.72	* 0.88	* 0.92	* 0.78	
NOP	* 0.64	* 0.82	* 0.82	* 0.83	
ArG	* 0.69	0.59			
ArF	* 0.77	* 0.73	* 0.65		
SiG	-0.58	-0.62		-0.61	
SiF				* 0.64	
Southern part					
Ba	0.39	* 0.64			-
SiO_2			*-0.61	-0.39	
Al_2O_3			* 0.64	* 0.45	
K_2O	0.43	* 0.56			* 0.68
Na_2O	0.36	* 0.64			* 0.63
POC			* 0.56		
NOP					
ArG			* 0.60		
ArF			* 0.49		
SiG		* 0.46	*-0.45		
SiF					

Sequential extractions performed in the coarse silt fraction can give a plausible explanation for this behaviour (Fig. 19). The results show that 80%-92% of the Cu and Pb are bound by oxi-hydroxides, forming coatings on the coarse grains, therefore modifying the geochemical partitioning of these metals.

The Guaiba system and the Camaquã river are likely to provide the lagoon with important amounts of oxi-hydroxides that constitute an important geochemical carrier of the lagoonal sediments. These carriers bind most of the metals, especially Cu, Pb and Zn, reducing their complexation with organic matter or sulphides. This is the probable reason for the lesser affinity that metals have for organic matter, yet this affinity is well established for other lagoonal environments (Laxen and Sholkovitz, 1981; Sweeney and Naidu, 1983; Tessier et al., 1985).

The behaviour of Zn and Cr shows a different picture, because, although these metals have a good correlation with oxi-hydroxides, their process of enrichment is quite different in the southern portion of the lagoon. Our data indicate that the inputs of anthropogenic Zn and Cr are originated in the Guaiba system.

The geochemical evolution of Zn and Cr throughout the lagoon is gradual. Starting at the Guaiba mouth, a reduction of the fraction associated with the oxi-hydroxides can be observed, simultaneous with the increase of the alumino-silicated (residual phase). Assuming that the source of these elements is the Guaiba, it can be supposed that there is a gradual process of spreading of Zn and Cr throughout the lagoon and in the southern portion they tend to be associated with the fine fraction.

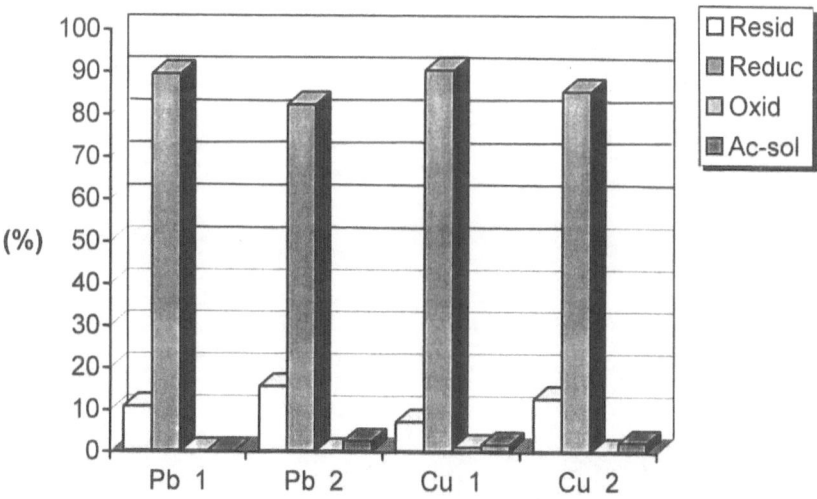

Fig. 19: Geochemical partitioning of Cu and Pb in the coarse silt fraction (16-64 µm) in two samples from km 119 and km 133 (southern portion of the lagoon)

Cr is very strongly bound to the residual phase in the Patos lagoon. This behaviour is described for a number of other similar environments (Gibbs, 1977; Förstner et al., 1982; Yetang and Förstner, 1984; Nirel, 1987). It is possible that the anthropogenic Cr binds with the alumino-silicates (residual phase), since the reducing conditions observed in the sediments of the lagoon (Baisch, 1994) can favour the incorporation of this metal into the structures of the clay minerals (Prohic and Kniewald, 1987).

It should also be underlined that the reducing conditions observed in the sediments can have an important role in the control of the metal contents, since the oxi-hydroxides become meta-stable and can release metals into the solution.

Conclusions

Figure 20 is a summary of the main results obtained in this work.

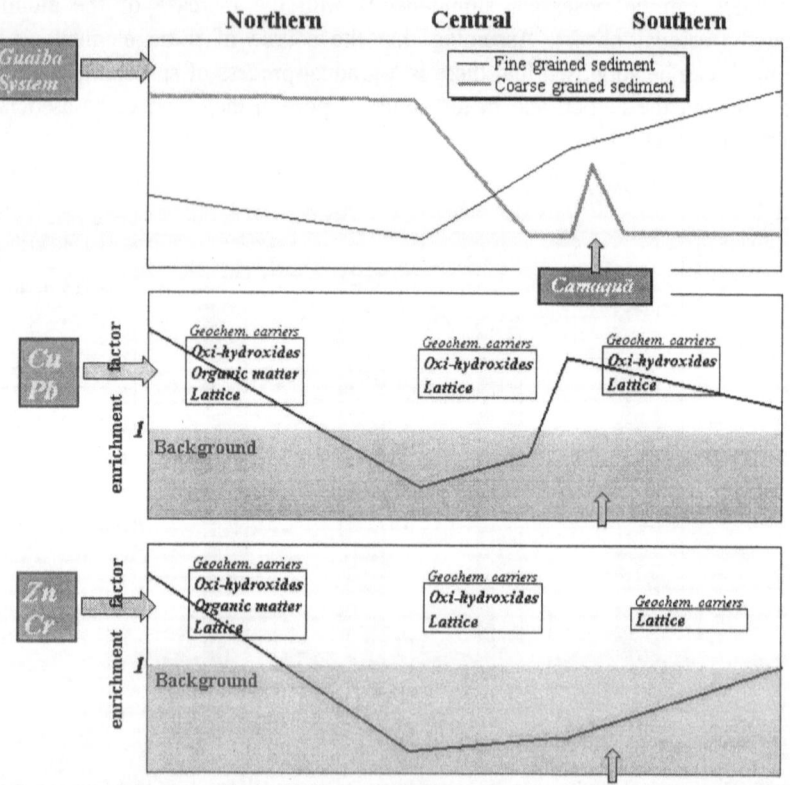

Fig. 20: Overview of the metal distribution and geochemical partitioning in the Patos lagoon

Although the northern portion is subjected to considerable detritic and anthropic inputs derived from the Guaiba system, the enrichment factors (Ef) do not show very high levels of contamination of the sediments. Southward, the metallic loads suffer a strong dilution/impoverishment, attributed to: 1) mixing with lagoonal sediments impoverished in metals; 2) dissolution of metals associated with oxi-hydroxides and organic matter, due to reducing conditions of the sediments. In the central portion of the lagoon, all metal concentrations fall close to background levels, being associated with the residual phase.

In the southern portion, metals can be divided into two groups according to their behaviour:

1. Cu and Pb appear strongly associated with the oxi-hydroxides. Two paths can be considered for these metals to arrive at the southern portion. The first is precipitation of colloidal oxi-hydroxides originated at the Guaiba system. The second is the direct discharge of detritic sediments rich in oxi-hydroxides that bear Cu and Pb. The formation of oxi-hydroxide coatings on coarse grained sediments was confirmed by sequential extraction of the sit fraction, yet no direct observation (electron microscope) was possible.

2. Zn and Cr constitute the second group. Although these metals are also associated with oxi-hydroxides, the dominant phase is the residual. The increase in the total concentration of these two metals in the southern portion is attributed to the processes of gradual settling of suspended material, which engenders an enrichment of Zn and Cr in the fine fraction of this portion of the lagoon.

The control of the oxi-hydroxides over the geochemistry of the metals in the sediments of the Patos lagoon is associated with the high contents of this geochemical carrier in the environment, preventing, through competition, their association with other geochemical carriers like organic matter and sulphides.

References

Baisch, P. (1985). Relação da distribuição geoquimica com as fontes antropogenicas dos metais pesados. Estuario da Lagoa dos Patos. *II Encon. Bras. Oceanol. Resumos.* p.45

Baisch, P. (1987). *Les oligo-éléments métalliques dans les sédiments de la Lagune dos Patos - Brésil.* Mémoire DEA Océanologie. Université de Bordeaux I. France. 62 p.

Baisch, P. (1989). *Geoquimica dos sedimentos (elementos tracos, maiores e materia organica) do sistema fluvio-lagunar-estuarino da Lagoa dos Patos.* Relatório técnico final. CIRM-FURG. Proj. Lagoa dos Patos. Parte I. 82 p.

Baisch, P. (1994). *Les oligo-éléments métalliques du Système Fluvio-lagunaire dos Patos (Brésil) - Flux et Devenir.* Ph.D. thesis. Université de Bordeaux I. no. 1136. 345p.

Baisch, P. (1996). Impact of heavy metals from mining activity in Camaquã river basin (south of Brazil). *II Intern. Symp. Environ. Geoch. Trop. Countries.* 3p, 1 tab, 2 fig. (CD-ROM).

Baisch, P. and Niencheski, L.F. (1985). Metais pesados nos sedimentos do estuario da Lagoa dos Patos. *Anais do I° Seminario sobre Pesquisa da Lagoa dos Patos,* 15 p.

Baisch, P. and Niencheski, L.F. (1986). *Metais Pesados nos sedimentos da Lagoa dos Patos. Fase II*. Final technical report - CIRM. 86 p.

Baisch, P., Almeira, M.T., Beaumord, C. and Kantin, R. (1982). Determinação de metais pesados nos sedimentos da Lagoa dos Patos. *XXXIV. Reunião Anual da Sociedade Brasileira para o Progresso da Ciência*. Resumo p.583 .

Baisch, P., Jouanneau, J.M. and Asmus, H. (1989). Chemical composition of sediments from de Patos Lagoon, Brazil. *XIII Intern. Geoch. Explor. Symposium et II Brazil. Geoch. Congress*. Vol.1. pp.11-20.

Baisch, P., Jouanneau, J.M. and Latouche, C. (1996). Flux du matériel en suspension, d'éléments traces et de la matière organique dans la lagune dos Patos (Brésil). *II Intern. Symp. Environ.Geoch. Trop.Countries*. 4p, 3 fig. (CD-ROM).

Baisch, P., Niencheski, L. F. and Lacerda, L. (1988). Trace metals distribution of Patos Lagoon estuary. *In: Metals in Coastal Environnments of Latin America*. Seeliger, U, Lacerda, L. D. and Patchineelam, S.R. (eds). Springer-Verlag. Berlin. pp 59-64.

Ballinger, D.G. and Mc Kee, G.D. (1971). Chemical characterization of bottom sediments. *J. Water Pollut. Con. Fed.*, 43(2):216-227.

Baumgarten, M.G., Klen, A.H. and Niencheski, L.F. (1990). Niveis de Cobre, Zinco e Chumbo dissolvidos na Lagoa dos Patos (RS). *II Simp. de Ecos. Costa Sul e Sudeste Brasileira*. Águas de Lindoia. São Paulo. v.II. pp.117-126.

Bremer, J.M. (1965). Total nitrogen. *Agronomy*. Vol.9:1149-1178.

Bruland, K.W., Bertine, K.K., Koide, M. and Goldberg, E.D. (1974). History of metal pollution in southern California coastal zone. *Environ. Sci. Technol*. 8:425-432.

Carvalho, C.E.V. and Lacerda, L.D. (1992). Heavy metals in the Guanabara Bay biota: Why such low concentrations. *Cien. Cult.*, 44(2/3):184-186.

CEEIG - Comite Executivo de Estudos Integrados da bacia do Rio Guaiba (1981). Report of management project-001/79, *Enquadramento dos mananciais*.140 p.

DMAE - Departamento Municipal de Agua e Esgotos (1983). *Inventário e classifição da ictiofauna do rio Guaiba - Pesquisa de metais e substancias toxicas*. Technical report no. 41, Centro de Estudos de Saneamento Basico-CESB. Porto Alegre. 84 p.

Duce, R.A., Hoffman, G.L., Ray, B.J., Fletcher, I.S., Wallace, G.T., Piotrowicz, S.R. Walsh, P.R., Hoffman, E.J., Miller, J.M. and Heffter, J.L. (1976). Trace metals in the marine atmosphere: sources and fluxes. In: Windom, H.L., Duce, R.A. (eds). *Marine polluant transfer*. Lexington Books, Lexington, MA.

Edenborn, H.M., Silverberg, N., Mucci, A. and Sundby, B. (1987). Sulfate reduction in deep coastal marine sediments. *Mar. Chem.*, 21:329-345.

El-Ghobary, H. (1983). *Diagenèse précoce en milieu littoral et Mobilité des éléments métalliques*. Thèse d'Etat. Université de Bordeaux I, 271 p.

Etcheber, H. (1981). Comparaison de diverses méthodes d'évaluation des teneurs en matières en suspension et en carbones organique particulaire des eaux marines du plateau continental aquitain. *J. Rech. Oceanogr.*, 6:37-42.

Filipek, L. and Owen, R. (1980). Early diagenesis of organic carbon and sulfur in outer shelf sediments from the Gulf of Mexico. *Amer. Jour. Sci.* 280:1097-1112.

Fiszman, M., W.C. Pfeiffer and L.D. Lacerda (1984). Comparison of Methods Used for Extraction and Geo-chemical Distribution of Heavy Metals in Bottom Sediments from Sepetiba Bay, R.J. *Environ. Technol. Lett.*, 5: 567-575.

Forstner, U. (1978). Metallanreicherungen in rezenten See-Sedimenten, geochemischer Background und zivilisatorische Einflüsse. *In: Mitt. Nationalkommitee der B.R. Deutschland für das Internationale Hydrologische Programm des UNESCO*, Vol.2. Koblenz, 66 p.

Forstner, U. (1982). Accumulative phases for heavy metals in limnic sediments. *Hydrobiologia* 91:269-284.

Forstner, U. and Muller, G. (1974). *Schwermetalle in Flüssen und Seen als Ausdruck der Umweltverschmutzung.* Springer, Berlin Heidelberg New York.225 p.

Forstner, U. and Salomons, W. (1980). Trace metals analysis on polluted sediments. Part I. Assessment of sources and intensities. *Environ. Technol. Letters,* 1:494-505.

Forstner, U. and Wittman, G. (1979). *Metal Pollution in the Aquatic Environment.* Springer. Berlin Heidelberg New York, 486 p.

Gaillard, J.F., Jeandel, C., Michard, G., Nicolas, E. and Renard, D. (1986). Interstitial water chemistry of Villefranche Bay sediments: trace metal diagenesis. *Mar. Chem.,* 8:233-247.

Gibbs, R.J. (1977). Transport phases of transition metals in the Amazon and Yukon rivers. *Geol. Soc. Am. Bull.,* 88:829-843.

Grant, A. and Middleton, R. (1990). An assessment of metal contamination of sediments in the Humber Estuary. U.K. *Est Coast. Shelf Sci,* 31:71-85.

Herz R. (1977). *Circulação das aguas de superficiais da Lagoa dos Patos.* Ph.D. thesis. Departamento de Geografia.Universidade de São Paulo.217 p.

Jouanneau, J.M. (1982). *Matières en suspension et oligo-éléments métalliques dans le système estuarien girondin. Comportement et Flux.* Thèse d'Etat. Univ. Bordeaux I. n°732. 150 p.

Kjerfve, B. (1986). Comparative oceanography of coastal lagoons. In: *Estuarine Variability.* D.A. Wolfe (ed.). Academic Press, New York, pp.:63-81

Lapaquellerie, Y. and Maillet, N. (1984). *Analyse chimique et minéralogique des roches par les rayons X (Fluorescence X et Diffraction X).* Rapport interne IGBA. Bordeaux. 130 p.

Laxen, D.P.H. and Sholkovitz, E.R. (1981). Adsorption (co-precipitation) of trace metals at natural concentrations on hydrous ferric oxide in lake water samples. *Environ. Technol. Lett.,* 2:561-569.

Laybauer, L. (1995). *Análise das transferências de metais pesados em águas e sedimentos fluviais na região das minas do Camaquã, RS.* M.Sc. thesis. UFRG. 164 p.

Loring, D.H. (1986). Intercalibration for trace metals in marine sediments. *Report Dep. Fish. and Oceans* Dartmouth, 60 p.

Martins, I.R., Martins, L.R. Toldo, Jr. E. and Gruber, N.L. (1987). Processos sedimentares na Lagoa dos Patos. *In: Anais do 1° Congresso Abequa.* pp.191-213.

Meguellati, N., Robbe, D., Marchandise, P., and Astruc, M. (1983). Intérêts des minéralisation sélectives pour le suivi des pollutions métalliques associées aux sédiments, *Journal Français d'Hydrologie,* 13(3):275-287.

Middleton, D. and Grant, A. (1990). Heavy metals in the Humber estuary : Scobicularia clay as a pre-industrial datum. *In: Proceedings of the Yorkshire Geological Society,* 48:75-80.

Nirel, P. (1987). *Evolution de la distribution particulaire des éléments chimiques en milieu estuarien.* Ph.D. thesis. Université de Paris VII,187 p.

Pestana, M. , Formoso, M. and Teixeira, E. (1995). Heavy Metals in Stream Sediments from Copper and Gold mining Areas in Southern Brazil. *In: Int. Conf. of Heavy Metals in the Environment.* Proceedings. Vol. 1:307-310.

Prohic, E. and Kniewald, G. (1987). Heavy metal distribution in recent sediments of the Krka River estuary–an example of sequential extraction analysis. *Mar. Chem.,* 22:279-297.

Quevauvillier, Ph., Lavigne, R. and Cortez, L. (1989). Impact of industrial and mine drainage wastes on heavy metals distribution in the drainage basin and estuary of the Sado river (Portugal). *Environ Poll.,* 59:267-286.

Salomons, W., Kerdjik, H., Van Pagee, H., Klomp, R. and Schreur, A. (1988). Behaviour and Impact Assessment of Heavy Metals in Estuarine and Coastal Zones. *In: Metals in Coastal*

Environments of Latin America. Seeliger, U, Lacerda, L.D. and Patchineelam, S.R. (eds.). Springer-Verlag, Berlin, pp.157-193

Sholkovitz, E.R., Boyle, E.A. and Price, N.B. (1978). The removal of dissolved humic acids and iron during estuarine mixing. *Earth Planet. Sci. Lett.*, 40:130-136.

Sweeney, M.D. and Naidu, A.S. (1983). Heavy metals in Arctic nearshore sediments, Northern Alaska: concentrations, extraction methodology and chemical associations. *In: International Conference of Heavy Metals in the Environment*, Heidelberg. pp.1094-1097.

Tessier, A., Carignan, B., Dubreuil, B. and Rapin, F. (1989). Partitioning of zinc between the water column and the oxic sediments in lakes. *Geochim.Cosmochim. Acta*, 53:1511-1522.

Tessier, A., Rapin, F. and Carigan, R. (1985). Trace metals in oxic lake sediments: possible adsorption onto iron oxyhydroxides. *Geochim. Cosmochim. Acta*, 49:183-194.

Travassos, M.; Baisch, P., Lacerda, L. (1993). Geochemistry distribution of heavy metals of the Patos Lagoon Estuary - Brazil. *In: Intern. Conf. Heavy Metals in the Environment*. Toronto. Vol. 1:185-188.

Trefrey, J.H., Metz, S. and Trocine, R.P. (1985). The decline in the lead transport by the Mississippi River. *Science.*, 230:439-441.

Vilas-Boas, D.F. (1990). *Distribução e comportamento dos sais nutrientes, elementos maiores e metais pesados na Lagoa dos Patos - RS*. M.Sc. thesis. Univ. de Rio Grande. 122 p.

Yetang, H. and Forstner, U. (1984). Chemical forms of some heavy metals in Huang He river sediments (China) and comparison with data from Rhine river sediments (West Germany). *Geochem.*, 3:37-44.

9 Mass Balance Estimation of Natural and Anthropogenic Heavy Metal Fluxes in Streams Near the Camaquã Copper Mines, Rio Grande do Sul, Brazil

Luciano Laybauer[1] and Edison D. Bidone[2]
[1] Curso de Pós-Graduação em Geociências, Instituto de Geociências, Universidade Federal do Rio Grande do Sul, Caixa Postal 15065, Porto Alegre, RS, Brazil
[2] Departamento de Geoquímica, Instituto de Química, Universidade Federal Fluminense, Outeiro de São João Batista, s/n, Centro, Niterói, RJ, Brazil

Abstract

The Camaquã Copper Mines were closed in 1996 after more than one century of active exploitation. The mines are located in the basin of João Dias creek in the upper course of the Camaquã River, Rio Grande do Sul State, Southern Brazil. The mine-generated effluents enriched in heavy metals were released into the João Dias creek. Both anthropogenic and natural inputs of heavy metals were mixed and added along the creek, and thus simple analysis of metal concentrations in the water is not sufficient to determine whether the inputs are from natural or anthropogenic sources. This study is an attempt to solve these questions based on mass balance studies of heavy metal fluxes between fluvial segments, and shows that this approach is appropriate to differentiate between natural sources and mining (anthropogenic) activities into the basin (total ~ 410 t yr^{-1}, Fe ~ 263 t yr^{-1}, Al ~ 130 t yr^{-1}, Cu ~ 16 t yr^{-1}, Zn ~ 3 t yr^{-1}). It also identifies the critical fluvial segments contaminated by heavy metals - about 98% of the total anthropogenic input occurred in the creek segment under direct influence of mining activities without any effluent control. The study also shows the low dilution capability of the João Dias creek and the tendency to "export" heavy metals out of the basin, rather than incorporate them in bottom sediments.

Introduction

Heavy metals, such as Cu, Pb, Zn, Hg and Cd, from anthropogenic sources are contaminants in superficial waters and generally cause adverse effects on the environment. The pollution associated with these elements is of great interest and has been extensively studied, due to their persistence in the environment as micro-pollutants of high toxicity (UNEP, 1992).

Geologic weathering is the source of background levels of heavy metals in natural superficial waters and leaches elements of the mineralogic assemblage from rocks and soils. Obviously, mineralised zones, such as this case study, an old mining region, show elevated concentrations of some metals in waters, soils and bottom sediments. When a mineral deposit is exploited to retrieve and process the ore, it is very common to increase the concentrations of metals in the area. Consequently, there is the general problem of how to distinguish natural geologic weathering from metal enrichment attributable to human activities, like mining, agriculture or manufacturing (Förstner and Wittmann, 1981).

The Camaquã Copper Mines are situated in the central part of Rio Grande do Sul State, Southern Brazil, and contain mainly copper sulphides (Fig. 1). The copper ore occurs as sulphide veins and as disseminated mineralisation, mostly in conglomerates and sandstones of Neo-proterozoic-Cambrian age (Licht, 1982; Benhome and Ribeiro, 1983; Laybauer, 1995).

Because copper sulphides are very soluble in the surface and groundwater environments and the exposure of these minerals to atmospheric oxygen results in one of the most acidic of all known weathering reactions (Förstner and Wittmann, 1981; Drever, 1982), special attention must be given to sulphide-mining operations and abandoned copper mines. The acidification process contributes to increase the metal charge of fluvial waters.

A very important problem in river water management is the identification of in which segments, in quantitative terms, the contaminant input is released by anthropogenic activities. If the considered water contaminants are heavy metals in a river basin naturally enriched in these metals, that target is made more difficult to identify because the anthropogenic and natural inputs of heavy metals are added along the entire length of the river, and the evidence of spatial concentration gradients in the river waters could be a result of the transfer of metals (geochemical conservative substances) from the upstream river segments (high hydrodynamic environment) to the downstream river segments (low hydrodynamic environment) (Laybauer, 1995; Travassos and Bidone, 1995; Laybauer et al., 1996). However, the simple analysis of metal concentrations in fluvial waters does not yield conclusive information about the metal discharge of mining operations, the natural and anthropogenic contributions to the final metal concentration measured in the stream, the dilution water capacity, the amount of metals retained in the hydrographic basin and the tendency to export metals out of the basin.

This chapter is an attempt to solve these questions, using an approach based on the mass balance of heavy metal fluxes between stream segments that drain the area.

Study Area

The study area is situated 45 km south from Caçapava do Sul City in the central region of Rio Grande do Sul State, Southern Brazil (Fig. 1).

The Camaquã Mines were the main copper mines of Southern Brazil and operated from the last century. The hydrographic sub-basin where the mining operations took place has an area of approximately 310 km². The main drainage of this basin, which received the mine effluents for over a century, is João Dias creek, which flows to the Camaquã River. The Camaquã River Basin is the most important drainage of the Sul-rio-grandense Shield to the Patos lagoon, an important Brazilian coastal lagoon (Baisch, 1991).

The Camaquã Copper Mines mined a sedimentary sequence formed mainly of conglomerates, sandstones and siltites, with only minor andesites of Neo-proterozoic age. The copper ore occurs as sulphide veins and disseminated mineralisation, mostly in conglomerates and sandstones of Neo-proterozoic-Cambrian age (Ribeiro, 1991). The main sulphide minerals observed in these deposits are chalcopyrite – $CuFeS_2$, bornite – Cu_5FeS_4, pyrite – FeS_2 and chalcocite – Cu_2S. Unmineralised rocks close to the mine region have anomalous copper concentrations, around 0.1% Cu, highlighting the elevated background of this element in the area. The disseminated ore itself has a copper concentrations varying between 0.3% and 1% (Ribeiro, 1991).

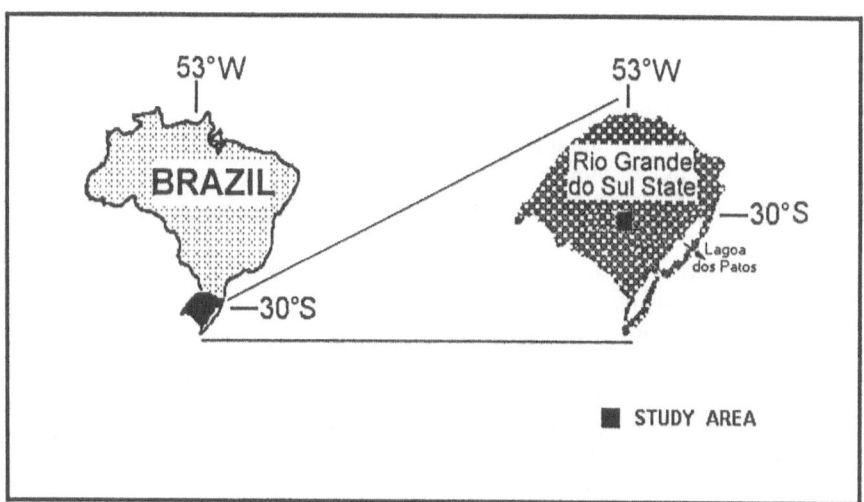

Fig. 1: Location of the study area

The exploitation process was carried out using both underground and open pit mining. During the eighties, when the CBC (Companhia Brasileira de Cobre) Water Monitoring Program was established and the tailing dam constructed, the main mining operation was developed as an open pit. The principal environmental control device used was a tailing dam that received mine effluents with a large amount of suspended solids, including fine grains, coming from the ore treatment (crushing, milling and flotation) and partially associated with mining operations (water pumping from underground and open pit mine) and also from the leaching of stock and tailing piles. A large proportion of these rejects and the effluents associated with these two last processes were dumped directly into João Dias creek.

Material and Methods

The methodological approach proposed in this study is based on a mass balance of heavy metal fluxes between fluvial segments to distinguish between anthropogenic and natural levels of pollution, using a chemical element as indicator of the natural component (Laybauer, 1995; Travassos and Bidone, 1995; Laybauer et al., 1996). This approach is synthesised in the following equation:

$$AC = (MeO - MeI) - [(EiO - EiI) \times (MeOB/EiOB)] \qquad (1)$$

where in the considered stream segment:

AC = anthropogenic component of the total metal flux changes,
MeO = metal flux output,
MeI = metal flux input,
$(MeO - MeI)$ = total metal flux increment,
EiO = indicator element flux output,
EiI = indicator element flux input,
$(EiO - EiI)$ = total indicator element flux increment,
$MeOB$ = metal flux output in the natural background fluvial segment (upstream control area),
$EiOB$ = indicator element flux output in the natural background fluvial segment (upstream control area), and
$[(EiO - EiI) \times (MeOB/EiOB)]$ = natural component of the total metal flux change.

These steady state mass balance calculations were performed considering the terms metal flux "input" and "output" as the multiplication of the total metal (particulate + dissolved) annual mean concentration in the input or output station by the average annual specific discharge (i.e., Q/A = annual mean of fluvial water flux/basin area of the considered sampling station). The resultant unit for the metal fluxes is tons km^{-2} yr^{-1}. Alternatively, to obtain metal fluxes in tons yr^{-1} it is sufficient to multiply the values in tons km^{-2} yr^{-1} by the basin area of the fluvial station.

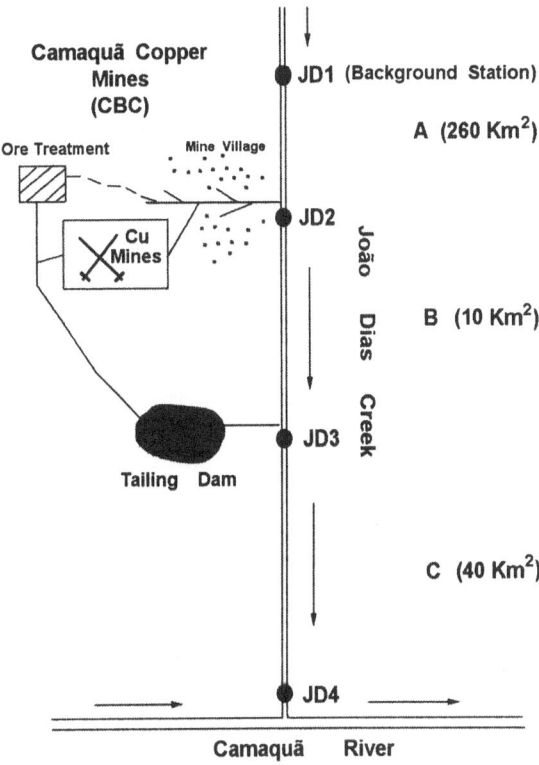

Fig. 2: Diagnostic representation of the mining area, João Dias creek and locations of the four sampling areas

In order to estimate the metal fluxes in the João Dias creek, four sampling stations were selected from upstream to downstream (Fig. 2): station 1 (JD1) represents the upstream control area or the background station; station 2 (JD2) and station 3 (JD3) are located in the region of mining activities, and station 4 (JD4) is at the creek's mouth. These sampling stations were taken as limits of the creek segments (Fig. 2): segment A, between the JD1 (input station) and JD2 (output station); segment B (input station = JD2 and output station = JD3) and segment C (input station = JD3 and output station = JD4).

Heavy metal (Cu, Zn, Fe, Al and Pb) concentrations in superficial water (particulate + dissolved) were obtained from the CBC Water Monitoring Program covering the period from 1980 to 1993, with an irregular monthly sampling. The metal concentration analyses were performed by atomic absorption spectrometric method (AAS), according to the procedures of the Standard Methods for Examination of Water and Wastewater, also followed during sampling and preservation of samples (APHA, 1980; 1985).

The available fluvial discharge data were insufficient and it was necessary to employ a hydrological model called IPHMEN II (Tucci et al., 1990) to estimate the fluvial discharges at all selected creek stations. A less than 5% difference was verified between the estimated fluvial discharges using the model and the available values obtained from the fluvial discharge measurement surveys in the study area (Table 1).

Table 1: Statistics from observed and calculated discharge series, using IPHMEN II (Tucci et al., 1990)

Determination coefficient	Volumes (m³ s⁻¹)		Difference (%)	Mean discharge (m³ s⁻¹)		Standard deviation
	observed	calculated		observed	calculated	
0.97	6048.20	5857.04	3.16	76.56	74.14	42.71

In the João Dias creek Basin, Pb was taken as the standard background element, because its abundance was very restricted in the water, i.e. the mean value was everywhere near 0.01 mg L⁻¹ with small standard deviation (Table 2). Besides Pb is homogeneously distributed (about 20 mg kg⁻¹) in all hydrographic basin materials, such as rocks, soils and uncontaminated bottom sediments (Licht, 1982; Laybauer, 1995).

Results and Discussion

Table 2 shows the database used in the proposed mass balance approach to the heavy metal fluxes in the João Dias creek.

The largest increment in metal concentrations (Zn ~ 150%; Fe ~ 200%; Cu and Al ~ 400%) occurred between the background control station (JD1) and the JD2 station, which is under the influence of the mining effluent. Furthermore, the metal concentrations showed a positive spatial gradient downstream, except for Pb which has a homogeneous distribution along the creek. However, considering the data between stations JD2 and JD4, the fluvial discharge increased ~ 20% and the metal concentrations changed ~ 20-33% (Table 2). These changes may be due to lower metal input (from natural and/or anthropogenic sources) downstream and the transfer of the metal concentrations from the upstream segments (high hydrodynamic environment) to the downstream segments (low hydrodynamic environment). The Al negative gradient (~ –20%), may be due to water dilution influence downstream without significant Al input into the river and/or removal of Al from the water by some hydrogeochemical process (e.g., sedimentation of Al bound to particulate suspended material). The proposed mass balance approach to the metal fluxes tries to solve these problems.

Table 3 shows the results using the proposed approach, where metal fluxes were estimated by multiplying the total metal (particulate + dissolved) annual mean concentration by the specific superficial water flux annual mean (Q/A); (i) input station and (o) output station; Increment = Me(o) – Me(i) total metal flux increment in the considered fluvial segment; Natural = [(Pb(o) – Pb(i) × (Me(o)B/Pb(o)B)] natural component of the total metal flux increment in the considered fluvial segment using Pb as a natural indicator. B refers to background control station JD1; AC = (Me(o) – Me(i)) – [(Pb(o) – Pb(i)) × (Me(o)B/Pb(o)B)] anthropogenic component of the total metal flux increment in the considered fluvial segment.

The metal flux estimates using the specific discharge (Q/A) instead of the simple discharge value (Q) is a better procedure for the analysis of the metal flux spatial gradients along the drainage, because it does not consider the weight of the discharge. In fact, if the fluvial water discharge (Q) increases strongly downstream, then it will show a clear positive gradient in that direction. This gradient naturally bears high energy, making evaluation of metal flux gradients along the river almost impossible. The Q/A procedure eliminates the natural tendency of the metal fluxes to simply reflect the downstream increase in discharge.

Table 2: Database used to estimate the heavy metal fluxes in the João Dias creek. A = basin area (km^2); Q = annual mean of the fluvial water discharge ($m^3 s^{-1}$); Metal concentration (dissolved + particulate) annual mean ± standard deviation ($mg L^{-1}$) and (n) = number of samples

Station	A	Q	Cu	Pb	Zn	Fe	Al
JD1	10	0.18	0.03 ± 0.02 (19)	0.01 ± 0.01 (14)	0.02 ± 0.02 (22)	1.26 ± 0.43 (28)	0.29 ± 0.27 (24)
JD2	260	5.82	0.12 ± 0.12 (40)	0.01 ± 0.01 (20)	0.03 ± 0.03 (38)	2.79 ± 2.26 (42)	1.10 ± 1.25 (40)
JD3	270	5.98	0.15 ± 0.10 (62)	0.01 ± 0.01 (36)	0.04 ± 0.03 (63)	3.21 ± 3.26 (58)	0.99 ± 1.20 (57)
JD4	310	6.97	0.16 ± 0.12 (62)	0.01 ± 0.01 (36)	0.04 ± 0.03 (63)	3.36 ± 3.01 (59)	0.87 ± 1.11 (58)

In the João Dias creek, the metal fluxes have a spatial gradient similar to the metal concentration distribution along the creek; i.e., the metal fluxes have a positive spatial gradient downstream. The greatest increase in metal fluxes occurs between the background station (JD1) and the JD2 station, the most directly influenced by the mining effluents. Pb is an exception, for it is present at the same flux values in all stations. This fact is in agreement with the suggestion that Pb is a natural element indicator in the basin. Al showed a tendency for a negative gradient from the JD2 station downstream.

Table 3: Metal fluxes (tons km^{-2} yr^{-1}) in the João Dias creek, Camaquã Copper Mines

Segment	Station	Cu	Zn	Fe	Al	Pb
	JD1 (i)	0.02	0.01	0.78	0.18	0.01
	JD2 (o)	0.08	0.02	1.94	0.76	0.01
A	Increment	0.06	0.01	1.15	0.58	0.00
	Natural	0.00	0.00	0.13	0.06	------
	AC*	0.06	0.01	1.02	0.52	------
	JD2 (i)	0.08	0.02	1.94	0.76	0.01
	JD3 (o)	0.10	0.03	2.22	0.68	0.01
B	Increment	0.02	0.01	0.28	-0.08	0.00
	Natural	0.00	0.00	0.00	------	------
	AC*	0.02	0.01	0.28	------	------
	JD3 (i)	0.10	0.03	2.22	0.68	0.01
	JD4 (o)	0.11	0.03	2.36	0.61	0.01
C	Increment	0.01	0.00	0.14	-0.07	0.00
	Natural	0.00	0.00	0.02	------	-----
	AC*	0.01	0.00	0.12	------	-----

* AC = anthropogenic contribution

The JD2 station is located downstream from mining facilities which released untreated effluents (contaminated by metals) into the creek (e.g., effluents from the ore treatment plant, pumped water from the underground mine, leaching of copper stock piles, mine village), while the JD3 station is just below the mine tailing dam into which the effluents from the open pit mine were released (Fig. 2). The observed difference in metal flux between stations JD2 and JD3 (between untreated effluents and effluents from the tailing dam) is < 15-30%, suggesting that the mine tailing dam was effective in reducing the impact of the effluents.

The metal flux increments in segment A are several times higher than those in segments B and C. Moreover, it was observed that the estimated total metal flux increment in the creek segments were essentially due to anthropogenic inputs. Therefore, the predominant anthropogenic metal inputs occurred in segment A.

The "negative values" obtained for Al fluxes in segments B and C may be associated with the incorporation of this element into the bottom sediments of the João Dias creek. However, this was not observed for the other metals considered in this study. These "negative values" were only 15% lower than the positive increment observed in segment A. As already verified, Al and Fe are associated with suspended solids originated in the mining effluents, whereas Cu and Zn are preferentially ionic species (Laybauer, 1995). Both finely particulate and ionic materials favour metal transferences down this stream. Therefore, the positive gradient of the metal fluxes in the fluvial sampling stations, associated with the hydrodynamic characteristics of the creek, suggest a predominant tendency for metals to be "exported" from the João Dias basin to the Camaquã River (Fig. 2).

Using Cu as an example and considering the ~ 600% negative difference value of Cu flux increments between creek segments A and C (Table 3), the ~ 30% positive difference value for Cu flux between stations JD2 and JD4 (Table 2); the ~ 25% positive difference value for Cu concentrations between stations JD2 and JD4 (Table 2); and the ~ 20% positive difference value for fluvial discharges between stations JD2 and JD4 (Table 2), it can be established that the João Dias creek presents a low ability to dilute and to retain this metal in the basin. Thus, there is a tendency to "export" Cu out of the João Dias basin. In fact, if anthropogenic Cu flux increments were great in the B and C river segments, the fluvial discharge values would be insufficient to reduce (by dilution) the Cu concentrations. Therefore, the negative gradient observed for the metal flux increments depends on the reduction of the charges of anthropogenic metal inputs, rather than on water dilution or the incorporation of copper into the stream sediments. Table 4 shows a synthesis of the anthropogenic component of the metal fluxes in the João Dias creek.

Table 4: Synthesis of estimated anthropogenic metal loads (tons yr^{-1}) in João Dias creek

Seg-ment	Station	Area (km^2)	Cu	Zn	Fe	Al	Total
A	JD1$_{(i)}$ - JD2 $_{(o)}$	260	15.0 (3.7)	2.5 (0.6)	255.0 (63.4)	130.0 (32.3)	402.5 (98.1)
B	JD2 $_{(i)}$ - JD3 $_{(o)}$	10	0.2 (6.5)	0.1 (3.2)	2.8 (90.3)	0.0 (0.0)	3.1 (0.8)
C	JD3 $_{(i)}$ - JD4 $_{(o)}$	40	0.4 (7.7)	0.0 (0.0)	4.8 (92.3)	0.0 (0.0)	5.2 (1.1)
Total tons yr^{-1} (%)			15.6 (3.7)	2.6 (0.6)	262.6 (64.0)	130.0 (31.7)	410.3 (100)

( ) Percentage of metal loads in each segment and whole basin.

Virtually all (98%) of the anthropogenic metal input into the drainage occurs in segment A, the largest fluvial segment and the most impacted by the mining activities. Fe and Al contributions are respectively 64% and 32% of the total metal anthropogenic inputs (~ 410 tons yr^{-1}).

Conclusions

The João Dias basin has rocks, soils and sediments naturally enriched in heavy metals. The natural and mining related inputs of heavy metals were mixed and added along the drainage, and anthropogenic components were "masked" by metal input from natural sources.

The largest increment in metal concentrations (Zn ~ 150%; Fe ~ 200%; Cu and Al ~ 400%) occurred between the background control station (JD1) and the JD2 station, which is under the influence of the mining effluent release. Approximately 98% of the anthropogenic metal input occurs in segment A, the largest fluvial segment and the most impacted by the mining activities. Fe and Al contributions are respectively 64% and 32% of the total anthropogenic metal inputs (~410 t yr^{-1}). The observed difference between the metal fluxes at stations JD2 and JD3 suggests that the mining tailing dam was effective in reducing metal concentrations in the effluents. The research showed that the João Dias creek has a significant capacity to "export" heavy metals to the Camaquã River.

Our approach succesfully estimated the metal charges released by the mining activities into the drainage, identify the critical fluvial segments contaminated by heavy metals and also estimate the natural and anthropogenic metal fluxes. Therefore, the proposed mass balance approach to metal fluxes, as employed in the João Dias creek, may be a useful methodological approach to the river water management.

References

APHA - American Public Health Association (1980). *Standard Methods for Examination of Water and Wastewater*, 15. ed., Washington, 1134 p.

APHA - American Public Health Association (1985). *Standard Methods for Examination of Water and Wastewater*, 16. ed., Washington, 1268 p.

Baisch, P. (1994). *Les oligo-éléments métalliques du Système Fluvio-lagunaire dos Patos (Brésil) - Flux et Devenir*. Ph.D. thesis. Université de Bordeaux I. no. 1136. 345p.

Benhome, M.G. and Ribeiro, M.J (1983). Datações K-Ar das argilas associadas a mineralização de cobre da Mina Camaquã e de suas encaixante, In: *Atas do I Simpósio Sul-Brasileiro de Geologia*, September 1983, SBG, Porto Alegre, pp.: 82-88.

Drever J.I. (1982). *The Geochemistry of Natural Waters*, Englewood Cliffs: Prentice-Hall, 388p.

Förstner U. and Wittmann G.T.W. (1981). *Metal Pollution in the Aquatic Environment*, Springer-Verlag, Berlin, 486p.

Laybauer L. (1995). *Análise das transferências de metais pesados em águas e sedimentos fluviais na região das Minas do Camaquã, RS*. Master of Science thesis in Geosciences, Universidade Federal do Rio Grande do Sul, Porto Alegre, 164 p.

Laybauer L., Bidone E.D. and Nardi L.V.S. (1996). Fluvial heavy metal contamination in the Camaquã copper mine region, southern Brazil: an approach based on natural and anthropogenic metal flux component segregation. *In*: *Encontro da Academia Brasileira de Ciências*, May 1996, Porto Alegre, 68 (2), 292-293.

Licht O.A.B. (1982). *Prospecção geoquímica aplicada a pesquisa de sulfetos não aflorantes associados a rochas sedimentares Eo-Paleozóicas na região da fazenda Santa Maria, Caçapava do Sul, RS*. Master of Sciences thesis in Geosciences, Universidade Federal do Rio Grande do Sul, Porto Alegre, 119p.

Ribeiro M. (1991). *Sulfetos em Sedimentos Detríticos Cambrianos do Rio Grande do Sul*. Doctoral thesis, Universidade Federal do Rio Grande do Sul, Porto Alegre, 2v.

Travassos M.P. and Bidone E.D. (1995). Avaliação da contaminação por metais pesados na bacia do rio Caí-RS através de uma análise dinâmica. In: *Desenvolvimento Sustentável dos Recursos Hídricos*. Sant'ana R. et al. (eds.), ABRH, 3, 205-211.

Tucci C.E.M., Damiani A.R.R. and Perrit R. (1990). Potenciais Impactos no Escoamento Devido a Modificação Climática: Avaliação Preliminar, RBE, *Caderno de Recursos Hídricos*, 8 (1), 65-79.

UNEP (1992). *Chemical Pollution: A Global Overview*. United Nations Environmental Program, Geneva, 106p.

10 A Model for Geochemical Partitioning of Heavy Metals in the Mar Del Plata Coastal Ecosystem, Argentina

Jorge E. Marcovecchio[1], Laura D. Ferrer[1], Alberto O. Barral[2], Marcelo O. Scagliola[3] and Adán E. Pucci[1]
[1] Lab. de Química Marina, Instituto Argentino de Oceanografía (IADO), Av. Alem 53, 8000 Bahía Blanca, Argentina
[2] Instituto Nacional de Investigación y Desarrollo Pesquero (INIDEP), C.C. 175, 7600 Mar del Plata, Argentina
[3] Obras Sanitarias Mar del Plata Sociedad de Estado (OSSE), Roca 1213, 7600 Mar del Plata, Argentina

Abstract

Geochemical partitioning of Zn, Cu, Fe and Mn was determined in the coastal marine sediments of Mar de Plata (Argentina), to assess the degree of contamination and the metal chemistry. The sequential extractions were tested to be included in monitoring programmes of the coastal environments. Five sediment samples were collected in the coastal area near to the city of Mar del Plata and analysed for total metal concentrations by atomic absorption spectrophotometry. Sequential extractions were also performed, partitioning the metals into five geochemical fractions. Considering the distribution trends as observed from the heavy metal concentrations, two main sources were identified in the considered area: that related to Mar del Plata harbour (Station 5), and the area affected by the pre-treatment plant of the urban and industrial sewage disposal of the city (Station 2). The method has shown itself to be an excellent tool for environmental evaluation, and it would be desirable to include it in future programmes directed to assessing the health quality of aquatic ecosystems.

Introduction

The role of heavy metals as environmental pollutants has been largely recognized. Several of these elements - i.e., cadmium, lead, mercury - have sometimes been found in anomalously high concentrations in the marine environment, apparently related to naturally occurring deposits (Giordano et al., 1992). However, anthropic activities still remain the cause of the increased amount of polluting substances - heavy metals among them - which have been dumped into aquatic systems. When contaminants such as heavy metals are introduced into marine or estuarine ecosystems, they often accumulate in the sediments (Rule and Alden, 1990). Thus, heavy metal pollution deserves special attention because of its high toxicity and persistence in the aquatic environment; particularly, ecosystems such as seaports or other industrialized coastal areas that have chronic inputs of metals are usually associated with highly contaminated sediments. This characteristic has led to concerns over the ecological effects that may be associated with sediment quality. Studying the levels and distributions of trace metals in marine or estuarine sediments has proved a practical way of assessing metal pollution in coastal ecosystems (Ryan and Windom, 1988). However, when the study is limited only to total metal contents, the result is a partial evaluation which must be complemented by additional information in order to assess the risk of metal contamination not only to the ecosystems but to human health (Lacerda et al., 1988). Thus, sediments may contain appreciably high concentrations of heavy metals, and the question arises as to what percentage of the total concentration of heavy metals will be available for release into the water or to be incorporated in aquatic organisms over a long period of time (Gibbs, 1973; 1977). Since the reactivity of metals depends on their chemical forms, and thus determines their capacity of interaction with biotic or non-biotic components of the ecosystem, and keeping in mind that the bio-availability and recycling of metals in the aquatic environment are dependent upon the corresponding associated geochemical fractions in the sediments, the importance of developing studies on geochemical partitioning of metals is obvious.

Even though studies on geochemistry and partitioning of trace metals have been largely carried out in many aquatic ecosystems (e.g., Lacerda et al., 1988; Elsokkary and Müller, 1989; Ponce-Velez and Vasques Botello, 1991; Giordano et al., 1992), up to this moment this kind of information is scarcely available for coastal or estuarine ecosystems from the south-western Atlantic Ocean (Marcovecchio et al., 1993).

The first aim of the present chapter is to determine the geochemical partitioning of some heavy metals (copper, iron, zinc and manganese) in coastal marine sediments from Mar del Plata littoral, in the south-eastern part of Buenos Aires Province, Argentina.

Materials and Methods

Samples of coastal marine sediments were collected in five sampling stations near Mar del Plata city (38° 00' S and 57° 02'W), located on the south-eastern coast of Buenos Aires Province, in Argentina (Fig. 1). In each station samples of the overlying layer of sediment (upper 5 cm) were collected from a depth of 20 m, on board the research vessel *Luisito* (Escuela Nacional de Pesca, Mar del Plata, Argentina).

All samples were kept in polyethylene bags, and stored in a freezer (at -20°C) until their treatment in the laboratory. Before chemical manipulation, all visible marine organisms and shell fragments were hand removed. Samples were dried at $40 \pm 5°C$, until constant weight. After that, samples were sieved, and the fraction

$40 \pm 5°C$, until constant weight. After that, samples were sieved, and the fraction smaller than 62 µm was separated. This fraction was analytically treated in order to determine the geochemical partitioning of copper, zinc, manganese and iron, following the method described by Megalatti et al. (1983) and modified by Maddock and Lópes (1988) and Lacerda et al. (1988). This method allows information to be obtained on the following geochemical fractions of metals:

- Fraction I : *Exchangeable adsorbed metals* (after treatment with ammonium acetate).
- Fraction II: *Oxidizable metal complexes* (after treatment with hydrogen peroxide, nitric acid and ammonium acetate).
- Fraction III: *Metals in carbonates* (after treatment with acetic acid and sodium acetate).
- Fraction IV: *Reducible complexes* (after treatment with hydroxylamine, hydrochloride and acetic acid).
- Fraction V: *Residual metals.*

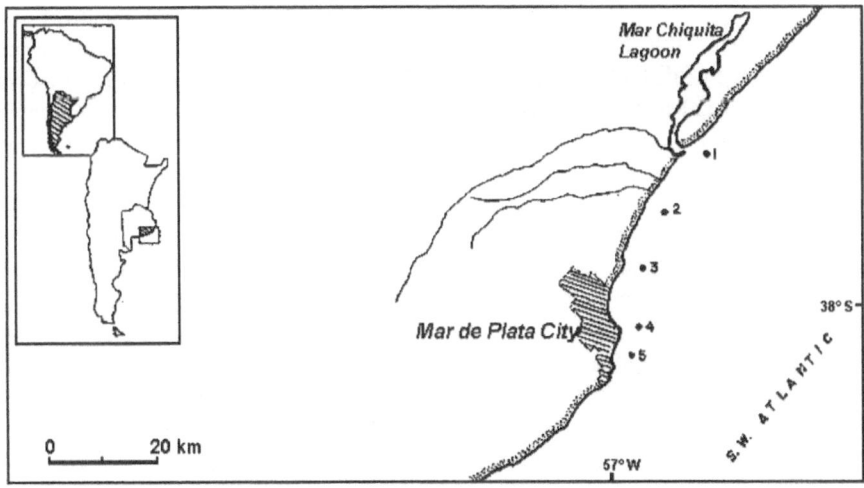

Fig. 1: Location of the sampling stations in the studied area

Furthermore, the total metal contents of the studied sediments were determined following the method described by Marcovecchio et al. (1988). A Shimadzu AA-640-13 atomic absorption spectrophotometer was used for the analyses, working with air-acetylene flame and with deuterium background correction (D_2BGC). Analytical grade reagents were used for the blanks and corresponding calibration curves. Comparisons between the metal concentrations obtained were performed through one way analyses of variance (ANOVA), using the SYSTAT software.

Results and Discussion

The geochemical partitioning of zinc, copper, manganese and iron was determined in coastal marine sediments from the littoral area neighbouring Mar del Plata city, in the Province of Buenos Aires, Argentina (Fig. 2). The studied area is part of an oceanographically steady system, with a strong littoral drift which flows from south to north, with an average current speed of less than one knot. Moreover, the system shows a clear seasonal variation in the values of the oceanographic parameters (i.e., salinity, temperature, or micro-nutrient concentrations - including nitrates, nitrites, ammonium, phosphates and silicates), which corroborate the productivity and environmental conditions of the ecosystem (Brandhorst and Castello, 1971). This characteristic of the ecosystem is an important starting point which must be taken into account in order to understand the occurrence and distribution of heavy metals in the area, obviously linked not only to the pollutant inputs but also to the oceanographic circulation.

Total Metal Concentrations

The total concentration of heavy metals as recorded in the sediments of the assessed area (Table 1) proved to be lower than values reported by other authors for sediments of different environments. The levels of trace metals as reported in the present paper appears to have a homogeneous distribution along the evaluated area, even though several sampling stations showed higher contents of some metals than of others for all the studied elements. In order to establish an adequate comparison, mean values of total copper as determined for Mar del Plata coastal ecosystem - which varies between 16.27 and 48.33 $\mu g\ g^{-1}$ - are compared with those reported by Lacerda et al. (1988) for sediments from Guarapina Lagoon (19.8 to 26.8 $\mu g\ Cu\ g^{-1}$), Estrela River (74.5 to 731.0 $\mu g\ Cu\ g^{-1}$), Iguaçu River (21.7 to 191.0 $\mu g\ Cu\ g^{-1}$), or São Francisco Canal (21.8 to 28.3 $\mu g\ Cu\ g^{-1}$) among others. In addition, Lyons and Gaudette (1979) have reported values ranging from 2.4 to 35.1 $\mu g\ Cu\ g^{-1}$ in sediments from Jeffreys Basin, Gulf of Maine, north-eastern USA.

On the other hand, values of zinc as determined for sediments from Mar del Plata (varying between 45.93 and 103.8 $\mu g\ Zn\ g^{-1}$) are compared with those reported by Jickells and Knap (1984) for sediments from a Bermuda coastal environment (1.4 to 250.0 $\mu g\ Zn\ g^{-1}$), or those reported by Patchineelam et al. (1988) in sediments from Guarapina (Brazil) coastal lagoon (35 to 115 $\mu g\ Zn\ g^{-1}$).

The concentrations of manganese as determined in sediments from Mar del Plata, and reported in the present paper, varied between 330.84 and 631.20 $\mu g\ Mn\ g^{-1}$. These manganese contents are comparable with those reported by Greenaway et al. (1988) in sediments from a Jamaican coastal environment (from 42 to 519 $\mu g\ Mn\ g^{-1}$), or by Patchineelam et al. (1988) in sediments from Guarapina coastal lagoon, in the State of Rio de Janeiro, Brazil (from 150 to 400 $\mu g\ Mn\ g^{-1}$).

Table 1: Total metal concentrations in coastal sediments from Mar del Plata ecosystem (Concentrations in $\mu g\ g^{-1}$, dry weight)

Studied metal	Station no. 1	Station no. 2	Station no. 3	Station no. 4	Station no. 5
Zinc	103.80	96.05	45.93	63.46	73.27
Copper	16.61	48.33	16.27	21.98	37.10
Manganese	631.20	545.70	330.84	400.93	439.76
Iron	7813.70	10931.70	8160.20	5434.80	27262.60

Finally, iron concentrations as reported in the present paper for Mar del Plata coastal sediments, which varied between 5.43 and 27. 26 mg Fe g^{-1}, are fully compared with previous reports from other authors, like those of Lyon and Gaudette (1979) for sediments from Jeffreys Basin, Gulf of Maine, north eastern USA (11.5 to 43.6 mg Fe g^{-1}); also the data of Giordano et al. (1992) for sediments from the Italian coasts, varying between 2.4 ± 0.7 mg Fe g^{-1} and 52.2 ± 29.0 mg Fe g^{-1}, are mentioned.

The concentration values of all the investigated heavy metals showed a slight dispersion, probably as a consequence of both the large sampling area and the location of the recognized local sources of pollutants into the system: the outlet of the pre-treatment plant for the urban and industrial sewage disposal of Mar del Plata city (Station 2), and Mar del Plata harbour area (Station 5). Nevertheless, total heavy metal levels as recorded in sediments of Mar del Plata littoral ecosystem are lower than those reported for coastal areas - including harbours - from other environments in the world (Catsiki and Arnoux, 1987; Fowler, 1990).

Partitioning of Heavy Metals

The study of the geochemical partitioning of zinc, copper, manganese and iron showed specific results for each of the evaluated metals (Fig. 2), but all of them followed the same distribution trend along the studied coastal ecosystem (Fig. 3).

In the case of zinc (Fig. 2) the partitioning showed that most of the metal (average value 91%) was associated with Fraction 5 (F_5), which corresponds to metals bound to the crystalline lattices of mineral particles. This metal fraction is essentially unavailable in the sedimentary environment (Salomons and Förstner, 1984). Moreover, zinc associated with Fraction 2 (F_2, oxidizable metal complexes) and Fraction 4 (F_4, metal reducible complexes) appears to be recorded in most of the studied stations, in percentages varying between 0.35 and 3.5 % for F_2, and between 1 and 4.6 % for F_4. Besides, percentages of zinc associated with Fraction 1 (F_1, exchangeable adsorbed metals) varied 0.25 and 6 %. This occurrence of Zn-F_1 was only recorded in Station 2 (close to Mar del Plata city sewage pre-treatment plant), Station 4 (Cabo Corrientes), and Station 5 (close to Mar del Plata harbour). Finally, zinc associated with Fraction 3 (F_3, metals in carbonates) was only recorded at Stations 2 and 5, at percentages of 2.5 and 4% respectively.

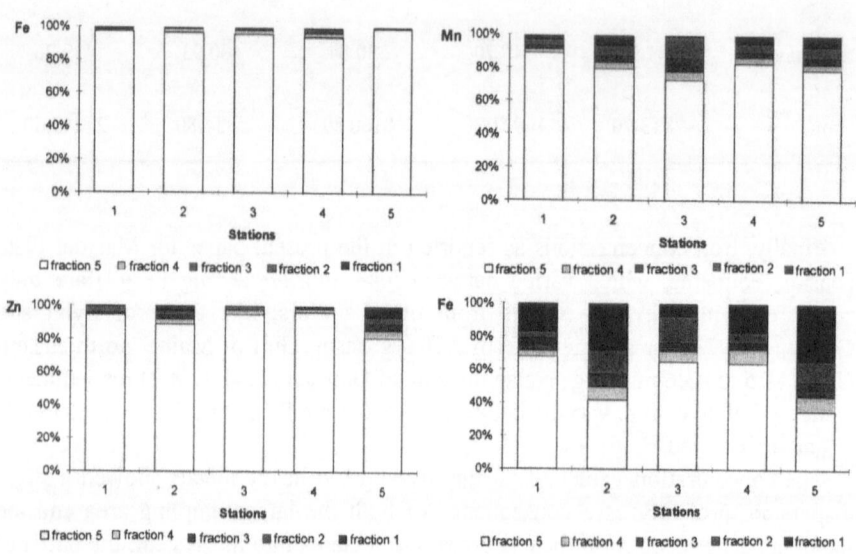

Fig. 2: Geochemical partitioning of zinc, copper, manganese and iron in the studied coastal marine sediments

When the geochemical partitioning of manganese was studied it was observed that the above mentioned five Fractions of metal were recorded in all the sampling stations (Fig. 2). Mn-F_1 varied between 1.7 % in Station 3 (Punta Iglesias) and 7 % in Station 5 (close to Mar del Plata harbour). Correspondingly Mn-F_2 varied from 3.2 % in Station 1 (outlet of Mar Chiquita Coastal Lagoon) to 12 % in Station 3 (Punta Iglesias). Furthermore, both the Mn-F_3 and Mn-F_4 showed percentages varying between 2.5 % and 2 % respectively in Station 1 (outlet of Mar Chiquita Coastal Lagoon), and 7.6 % and 6 % respectively in Punta Iglesias (Station 3). Finally, Mn-F_5 was highest, with percentages ranging from 72.7 % in the Station 3 (Punta Iglesias) to 89.8 % close to Mar Chiquita Coastal Lagoon (Station 1).

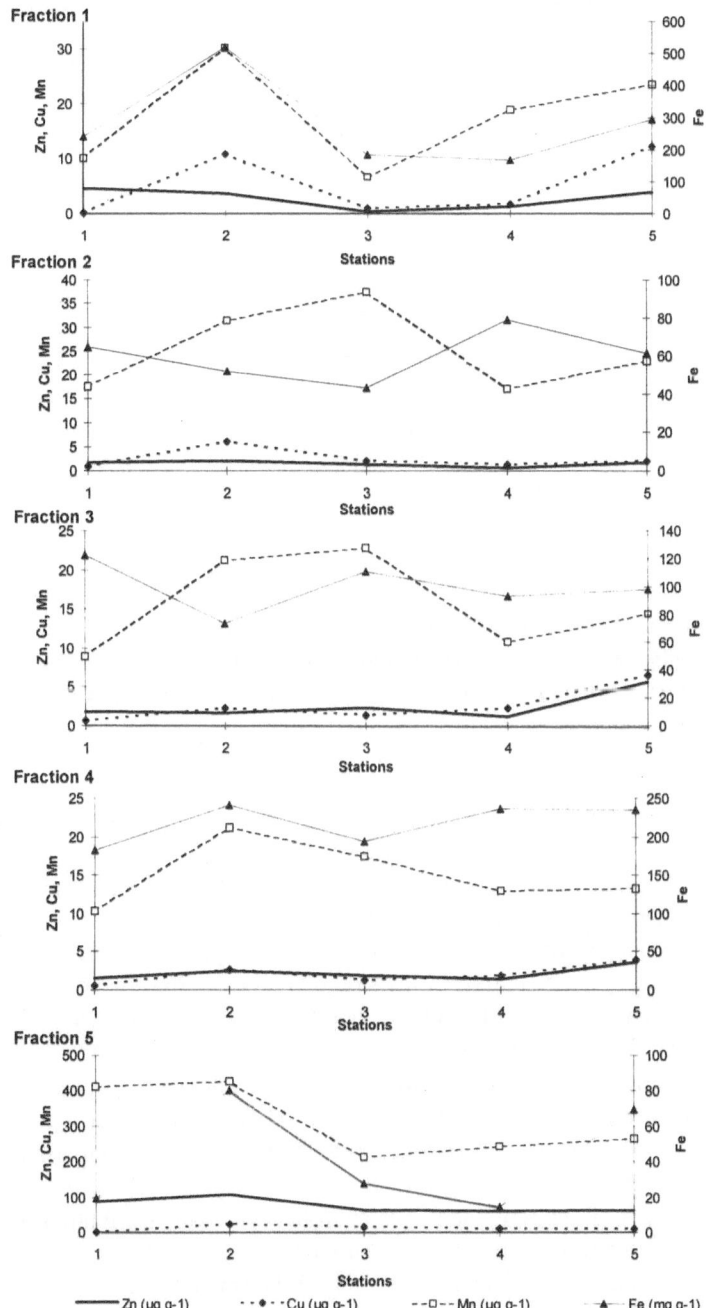

Fig. 3: Geochemical distribution of trace metals in coastal sediments of the Mar del Plata area

The variation of percentages between geochemical fractions was largest in the case of copper (Fig. 2). $Cu-F_1$ showed variations from 7.5 % in Station 3 (Punta Iglesias) to 35 % near Mar del Plata harbour (Station 5), while $Cu-F_2$ varied from 4.4 % in the outlet of Mar Chiquita Coastal Lagoon (Station 1) to 15 % in Station 3 (Punta Iglesias). The percentage of $Cu-F_3$ was lowest in Cabo Corrientes - Station 4 (4.6 %) - and highest close to Mar del Plata harbour - Station 5 (9%) - while $Cu-F_4$ showed percentages ranging from 6.8 % close to Mar Chiquita Coastal Lagoon (Station 1) to 9 % near Mar del Plata harbour (Station 5). Finally, in Fraction 5 the percentage of $Cu-F_5$ varied from 34.9 % in Station 5 (Mar del Plata harbour) to 67.8 % in the outlet of Mar Chiquita Coastal Lagoon (Station 1).

The last metal analysed was iron, and its geochemical partitioning (Fig. 2) showed that $Fe-F_1$ percentages varied between 0.7 % (Stations 2, 3 and 5) and 1.2 % (Stations 1 and 4), while those of $Fe-F_2$ varied between 0.1 % (Stations 2 and 5) and 0.6 % (Station 4). Moreover, Stations 2 and 5 showed the lowest percentages of $Fe-F_3$ (0.1%), while Station 4 showed the highest (0.8 %). The percentage of $Fe-F_4$ was lowest in Stations 2 and 5 (0.3 %) and highest in Station 1 (1%), while $Fe-F_5$ ranged between 96.1 % (Station 4) and 98.8 % (Stations 2 and 5).

Even though the absolute concentrations of the evaluated trace metals here reported were different to the corresponding ones from other authors and other environments, not only the distribution trend of all the geochemical fractions studied, but also their respective percentages of occurrence clearly agreed with corresponding data reported internationally: those of Lacerda et al. (1988) for sediments from the Frade River, Ingaíba River, Guarda River, São Francisco Canal, Estrela River, Iguaçú River, and Guarapina Lagoon (all of them in Brazil); of Rauret et al. (1989) for sediments from the coast off Barcelona, in Spain; of Ponce-Velez and Vasques Botello (1991) from sediments from the Términos Lagoon, at Campeche, México; or, of Giordano et al. (1992) from sediments from different Italian coastal environments.

Moreover, considering the distribution trends as observed from the evaluated heavy metals, two main sources of the studied pollutants have been identified for the considered area: Mar del Plata harbour (Station 5), and the area affected by the pre-treatment plant for the urban and industrial sewage disposal of the city (Station 2). As has been mentioned, one of the central points of scientific research linked with environmental assessment is the determination of the biological availability of sediment-bound trace elements (Luoma, 1989). One of the most important uses of the *"assessment of geochemical partitioning of trace metals"* is to identify the major influences of anthropic activities upon heavy metals distribution in natural ecosystems. In the results presented here it could be observed that the magnitudes of Fractions 1 and 2 ($Me-F_1$ and $Me-F_2$) for all the studied metals are remarkably increased; this fact is an extremely important point, keeping in mind that this two fractions constitute the *"potentially available fraction"* for the evaluated environment (Marcovecchio et al., 1993). This matter has been widely emphasized by many authors (e.g., Calmano and Förstner, 1983; Salomons and Förstner,

1984), who have pointed out that the decreasing importance of the residual fraction of metals ($Me-F_5$), or conversely, the increase of the other geochemical fractions ($Me-F_1$, $Me-F_2$, $Me-F_3$, or $Me-F_4$), are strongly associated with pollution processes.

Thus, method used in this study has shown itself to be an excellent tool for environmental evaluation, and it would be desirable to include it in future programmes directed to assess the *health quality* of aquatic ecosystems.

References

Brandhorst, W. and J.P.Castello (1971). Evolución de los recursos de anchoíta (*Engraulis anchoita*) frente a la Argentina y Uruguay. I. Las condiciones oceanográficas, sinópsis del conocimiento actual sobre anchoíta y el plan para su evaluación. *Proy.Des.Pesq.*, Ser. Inf. Técn., 29:63 pp.

Calmano, W. and U.Förstner (1983). Chemical extraction of heavy metals in polluted river sediments in Central Europe. *Sci.Tot.Environ.*, 28:77-90.

Catsiki, A.V. and A.Arnoux (1987). Etude de la variabilité des teneurs en Hg, Cu, Zn, et Pb de trois espèces de mollusques de l'Etange de Berre (France). *Mar.Environ.Res.*, 21:175-187.

Elsokkary, H. and G.Müller (1989). Geochemical association of heavy metals in sediments of the Nyle River, Egypt. In: *"Heavy Metals in the Environment"*, J-P.Vernet (ed), CEP Cons.Ltd., Edinburgh, UK, pp. 134-137.

Fowler, S.W. (1990). Critical review of selected heavy metal and chlorinated hydrocarbon concentrations in the marine environment. *Mar.Environ.Res.*, 29:1-64.

Gibbs, R.J. (1973). Mechanisms of trace metals transport in rivers. *Science*, 180 : 71-73.

Gibbs, R.J. (1977). Transport phases of transition metals in the Amazon and Yukon rivers. *Geol.Soc.Am.Bull.*, 88:829-843.

Giordano,R., L.Musmeci, L.Ciaralli, I.Vernillo, M.Chirico, A.Piccioni and S.Constantini (1992). Total contents and sequential extraction of mercury, cadmium and lead in coastal sediments. *Mar.Pollut.Bull.*, 24 (7):350-357.

Greenaway, A.M., R.J.Lancashire and A.I.Rankine (1988). Metal ion concentration in sediments from Hellshire, a Jamaican coastal environment. In : *"Metals in coastal environments of Latin America"*, U.Seeliger, L.D.Lacerda and S.R.Patchineelam (eds.), Springer-Verlag, Heidelberg, pp.77-85.

Jickells, T.D. and A.H.Knap (1984). The distribution and geochemistry of some trace metals in the Bermuda coastal environment. *Estuar. Coast. Shelf Sci.*, 18:245-262.

Lacerda, L.D., C.M.M.Souza and M.H.D.Pestana (1988). Geochemical distribution of Cd, Cu, Cr and Pb in sediments of estuarine areas along the southeastern Brazilian coast. In: *"Metals in coastal environments of Latin America"*, U.Seeliger, L.D.Lacerda and S.R.Patchineelam (eds.), Springer-Verlag, Heidelberg, pp.86-99.

Luoma, S.N. (1989). Can we determine the biological availability of sediment-bound trace elements ? *Hydrobiologia*, 176/177:379-396.

Lyons, W.B. and H.E.Gaudette (1979). Sediment geochemistry of Jeffreys Basin, Gulf of Maine: inferred transport of trace metals. *Oceanol.Acta*, 2(4):477-481.

Maddock, J.E.L. and C.E.A.Lópes (1988). Behaviour of pollutant metals in aquatic sediments. In: *"Metals in coastal environments of Latin America"*, U.Seeliger, L.D.Lacerda and S.R.Patchineelam (eds.), Springer-Verlag, Heidelberg, pp.100-105.

Marcovecchio,J.E., V.J.Moreno and A.Pérez (1988). Determination of some heavy metal baselines in the biota of Bahía Blanca, Argentina. *Sci.Tot.Environ.*, 75:181-190.

Marcovecchio,J.E., A.O.Barral, M.S.Gerpe, M.O.Scagliola and L.D.Ferrer (1993). A model of geochemical distribution of trace metals in a coastal ecosystem from Buenos Aires Province, Argentina. *In*: *"Perspectives on Environmental Geochemistry in Tropical Countries"*, J.J.Abrao, J.C.Wasserman and E.V.Silva Filho (eds.), UFF, Niteroi, Brazil, pp. 333-336.

Megelatti, N., O.Robbe, P.Marchandise and M.Astruk (1983). A new chemical extraction procedure in the fractionation of heavy metals in sediments. *Proc. Internat. Conf. Heavy Metals in the Environment*, 4:1090-1093

Patchineelam, S.R., C.M.Leitão Filho, K.Kristotakis and H.J.Tobschall (1988). Atmospheric lead deposition into Guarapina Lagoon, Rio de Janeiro State, Brazil. *In*: *"Metals in coastal environments of Latin America"*, U.Seeliger, L.D.Lacerda and S.R.Patchineelam (eds.), Springer-Verlag, Heidelberg, pp.65-76.

Ponce-Velez, G. and A.Vasques-Botello (1991). Aspectos geoquímicos y de contaminación por metales pesados en la Laguna de Términos, Campeche. *Hidrobiológica*, 1(2):1-10.

Rauret, G., R.Rubio, J.F.López-Sánchez and C.Samitier (1989). Metal solid speciation in anoxic marine sediments off the coast of Barcelona (Spain). *In*: *"Heavy Metals in the Environment"*, J-P.Vernet (ed), CEP Cons. Ltd., Edinburgh, UK, pp.590-593.

Rule, J.H. and R.W.Alden III (1990). Cadmium bioavailability to three estuarine animals in relation to geochemical fractions to sediments. *"Arch.Environ.Contam. Toxicol."*, 19 : 878-885.

Ryan, J.D. and H.L.Windom (1988). A geochemical and statistical approach for assessing metal pollution in coastal sediments. *In*: *"Metals in coastal environments of Latin America"*, U.Seeliger, L.D.Lacerda and S.R.Patchineelam (eds.), Springer-Verlag, Heidelberg, pp.47-58.

Salomons, W. and U.Förstner (1984). *"Metals in the hydrocycle"*, Springer-Verlag, Heidelberg, 349 pp.

11 Mangrove Swamps of the Subaé and Paraguaçú Tributary Rivers of Todos os Santos Bay, Bahia, Brazil[1]

J.F.Paredes, A.F.S. Queiroz, I.G. Carvalho, M.A.S.B. Ramos, A.L.F. Santos and C. Mosser
Instituto de Geociência, UFBA, Salvador, Brazil

Abstract

Different areas from the Todos os Santos Bay were monitored and the fate of heavy metals in these environments was determined. From 1982 to 1992, a significant worsening in the availability of heavy metals in the different sites of the Bay was observed. However, it must be emphasised that the comparison between the present results and those obtained from other works rests on a gap of at least three years. The determinations of heavy metal concentrations were based on the analysis of surface sediments of the different stations selected and the observed increase is probably due to the transport of the elements by tidal dynamics and upstream riverine transport of industrial effluents.

Introduction

The Todos os Santos is the largest bay in Brazil. It lies in the State of Bahia and extends over 750 km^2. The bay is part of the Reconcavo physiographic region, it has a relatively high population density and also sustains a traditional variety of industrial, agricultural and cattle activities. Moreover, this region has for the last 20 years see great urban and industrial development, including the industrial complexes of Subaé, near the city of Feira de Santana, Aratu and Camaçari, near the city of Salvador.

The purpose of this chapter is to examine the dispersion of Cr, Zn, Cu, Cd and Hg in the bed sediments and mangrove swamps located in two main estuaries in Todos os Santos Bay: the Paraguaçú River, Maragogipe Region and the Subaé River, Santo Amaro Region (Fig. 1 and Fig. 2), with the objective of studying the behaviour of these metals and their influences on that region. With this in mind, a comparison of these data with earlier data collected at the Paraguaçú and Subaé estuaries by the governmental agency Bahia-CEI (1987) will be discussed.

---

[1] Also published in *Tecnologia Ambiental*, CETEM No. 9, 1995, 15 pp.

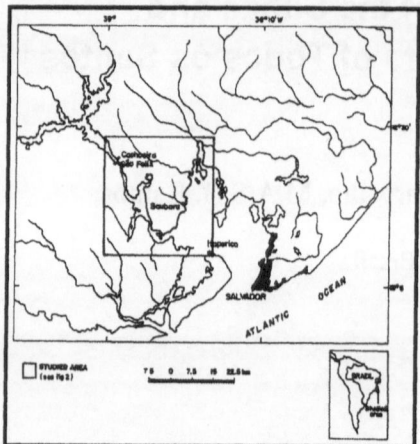

Fig. 1: General location of the study area

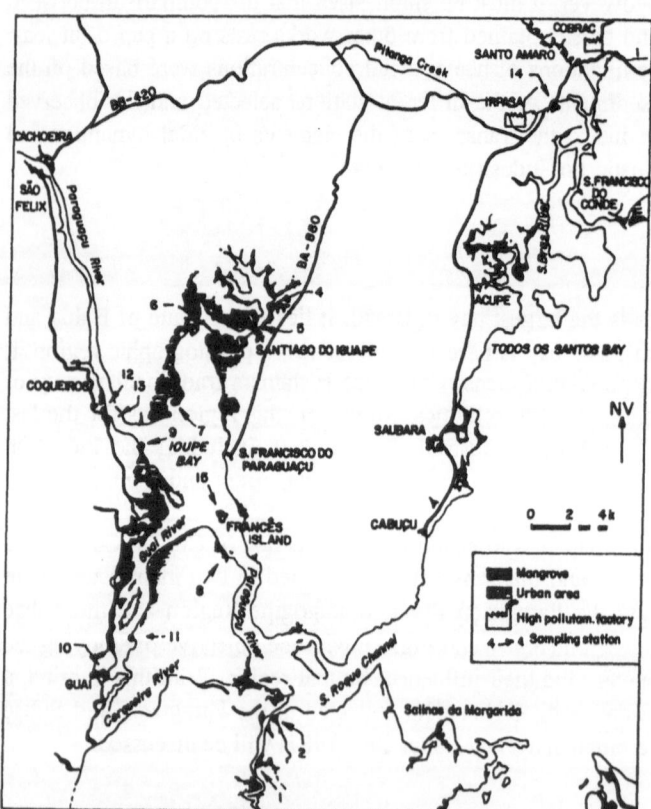

Fig. 2: Specific location of the sampling stations in Todos os Santos Bay

Both the Paraguaçú and the Subaé estuarine areas are characterised by marine influences. In the case of the Paraguaçú estuary, these influences are mainly observed during the low discharge period, which is controlled by the Pedra do Cavalo Dam, 20 km upstream of the estuary. Ramos (1993) showed that the flood-tide exhibited a larger influence in the Paraguaçú estuary than the ebb-tide, which ran only for about two hours, at least in the low discharge period. The flood-tide peaks usually occurred at the second and third hour of the flood. However, the faster currents observed were at the surface, at the beginning of neap-tide (August/1992), as shown in Fig. 3. All this estuarine complex (Iguape Bay plus São Roque Channel) is characterised by intense biological activity and conspicuous organic matter corresponding to the eutrophication of natural conditions. Indeed the Iguape Bay is bordered by luxuriant mangrove vegetation responsible for the great amount of organic matter.

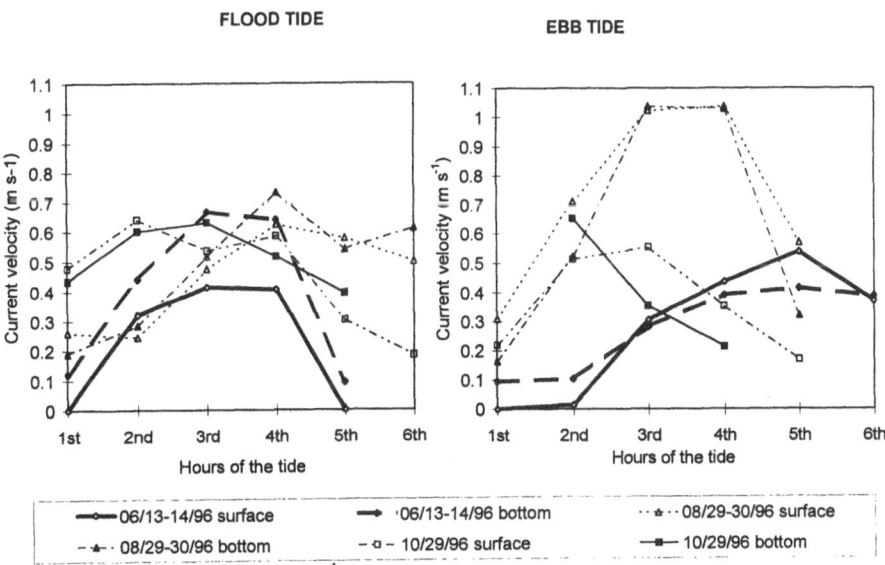

Fig. 3: Diagrams of current intensity during tide cycles monitored in the estuary of the Paraguaçú River

Methodology

Fourteen sound carrots of about 1 m length and diameter of approximately 70 mm were collected from mangrove sediments at the Maragogipe lagoon estuary, the São Roque Channel and at the narrow Subaé estuary, according to the sampling points plotted in Fig. 2 (stations 1 to 14).

The sediment carrots were then sectioned into 25-cm-thick slices to model or pattern the heavy metals in question, according to the sediment depth (Queiroz, 1992). However, for the present study, only the uppermost slice (first 25 cm) was considered.

Cr, Zn, Cu, Fe and Mn levels were measured at the Spectro-chemical Analysis Laboratory of the Surface Geochemistry Centre, Strasbourg, France, using the method described by Samuel et al. (1983). Hg, Cd and Pb were analysed in the Atomic Absorption Spectro-chemistry Laboratory of the ORSTOM Centre, Bondy, France, using the method described by Pinta (1985).

Five bottom sediment samples were collected in front of the Francês Island, between the lagoon estuary of Paraguaçú (Iguape Bay) and the São Roque Channel which connects the estuary with Todos os Santos Bay (station 15, Fig. 2). These samples were obtained during the low effluent discharges period (June through October, 1992). The samples were taken from a depth of 10 m below surface water, either in spring tides or neap tides with a Van Lee sampler. These five samples were then subjected to mineralogical study and chemical analysis. Zn, Cd, Pb and Cu were analysed at the Atomic Absorption Laboratory of the Geochemistry Department, Federal University of Bahia (UFBA), using the method described by Fletcher (1981). Hg was measured in the Atomic Absorption Laboratory of the Bahia State Research and Development Centre (CEPED), according to the method described by the US Environmental Protection Agency (US EPA, 1974).

Results and Discussion

The results are presented in Table 1 and Table 2.

Table 1: Heavy metal concentrations in surface sediments in Stations 1 to 14

Station	Fe_2O_3 (%)	Mn ($\mu g\ g^{-1}$)	Zn ($\mu g\ g^{-1}$)	Cu ($\mu g\ g^{-1}$)	Pb ($\mu g\ g^{-1}$)	Cd ($\mu g\ g^{-1}$)	Hg ($\mu g\ g^{-1}$)	Cr ($\mu g\ g^{-1}$)
01	0.50	78.1	15.3	6.8	5.3	0.01	<0.05	23.3
02	6.84	400.0	83.5	21.5	16.5	0.12	<0.05	109.0
03	5.08	186.0	59.5	14.6	12.0	0.06	<0.05	74.5
04	2.92	173.0	21.1	8.2	2.5	<0.01	0.30	36.1
05	7.09	337.0	78.9	31.9	15.5	0.13	<0.05	108.0
06	6.88	376.0	67.0	25.0	19.0	0.16	<0.05	112.0
07	4.43	206.0	44.6	8.6	4.0	<0.01	<0.05	33.9
08	3.17	175.0	28.0	10.1	5.5	0.05	<0.05	34.3
09	5.28	487.0	45.6	9.4	5.0	0.07	<0.05	107.0
10	2.05	122.0	39.1	9.5	4.0	0.08	<0.05	22.0
11	4.84	228.0	52.2	20.5	7.0	0.14	<0.05	67.7
12	3.84	230.0	47.5	10.4	8.5	0.04	<0.05	57.4
13	4.18	220.0	40.3	8.6	10.0	0.08	<0.05	49.0
$\overline{X}$ (n=13)	4.39	250.6	47.8	36.6	8.8	0.07	<.05(n=12)	64.2
14	5.72	207.0	102.0	51.0	156.0	11.40	<0.05	73.4

Table 2: Heavy metals concentrations in surface sediments in Station 15

Sampling date	Fe$_2$O$_3$ (%)	Mn (μg g^{-1})	Zn (μg g^{-1})	Cu (μg g^{-1})	Pb (μg g^{-1})	Cd (μg g^{-1})	Hg (μg g^{-1})
06.14.92	0.22	355.79	48.73	38.80	62.58	<1.3	0.173
08.30.92	0.20	233.57	43.13	45.11	50.90	<1.3	0.232
09.20.92	0.22	289.03	44.83	37.87	52.90	<1.3	0.208
10.20.92	0.20	240.67	43.24	33.37	44.70	<1.3	0.200
10.26.92	0.22	221.62	43.55	45.56	44.40	<1.3	0.208
$\overline{X}$	0.21	268.13	44.70	40.14	51.10	<1.3	0.203

The results of the study carried out in 1982 by Bahia-CEI are presented in Table 3. Although other areas of the Todos os Santos Bay were studied by that governmental agency, only the data from the sediments of the mouth of Paraguaçú River and from the estuarine area of the Subaé River will be reported here.

Table 3: Average contents of heavy metals in surface sediments from mangroves (M) and estuaries (E) of the studied area, to compare with those from the Jacuipe River and standard shales

Source	Fe$_2$O$_3$ (%)	Mn (mg kg^{-1})	Zn (mg kg^{-1})	Cu (mg kg^{-1})	Pb (mg kg^{-1})	Cd (mg kg^{-1})	Hg (mg kg^{-1})	Cr (mg kg^{-1})
This study (1992-1993)	4.39	250.6	47.8	36.6	8.8	0.07	<0.05	64.2
Paraguaçú Estuary (Francês Isl.) (E) Ramos (1993)	0.21	268.1	44.7	40.1	51.1	<1.3	0.203	-
Paraguaçú Estuary (Iguape Bay) (M) Queiroz (1992)	4.39	250.6	47.8	13.6	8.8	0.07	<0.05	64.2
Subaé Estuary (Santo Amaro) (M and E) Queiroz (1992)	5.72	207.0	102.0	51.0	156.0	11.4	<0.05	73.4
Former studies Subaé Estuary (1982)(E) Bahia-CEI (1987)	-	-	38.7	20.0	12.6	2.7	0.019	-
Paraguaçú Estuary (1982) (E) Bahia-CEI (1987)	-	-	37.0	17.3	12.0	0.4	0.029	-
Other studies Jacuipe Estuary (1989) (M and E) Queiroz (1989)	2.35	104.6	44.95	15.2	12.6	0.05	-	57.1
Standard shales Turekian and Wedepohl (1961)	4.7	850.0	95.0	45.0	20.0	0.3	0.4	90.0

Table 3 also shows the average values of heavy metal concentrations in surface sediments from the mangroves of the Paraguaçú (stations 1 to 13) and Subaé (station 14) estuaries and in bottom sediments from the Paraguaçú River (station 15), as well as the average values obtained by Queiroz (1989) for the mangrove sediments from the estuary of the Jacuipe River, and those values indicated by Turekian and Wedepohl (1961) for standard shales.

All observations in the Paraguaçú lagoon estuary were performed in the mean salinity of 30% (28-34%). This high salinity accounts for the halite crystal formation in the sediments collected due to the natural evaporation of water in the laboratory. The mineralogical composition of these sediments (quartz, kaolinite, goethite and minor chlorite and illite) (Queiroz, 1992; Ramos, 1993) reflects the dominant lithologies of the region, represented by the crystalline basement and the sedimentary Brotas group and the Barreiras formation (Inda and Barbosa, 1978).

The concentrations of Cr, Zn, Cu, Pb and Cd in the studied sediments are lower than the respective values indicated by Turekian and Wedepohl (1961) for standard shales (Table 3). The only exception is observed in the sediments from the mangrove of Santo Amaro, in the Subaé estuary (station 14), and also for Pb in the sediments collected at station 15 (Francês Island). Besides these anomalous values, the high Hg concentration registered in the sediments from Francês Island may be included here. They reveal a critical situation in the estuarine ecosystem of the Subaé River. On the other hand a significant worsening of the environmental quality from 1982 to 1992 must be emphasised, as must the fact that the values found in this study, excluding Fe in Francês Island and Cu and Pb in Iguape Bay are higher than the average values found by Queiroz (1989) in sediments collected in the estuarine area of the Jacuípe River, east of the Recôncavo Region (Fig. 1). This basin drains the main industrialised area of the Recôncavo.

The concentrations of Cu, Pb and Hg found in the samples from Francês Island, are significantly higher when compared with those obtained in the mangroves of the Iguape Bay. On the other hand, if these results are compared with the concentrations obtained in 1982 by Bahia-CEI in the São Roque Channel, it can be concluded that in the Francês Island, anthropic sources are decreasing the environmental quality. This rises the following question: *What is the direction of the transport of heavy metals?* We believe that the main transport of these elements is associated with tidal dynamics, being related with the hydrographic conditions in Todos os Santos Bay. The action of the atmospheric plume from Mataripe Oil-Refinery must also be considered (Fig. 1). Fossil fuels bear most of the heavy metals included in this study, and under the action of the regional winds which blow from SE and from E the Mataripe plume will affect the estuaries studied here.

Conclusions

In this chapter critical conditions were identified in both areas.

A significant aggravation of the environmental quality can be observed between 1982 and 1992 (Table 3), particularly on the estuarine waters of Subaé River, where all concentrations of heavy metals sharply increased in the sediments such that all of them largely overtook the concentrations of the standard shales, with the single exception of Hg. The conditions mentioned relatively to the Maragogipe area can be related to the transport of these elements by tidal dynamics, while the influence in the Subaé estuary must be mainly associated with the riverine transport of upstream industrial effluents.

Comparing the concentrations of heavy metals exhibited in sediments obtained in front of Francês Island (Iguape Bay), we may see that Pb largely exceeds the corresponding shale pattern, while Cu almost reaches it. Regarding Cd we may also conclude that the situation is becoming critical, since 1982, the Bahia-CEI data concerning the São Roque Channel, exceeded not only the permissible concentrations for Cd (0.01 - 0.005 mg kg^{-1}) but also is above the standard shales (0.3 mg kg^{-1}).

Regarding the lagoon estuary of the Paraguaçú River, the transport directions which give rise to the heavy metal concentrations mentioned in this chapter remain unknown. Therefore, further studies are recommended on this area in order to find out the lay-out of these transports, the accumulation of these heavy metals, their effects on the environment and their predicted evolution.

References

Bahia-CEI (1987). *Qualidade Ambiental na Bahia - Recôncavo e Regiões Limítrofes*. Secretaria do Planejamento, Ciência e Tecnologia-Centro de Estatística e Informações, Salvador, Bahia, Brazil, 48p.

Fletcher, W.K. (1981). Analytical Methods in Geochemical Prospecting. ChapterVI: Atomic Absorption Spectrophotometry. In: *Handbook of Exploration Geochemistry*, Vol. I, G.J.S. Govett (Ed.). Elsevier Sci. Pub. Co., p.109-136.

Inda, H.A. and Barbosa, J.F. (1978*). Texto explicativo para o Mapa Geológico do Estado da Bahia - Escala 1:1.000.000*. Governo do Estado da Bahia, Secretária das Minas e Energia, Coordenação da produção Mineral, Salvador, Bahia, Brazil, 137p.

Pinta, M. (1985). *Spectrométrie d'Absorption Atomique*. Techniques de l'Ingénieur, 10, P 2825 91-24.

Queiroz, A.F.S. (1989*). Estudos Geoquímicos e Sedimentológicos no Manguezal do Estuário do Rio Jacuipe, Camaçarí, Bahia*. M.Sc. Thesis, Graduate Course in Geociences, Federal University of Bahia, Brasil, 227p.

Queiroz, A.F.S. (1989). *Mangroves de la Baia de Todos os Santos - Salvador - Bahia - Brésil: ses caractéristiques et l'influence anthropique sur sa géochimie*. D.Sc. Thesis, Louis Pasteur University, Strasbourg, France, 148p.

Ramos, M.A.S.B. (1993). *Estudos Geoquímicos Relativamente à Dinâmica de Marés no Estuário Lagunar do Rio Paraguaçú - Bahia - Brasil*. M.Sc. Thesis, Graduate Course in Geociences, Federal University of Bahia, Brasil, 84p.

Samuel, J. ; Rouault, R.; Balouka, I.; Staub, P. et Besnus, Y. (1990*). Les Méthodes d'Analyses des Matériaux Géologiques Pratiquées au Laboratorie d'Analyses Spectrochimiques*. Notes Techniques de l'Institut de Géologie, n° 16, Louis Pasteur University, Strasbourg, France, 46p.

Turekian, K.K. and Wedepohl, K.H. (1961). Distribution of the Elements in Some Major Units of the Earth's Crust. *Geol. Soc. of Am. Bull.*, 72:175 - 192.

US EPA (1974). *Methods for Chemical Analysis of Water and Wastes*. US Environmental Protection Agency, Office of Technology Transfer, Washington DC, USA, 20460. p.134-197.

12 Metal Scavenging and Cycling in a Tropical Coastal Region

Christovam Barcellos[1], Luiz Drude de Lacerda[1] and Sergio Ceradini[2]
[1] Departamento de Geoquímica, Universidade Federal Fluminense, Niterói, RJ 24020-150, Brazil
[2] Divisione Ambiente, CISE, Via Reggio Emilia 39, Segrate, Milan, MI 20090, Italy

Abstract

The retention of terrigenous metals in coastal regions plays an important role in the oceanic global mass balance of these elements. In Sepetiba Bay (Rio de Janeiro), efficient Cd and Zn removal from the water column was observed due to processes influenced by sediment resuspension and deposition. These mechanisms involve adsorption onto particle surfaces, which is independent of matrix composition and the origin of the suspended particles. Cadmium can be concentrated along the bay shore where sediment and organic matter cyclings are intensive.

Introduction

Zinc and cadmium have been recognized to behave as nutrient-type metals in oceanic (Boyle et al., 1976; Simpson, 1981) and estuarine (Yeats, 1988; Windom et al., 1991) waters. In coastal regions these metals can be scavenged by coagulation, adsorption and incorporation into the particulate matter. The production and degradation of organic matter cause changes in water and particulate matter qualities which may affect these processes. Cadmium has a paired distribution with P in the water column due to their simultaneous incorporation and release by phytoplankton (Boyle et al., 1976; Nolting et al., 1991; Lee and Fisher, 1992). Zinc has been reported to change from dissolved to particulate form in response to the photosynthetic activity of the phytoplankton (Salomons and Forstner, 1984).

The aim of this study was to establish the mechanisms acting on Cd and Zn scavenging in the coastal waters rich in organic matter, by using the records of recent (from sediment traps) and old deposits (from bottom sediments) sediments in a tropical bay. The origin of sediments, using both sampling strategies, was established based on major element composition (Barcellos and Lacerda, 1997).

The use of Sepetiba Bay as a field model for metal fate in coastal waters is favourable due to three different factors: 1) relatively high input of metals; 2) singular and separated sources of sediments (fluvial), Cd and Zn (industrial effluents to the bay) and organic matter (from urbanization along the southeast shore) which allows comprehension of the relations among these materials in coastal environments; 3) convenience due to the proximity to Rio de Janeiro City.

Previous studies have reported Cd and Zn enrichment in the bottom sediments of the bay (Souza et al., 1986; Patchineelam et al., 1989; Barcellos and Lacerda, 1993) but this enrichment cannot be explained only by the inputs of metal-rich industrial solid wastes. Geochemical exchange between coastal waters and mangroves was also identified as an important mechanism influencing heavy metal fate in the land-sea interface (Lacerda and Rezende, 1991). The Sepetiba shore regions were found to be a sink for river-borne sediments (Argento and Vieira, 1988) and for authigenic organic matter (Rezende et al., 1990) due to the current circulation pattern of the bay. In the present work natural tracers were used for the investigation of the production and transport of sediments and their relationships with Cd and Zn cycling in Sepetiba Bay.

Methodology

Site Description

Sepetiba Bay is a semi-enclosed environment (with 447 km^2 of area) located at circa 60 km west of Rio de Janeiro City. Its drainage basin is inhabited by 1.2 million people and harbours some of the main regional industrial activities, comprising a total area of 2,065 km^2. Earlier surveys showed that local sediments and biota are contaminated by Pb, Cr, Zn and Cd from industrial sources (Lacerda, 1983). Water depth ranges from 2 m in the inner zone to 20 m in the broad sea interface. A clockwise circulation pattern promotes the permanent exchange of water with the sea (Signorini, 1980) but creates extensive mud flats along the eastern coast of the bay. Water quality studies showed vertical water column mixing and important horizontal and temporal inequalities (Azevedo et al., 1994) due to the riverine or marine influence, the existence of intense primary production cells and the extent of mangroves which cover 35% the of bay perimeter.

Sample Collection and Analysis

Figure 1 shows major geographical features and sediment sampling sites along the Sepetiba Bay shore region. Other bottom sediment samples (M1, M2, M3) were collected in the continental shelf in front of the Paraíba do Sul estuary located on the coast of Rio de Janeiro. These samples were considered as typical marine end-members. Other sediment samples were also collected in the Itaipú, Saquarema, Maricá and Araruama lagoons, along the Rio de Janeiro coast, as well as in the São Francisco and Doce River estuaries. These sediments were used to estimate the regional background concentrations of Cd and Zn and as typical end-members of river estuaries and lagoons without significant industrial sources.

Sediment traps (85 mm diameter and 420 mm height) were disposed in four stations in the transition zone of the bay, exposed to marine and riverine influences for 4 days. The traps were recovered by divers and the collected particulate matter was separated by centrifugation at 3000 RPM for 10 minutes.

Bottom sediment samples were collected from the upper 3 cm of mud deposits along the bay's shore and immediately wet sieved (64 μm). All samples were dried at 50°C for 48 hours and about 100 mg aliquots were digested in 20 ml of HNO_3:HF:HCl (1:6:3). Weakly bound metals were extracted from the trapped material and bottom sediments using 0.3N HCl with agitation for 3 hours.

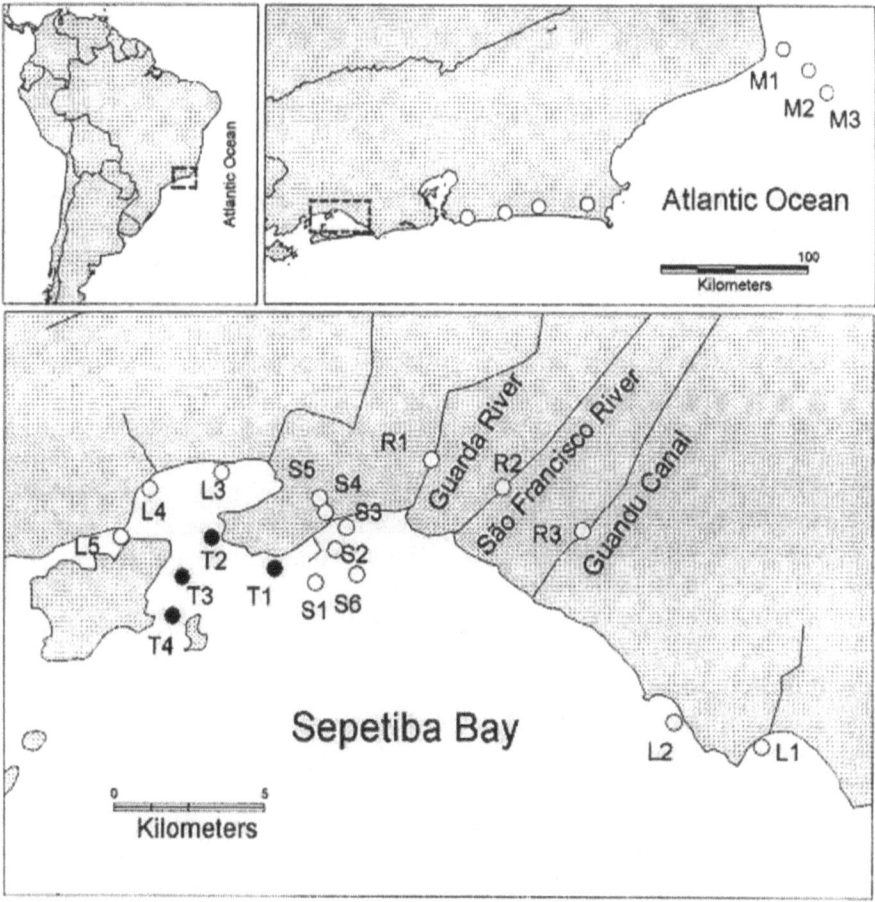

Fig. 1: Map of Rio de Janeiro coast and detail of Sepetiba Bay. *Open circles* indicate the bottom sediment sampling stations and *dark circles* indicate the sediment trap installation sites

Zinc analysis was performed by inductively coupled plasma atomic emission spectrometry, and Cd with a graphite furnace atomic absorption spectrometry. Determinations of Zn and Cd concentrations in samples L1, L2, L4 and L5 were simultaneously analysed in two laboratories, in order to check methodological and equipment precision. Comparison showed consistent results between the analysis performed at CISE (Milan, Italy) and Geochemistry Department, Universidade Federal Fluminense (Niterói, Brazil).

Phosphorus analysis was carried out by the ammonium molybdate colorimetric method (Grasshoff et al., 1983). Dilution and standard addition test were performed in order to evaluate potential matrix interferences.

Macro-element concentrations in the sediments were obtained by X-ray fluorescence technique using lithium tetraborate fusion and secondary regional certified samples as reference standards. At least 2 replicates of standards and samples were analysed simultaneously, presenting errors of 2-5%.

Sediment Deposition, Production and Transport

Sediment deposition rates measured by means of sediment traps varied from 8.6 (in the westernmost site) to 23.0 mg cm^{-2} d^{-1} (in the eastern site). Sedimentation rates in the bay were estimated by previous works (Japenga et al., 1988; Barcellos et al., 1991; Quevauviller et al., 1992) leading to a range of results between 0.8 and 1.3 cm y^{-1}. An average particle burial rate of 1 to 2 mg cm^{-2} d^{-1} is calculated assuming 0.4 to 0.6 g cm^{-3} for the bulk sediment density for non-compacted clay sediments (Hakanson, 1980; Barcellos et al., 1991). One order of magnitude difference between sediment deposition and burial rates is probably due to frequent resuspension episodes and the intensive sediment load to the water column (Hakanson et al., 1984). On the other hand, river transported sediments (300 to 590 x 10^3 t y^{-1}; IFIAS, 1988 and Rezende, 1993) corresponds to a range of 30 to 60% of the total estimated sediment burial. Authigenic and marine matters are thus possibly important sources of sediments to the bay.

The origin and fate of the sediments were investigated using concentration data of the major components as natural mineralogic tracers (Salomons and Mook, 1988). Table 1 presents the concentration values in the marine (M1, M2 and M3), riverine (R1, R2 and R3), shore (L1, L2, L3, L4 and L5), bottom sediment (S1, S2, S3, S4, S5 and S6) and trapped (T1, T2, T3 and T4) materials. Some of the bottom sediment samples were collected in an industrial waste contaminated embayment which constitutes a secondary source of sediments for the bay. The spreading of sediments from this source represents a small fraction of total sediment input, however important from the pollution point of view because of the high metal concentrations observed in such sediments (Barcellos et al., 1991).

The range of major element concentrations indicates a dominance of clay aluminium-silicate matrix. This dominance however decreases from the riverine and shore to the marine end-member as verified by the increase of Si and the decrease of Fe and Ti concentrations. Variations in the major element concentrations suggest the existence of diverse sources for the particulate matter and their mixture in the coastal region. Higher Al and K and, to a lesser extent Ca concentrations are observed in respect to other estuarine sediments such as in the Rhine and Meuse Estuary sediments, which are under the influence of North Sea sediment deposition and dredging (Nolting et al., 1990).

Correlation analysis using major elements data showed significant relationships between Ca, K and Si, and between Al and Ti. Iron presented negative correlation with the first group of elements. These groups of elements have been used as tracers respectively for the marine and terrigenous influences in the Brazilian coastal shelf (Carvalho and Lacerda, 1993).

Cluster analysis was performed using major elements data and average method. Six clusters of samples were identified (Fig. 2).

Table 1: Major element concentration in the sediments from the Sepetiba Bay, Rio de Janeiro. Values in % with exception of Mn and P in $\mu g \ g^{-1}$ (d.w.)

Sample	Fe	Mn	Ti	Ca	K	Si	Al	Mg	P
R1	4.93	793	0.60	0.30	1.07	19.0	7.99	0.79	-
R2	5.41	910	0.62	0.31	1.11	15.2	8.41	0.82	313
R3	4.34	705	0.82	0.34	1.67	18.4	6.45	0.61	244
L1	4.39	2,327	0.66	0.50	1.38	18.5	6.51	1.66	399
L2	4.68	3,310	0.63	0.39	1.38	17.6	7.37	1.23	624
L3	3.79	463	0.68	0.50	1.67	17.1	6.68	1.33	456
L4	3.60	367	0.63	0.50	1.74	19.1	6.47	1.35	361
L5	3.11	305	0.57	0.44	1.66	18.0	5.22	0.79	187
S1	5.25	520	0.76	0.30	1.30	19.1	8.43	0.84	-
S2	4.47	380	0.77	0.28	1.31	18.6	7.85	0.82	-
S3	3.90	320	0.61	0.31	1.51	20.9	6.09	1.32	-
S4	7.31	320	0.56	0.32	1.16	16.3	5.60	1.29	334
S5	6.27	280	0.67	0.30	1.33	16.9	6.88	1.16	100
S6	4.48	448	0.75	0.35	1.34	18.4	7.81	0.81	205
T1	3.43	769	0.55	0.50	1.56	19.7	6.03	1.23	580
T2	3.88	570	0.67	0.39	1.79	21.1	7.09	1.12	650
T3	3.54	439	0.56	0.52	2.17	21.4	6.12	0.98	740
T4	4.83	592	0.62	0.57	1.64	20.0	6.43	0.92	448
M1	1.64	390	0.37	0.44	1.54	31.3	3.00	0.30	165
M2	1.47	350	0.43	0.35	1.13	32.9	2.42	0.23	230
M3	1.79	400	0.78	0.42	0.91	32.7	1.92	0.29	116

The last two clusters (T1, T2, T3, T4 and L3, L4, L5) have the above cited marine influence feature composition. Slight differences between trapped material (T1, T2, T3 and T4) and western shore sediment deposits (L3, L4 and L5) can be observed. The lower Fe, Mn and P concentrations are the main differences between mud shore deposit and trapped material. The similarity between these groups implies the presence of resuspended sediments in trapped material.

The marine end-member samples (M1, M2 and M3) formed a separated cluster with significant compositional differences (e.g., higher Si and lower Fe).

The spreading of industrial waste material is restricted to the inner contaminated zone (S4 and S5), while the riverine influence was observed even for some of the sediments from the industrial contaminated zone (S1, S2 and S6). The isolation of the eastern shore deposited sediments (L1 and L2) from the rest of the clusters is statistically explained by its significantly higher contents of Mn and Mg. This area has been identified as an almost independent hydrodynamical cell with low salinity and high organic matter contents in the water column (Azevedo et al., 1994).

Zinc, Cadmium and Phosphorus Distribution Patterns

Table 2 presents the average concentrations of Zn, Cd and P in bottom sediments of different origins in the Sepetiba Bay and other Brazilian coastal regions.

Table 2: Zn, Cd and P concentration (in $\mu g\ g^{-1}$, d.w.) in the deposited sediments from river mouths, shore and traps from Sepetiba Bay and other Brazilian coastal regions

		Zn	Cd	P
Sepetiba Bay				
	River estuaries	204 ± 32	0.52 ± 0.11	278 ± 34
	Shore	824 ± 129	4.24 ± 0.43	405 ± 71
	Traps	1087 ± 106	4.10 ± 0.42	605 ± 48
Non-contaminated coastal regions				
	Rivers estuaries	61 ± 10	0.15 ± 0.05	174 ± 41
	Coastal lagoons	86 ± 9	0.23 ± 0.01	231 ± 57

Cd, Zn and P enrichments are observed in the Sepetiba Bay sediments compared to other coastal regions. However, these enrichments are by a factor of 20 for Cd, 15 for Zn and only 2 for P. This demonstrates a severe metal contamination and only a slight increase in the organic matter accumulation.

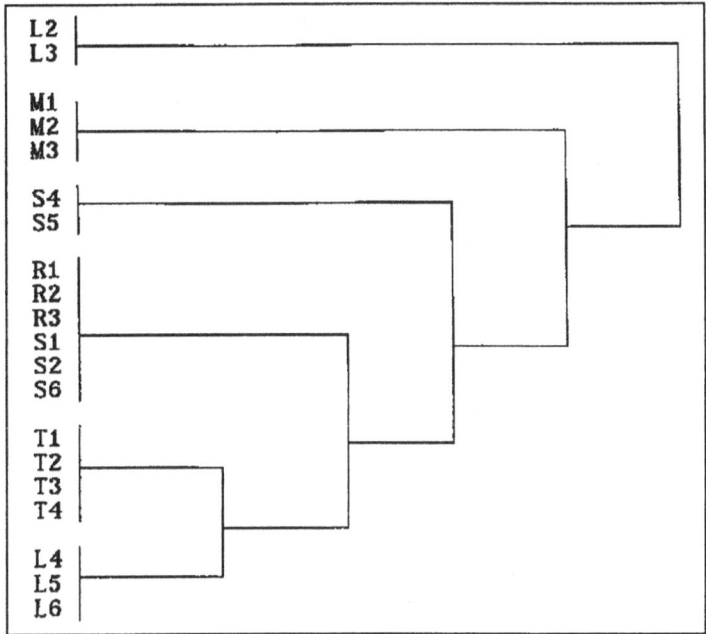

Fig. 2: Dendrogram of clusters for Sepetiba Bay sediment samples based on major elements concentrations. Arbitrary horizontal scale

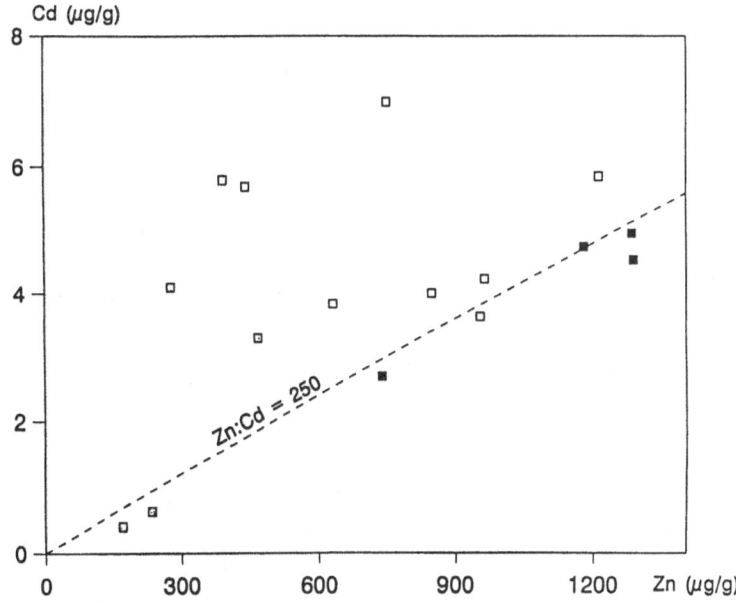

Fig. 3: Cadmium and Zn concentrations in the Sepetiba Bay sediments. *Dark squares* represent trapped sediment samples

Despite the different distances from the contamination sources, and the variability of geochemical matrixes and environmental conditions, Cd concentration values stand at a high stable plateau (above 2.5 µg g^{-1}) for all of the Sepetiba Bay sediment samples.

Zinc, Cd and P present significant paired correlations for Sepetiba Bay sediment samples. Figure 3 compares Zn and Cd concentrations in sediments. Zn and Cd have higher concentrations in trapped material than in the main contributing rivers. For this reason, Cd and Zn accumulation in the bay cannot be assumed to be a consequence of the deposition of river transported sediment alone. On the other hand, this is evidence of specific mechanisms working in the bay. The trend for Cd and Zn enrichment in particulate matter was also observed in the North Sea off the estuarine mixing zone (Kersten et al., 1988) and in the Ligurean open sea (Fabiano et al., 1988). In these areas the mechanism responsible for this behaviour was the phytoplankton metabolism, the so called "nutrient pump".

The mean Cd and Zn concentrations in phytoplankton from Sepetiba Bay are 2.0 and 422 ug g^{-1}, respectively (Andrade et al., 1990) which are below the observed concentrations in trapped and deposited sediments. However, these concentrations are high when compared to non-contaminated reference values, showing the efficiency of the "nutrient pump".

The trapped sediments conserve a Zn:Cd ratio of 265 ± 6. Significantly lower values are verified for the bay shore bottom sediments (190 ± 16). The ratio of 250:1 has been identified in the bay and several geological matrixes as an average abundance proportion of these elements (Suzuki et al., 1979; Simpson, 1981; Barcellos and Lacerda, 1993). Zn:Cd ratios are higher and present a wider range of variation for sediments from other Brazilian coastal regions (Table 2). The prevalence of Zn:Cd ratio within the metals source range may indicate that Cd uptake by resuspended and newly formed particles is regulated by its relative abundance, rather than by active biogeochemical controls. After deposition a relative enrichment of Cd in sediment (35% in respect to Zn) is observed. A large quantity of Cd remains linked to sediment and its relative concentration can be further increased.

Figure 4 shows the concentration of Cd in sediment as a function of P. Cadmium and P exhibit a steeply increasing tendency from the riverine samples to the trapped sediments. This enrichment is however stronger for Cd than for P. The observed Cd:P molar ratio varies from 0.29 to $0.65.10^{-3}$ for low impacted Brazilian coastal region. This range is within the expected values for estuarine and oceanic water samples (Boyle et al., 1976; Windom et al., 1991; Nolting et al., 1991). For the Sepetiba Bay trapped material, this ratio stands one order of magnitude above the oceanic mean. The similarity in the behaviours of Cd and P has been pointed out to be controlled by the aquatic biota uptake (e.g., Walsh and Hunter, 1992) or by mineralogic inclusion (Sadiq, 1989). Whatever the mechanism, the oversaturation of Sepetiba Bay sediments of Cd with respect to possible controlling P actions seems obvious.

After deposition, slight Cd and P losses are observed. From freshly formed sediments to shore deposits, the release of 0.8 mM of Cd for every Mol of P is calculated, which is similar to the proportion of paired enrichment in the estuarine waters (Windom et al., 1991). The aging of sediments, corresponding to organic matter degradation and lower redox potentials, can be responsible for the release of Cd and P from labile positions. The sampling strategy of superficial sediments avoids the influence of extreme anoxic conditions and consequent precipitation of Cd sulphides. Large quantities of P are released from coastal sediments by organic matter degradation as verified by the comparison between burial and depositional fluxes (Sundby et al., 1992). For Sepetiba Bay, the organic matter degradation and PO_4 production do not correspond to the complete Cd release from sediments. High baseline levels (represented in Fig. 4 as the vertical intercept) could be observed even with the complete elimination of organic matter.

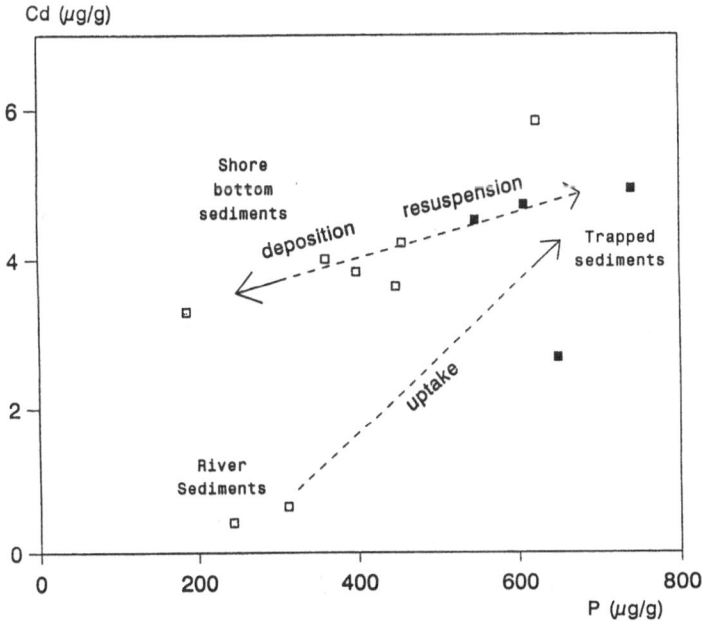

Fig. 4: Cadmium and P concentrations in the Sepetiba Bay sediments. *Dark squares* represent trapped sediment samples

The decoupling of Cd and P cycles in the Mediterranean Sea was explained by the differential efficiency of biological removal for dissolved P (99%) and Cd (9 to 32%) from the water column (Ruiz-Pino et al., 1990). The observed differences of Cd, Zn and P behaviours during uptake and release stages must be considered, where reaction reversibility and velocity may play a major role. The complete reversibility of Cd adsorption onto clay particles was verified by Commans (1987) in laboratory experiments. Zinc, however, can be stored occluded into resistant particle sites (Hegeman et al., 1992). The validity of these findings is restricted by the minor role played by clay minerals in the adsorption of metals in water. The presence of organic ligands and hydrous Mn and Fe oxides on estuarine particle surfaces creates powerful adsorption sites for metals (Davies-Colley et al., 1984; Fu and Allen, 1992). These coatings represent a barrier for metal absorption or at least a necessary intermediate step.

Table 3 presents the results of leaching tests performed with the trapped material of the bay.

Table 3: Solubilized fraction of Zn, Cd and P from trapped sediment after 0.3N HCl leaching, in percent over the total content

Station	Zn	Cd	P
1	46	54	41
2	58	70	10
3	65	70	27

Leaching treatment removed a large portion of Cd, followed by Zn and P from sediments. Cadmium, followed by Zn, is located in more mobile or easily degradable geochemical sites of the suspended matter. Phosphorus extractability was found to be low, due to its structural organic nature. Resistant forms of P and Cd can be stored during the algae minimum growth stage in nutrient rich waters. Cadmium concentration in algae is highly dependent on its availability in water and it is controlled by the P uptake and storage (Walsh and Hunter, 1992). Particle organic coatings were pointed out by Bruland (1992) as the main Cd carriers in coastal waters and can explain the large labile fraction of Cd and Zn concentrations in the suspended matter. The retention of Cd in the deposited sediments necessarily involves the modification of its chemical form into more resistant ones. The conflicting process of organic matter degradation and Mn oxides precipitation was described as a trap for metals in the Mediterranean Sea surface sediments (Fernex et al., 1992).

Conclusions

Sedimentation in the Sepetiba Bay is controlled by the mixture of riverine, marine and autochthonous sediments. The preferential sink of river-borne sediments is located in the eastern part of the bay, while in the western shore, deposits of marine origin are dominant. Despite the different sediment compositions and origins, high and stable Cd and Zn concentrations were observed in the Sepetiba Bay samples.

The production and transport of organic matter in the water column is able to scavenge large quantities of available Cd and Zn. The Cd contents in the sediment are not only explained by transport and mixing, but also uptake and adsorption onto suspended particle surfaces. These metal scavenging mechanisms are controlled by their relative abundance in the water column. The posterior Cd enrichment in deposits is a consequence of the differential release dynamics of metals and organic matter. Organic matter degradation can release P without the corresponding release of bound metals. The increasing of the exchange phase is a probable post-depositional mechanism explaining Cd retention in the sediments.

Acknowledgements

The authors are grateful to Dario P. Carvalho for providing field facilities during the sampling campaign. Representative Brazilian coastal sediment samples were kindly furnished by C.E. Rezende, M. Fernandez and C.E.V. Carvalho. All analyses were performed in the CISE laboratories during 1993. Particular thanks are deserved by A. Zoboli and R. Barban for their attention during XRF and AAS analysis. This work was undertaken with the support of the ICTP Programme for Training and Research in Italian Laboratories (Trieste, Italy) and the Conselho Nacional de Desenvolvimento Científico e Tecnológico (CNPq-Brazil).

References

Andrade, P.P.; Andrade, L.P.L.S.; Pfeiffer, W.C.; Azevedo, S.M.F.O.; Karez, C.S.; Rezende, C.E. (1990). Níveis de metais pesados no fitoplancton da Baía de Sepetiba - avaliação preliminar. *3° Congresso Brasileiro de Limnologia* (proceedings), Porto Alegre, Brazil. p. 229.

Argento, M.S.F.; Vieira, A.C. (1988). Impacto ambiental na Praia de Sepetiba. *Congresso Brasileiro de Sensoriamento Remoto* (proceedings), Natal, Brazil. pp. 1-7.

Azevedo, F; Patchineelam, S.R. (1996). Massas de água e material em suspensão na Baía de Sepetiba, *Ciência e Cultura* (in press)

Barcellos, C, Rezende, C.E. and Pfeiffer, W.C. (1991). Zinc and cadmium production and pollution in a Brazilian coastal region, *Marine Pollution Bulletin*, 22 (11):558-561.

Barcellos, C.and Lacerda, L.D (1993). Cadmium behaviour in a tropical estuary. *Heavy Metals in the Environment*, CEP Consultants, Toronto, Canada. Vol. 1. pp.:169-172.

Barcellos, C.and Lacerda, L.D (1997). Sediment origin and budget in Sepetiba Bay (Brazil) - an approach based on multielemental analysis, *Environmental Geology* (in press).

Boyle, E.A.; Scalter, F. and Edmond, J.M. (1976). On the marine geochemistry of cadmium. *Nature*, 263:42-44.

Bruland, K.W. (1992). Complexation of cadmium by natural organic ligands in the central north Pacific. *Limnology Oceanography*, 37(5):1008-1017.

Carvalho, C. and Lacerda, L.D. (1993). Utilização de Ca e Ti como traçadores de origem de sedimento na plataforma continental brasileira, *Congresso Brasileiro de Ecossistemas Costeiros*.

Comans, R.N.J. (1987). Adsorption, desorption and isotopic exchange of cadmium on illite: evidence for complete reversibility. *Water Research*, 21:1573-1576.

Davies-Colley, R.J.; Nelson, P.O., and Wiliamson, K.J. (1984). Copper and cadmium uptake by estuarine sedimentary phases. *Environmental Science Technology*, 18 (7):491-499.

Fabiano, M.; Baffi, F.; Povero, P.; Frache, R. (1988). Particulate organic matter and heavy metals in Ligurean open sea. *Chemistry and Ecology*, 3:313-323.

Fernex, F.; Fevrier, G.; Benaim, J.; Arnoux, A. (1992). Copper, lead and zinc trapping in the Mediterranean deep-sea sediments: probable coprecipitation with Mn and Fe. *Chemical Geology*, 98:293-306.

Fu, G. and Allen, H.E. (1992). Cadmium adsorption by oxic sediment. *Water Research*, 2:225-233.

Grasshoff, K., Ehrardt, M., Kremling, K. (1983). *Methods of Seawater Analysis*. Verlag Chemie, Weinhein. 419 p.

Hakanson, L. (1980). An ecological risk index for aquatic pollution control: a sedimentological approach. *Water Research*, 14:975-1001.

Hakanson, L.; Floderus, S. and Wallin, M. (1984). Sediment trap assemblages - a methodological description. *Hydrobiologia*, 176/177:481-490.

Hegeman, W.J.M.; van der Weijden, C.H.; Zwolsman, J.J.G. (1992). Sorption of zinc on suspended particles along a salinity gradient: a laboratory study using illite and suspended matter from the River Rhine. *Netherlands Journal of Sea Research*, 28(4):285-292.

IFIAS (1988). *Sepetiba Bay Management Study: Workplan*. The International Federation of Institutes for Advanced Study, Rio de Janeiro, report. 72 p.

Japenga, W; Wagenar, W.J.; Salomons, W.; Lacerda, L.D.; Patchineelam, S.R.; Leitão Filho, C.M. (1988). Organic micropollutants in the Rio de Janeiro coastal region, Brazil. *The Science of the Total Environment*, 75:249-259.

Kersten, M.; Dike, M.; Kriews, M.; Naumann, K.; Schmidt, D.; Schulz, M.; Schwikowski, M., and Steiger, M. (1988). Distribution and Fate of Heavy Metals in the North Sea. In: W. Salomons; B. Bayne; E. Duursma and U. Forstner (eds). *Pollution of the North Sea. An Assessment*. Springer, Berlin. pp.:300-347.

Lacerda, L.D. (1983). *Aplicação da metodologia de abordagem pelos parâmetros críticos no estudo da poluição por metais pesados na Baía de Sepetiba, Rio de Janeiro*. Ph.D. Thesis. Instituto de Biofísica Carlos Chagas Filho. UFRJ, Rio de Janeiro, Brazil. 125 p.

Lacerda, L.D.; and Rezende, C.E. (1991). Heavy metal biogeochemistry in mangrove ecosystems. In: *Global Perspectives on Lead, Mercury and Cadmium Cycling in the Environment*, New Dehli. pp.:289-297.

Lee, B.-G.; Fisher, N.S. (1992). Degradation and elemental release rates from phytoplankton debris and their geochemical implications. *Limnology and Oceanography*, 37(7):1345-1360.

Nolting, R.F.; Sundby, B.; Duinker, J.C. (1990). Behaviour of minor and major elements in the suspended matter in the Rhine and Meuse Rivers and estuary. *The Science of the Total Environment*, 97/98:169-183.

Nolting, R.F.; de Baar, H.J.W.; van Bennekom, C.; Masson, A. (1991). Cadmium, copper, and iron in the Scotia Sea, Weddell Sea and Weddell/Scotia confluence (Antarctica). *Marine Chemistry*, 35:219-243.

Patchineelam, S.R., Leitão-Filho, C.M., Azevedo, F.V. and Monteiro, E.A. (1989). Variation in the distribution of heavy metals in surface sediments of Sepetiba Bay; Rio de Janeiro, Brazil. In: CEP Consultants, *Heavy Metals in the Environment* (proceedings), Geneva, vol. 2, pp.:424-427.

Quevauviller, P.; Donard, O.F.X.; Wasserman, J.C.; Martin, F.M.; Schneider, J. (1992). Occurrence of methylated tin and dimethyl mercury compounds in a mangrove core from Sepetiba Bay, Brazil. *Applied Organometallic Chemistry*, 6:221-228.

Rezende, C.E.; Lacerda, L.D.; Ovalle, A.R.C.; Silva, C.A.R. (1990). Nature of POC transported in a mangrove ecosystem: a carbon stable isotopic study. *Estuarine Coast and Shelf Science*, 30:641-646

Rezende, C.E. (1993). *Origem, transporte e destino da matéria orgânica na interface fluvio-marinha, sob diferentes condições de uso do solo e sua relação com o trânsito de poluentes metálicos na Baía de Sepetiba - RJ.* Doctoral thesis, Instituto de Biofísica, UFRJ, Rio de Janeiro, 193 p.

Ruiz-Pino, D.P., Nicolas, E., Bethoux, J.P. and Lambert, C.E. (1991). Zinc budget in the Mediterranean Sea: a hypothesis for non-steady-state behavior. *Marine Chemistry*, 33:145-169.

Sadiq, M. (1989). Marine chemistry of cadmium: a comparison of theoretical and field observations. *Environmental Technology Letters*, 10:1057-1070.

Salomons, W. and Forstner, U. (1984). *Metals in the Hydrocycle*. Springer-Verlag, Berlin. 349 p.

Salomons, W. and Mook, W.G. (1987). Natural tracers for sediment transport studies. *Continental Shelf Research*, 7(11/12):1333-1343.

Sanford, L.P. (1992). New sedimentation, resuspension, and burial. *Limnology and Oceanography*, 37(6):1164-1178.

Signorini, S.R. (1980). A study of the circulation in Bay of Ilha Grande and Bay of Sepetiba: Part I. A survey of the calculation based on experimental field data. *Boletim do Instituto Oceanografico*, 29(1):41-55.

Simpson, W.R. (1981). A critical review of cadmium in the marine environment. *Progress in Oceanography*, 10:1-70.

Souza, C.M.M.; Pestana, M.H.D. and Lacerda, L.D. (1986). Heavy metal partitioning in sediments of three estuaries along the Rio de Janeiro coast. *The Science of the Total Environment*, 58:63-72.

Sundby, B; Gobeil, C.; Silverberg, N.; Mucci, A. (1992). The phosphorus cycle in coastal marine sediments. *Limnology and Oceanography*, 37(6):1129-1145.

Suzuki, M. , Yamada, T., Miyazaki, T. and Kawazoe, K. (1979). Sorption and accumulation of cadmium in sediment of the Thama River. *Water Research*, 13:57-63.

Walsh, R.S. and Hunter, K.A. (1992). Influence of phosphorus storage on the uptake of cadmium by the marine alga Macrocystis pyrifera. *Limnology and Oceanography*, 37(7):1361-1369.

Windom, H.; Byrd, J.; Smith, R.; Hungspreugs, M.; Dharmvanij, S.; Thumtrakul, W.; Yeats, P. (1991). Trace metal-nutrient relationships in estuaries. *Marine Chemistry*, 32:177-194.

Yeats, P.A. (1988). The distribution of trace metals in ocean waters. *The Science of the Total Environment*, 72:131-149.

13 Mercury Behaviour in Sediments from a Sub-Tropical Coastal Environment in SE Brazil

Paulo Rubens G. Barrocas[1] and Julio Cesar Wasserman[2]

[1] Visiting Researcher, CESTEH, Fundação Oswaldo Cruz, Rua Leopoldo Bulhões 1480, Rio de Janeiro, RJ, Brazil (Cooperation FIOCRUZ/FAPERJ)

[2] Departamento de Geoquímica, UFF, Outeiro de São João Batista s/n°, Niterói, RJ, Brazil

Abstract

The geochemical behaviour of mercury in sediments of the São João de Meriti system (Guanabara bay watershed, Rio de Janeiro, Brazil) was studied. Total concentrations of mercury were compared to sediment contents of clay-silt, organic/inorganic carbon, total hydrolyzable amino acids (THAA) and specific amino acids, total sulphur and iron. A sequential extraction was also used to clear up some questions arising from the total concentration correlations. Results (0.32-3.38 mg Hg kg^{-1}) show a clear contamination of the studied area by mercury, although the levels are not high when compared to other impacted areas around the world. The behaviour of mercury in the area is closely linked to organic matter. Good correlations were found between mercury and organic indexes (organic carbon, nitrogen, amino acids), 95% of the mercury present in the sediment was shown to be linked to organic matter (as indicated by geochemical partitioning). On the other hand, no correlation was found between mercury and sulphur, which is explained by the very stable bonds between mercury and organic matter.

Introduction

In the recent years, environmental researchers interested in metal pollution have dedicated considerable efforts to the study of coastal areas, mainly tropical estuaries. This concern is due to the lack of environmental information in these areas and also due to their high rates of contamination. Among the compartments studied to assess coastal environmental contamination by metals, sediment has been the most largely used (e.g; Wasserman et al., 1991; Wasserman et al., 1992; Fernandez, 1994). This is due to the capacity of the sediments in integrating the very low and unstable concentrations of water dissolved pollutants. Furthermore, the advantage of sediments when compared to organisms is that the metabolic processes of the latter may mask the environmental ones.

Among the heavy metals, mercury has proved to be the most harmful to human health. It is the only metal that, due to its biomagnification on the trophic chain and organification to the most toxic form, methylmercury, has been responsible for a considerable number of human deaths (Fujiki, 1972; Takeuchi, 1972; Förstner and Wittmann, 1983).

A number of studies (reviewed in deGroot, Salomons and Allersma, 1976; Förstner and Wittmann, 1983) have indicated that the estuarine environment is the most sensitive to heavy metal pollution. This is explained by the 1) depositional characteristics, 2) very reducing conditions and 3) high concentrations of organic matter, that cause the trapping of the metals, mainly in the sediments that commonly contribute more than 90% of these pollutants present in the environment (Eysink et al., 1988; Elsokkary, 1989, Wasserman et al., 1992; Wasserman et al. 1991). These increasing concentrations in the estuarine environment directly affect concentrations in the organisms since they use this environment as a nursery for juveniles.

In Brazil, the problem of environmental mercury contamination has been widely studied, mainly in the amazonian environment, where gold mining activities are widespread (Hacon et al., 1990; Veiga and Fernandes, 1991; Farid, 1992; Lacerda and Salomons, 1992, reviewed the subject).

On the other hand, in the Brazilian coastal environment, the available studies are restricted to the distribution of mercury and some monitoring programmes (Mósca, 1980; CETESB, 1986a; CETESB, 1986b; Vargas-Boldrini and Navas-Pereira, 1987; Eysink et al., 1988; Vargas-Boldrini, 1990; FEEMA, 1990); their approaches do not regard processes. Some recent studies have attempted a different approach, regarding the processes (Quevauviller et al., 1992; Barrocas and Wasserman, 1993; Barrocas et al. 1993). Even though these works have not reached any conclusive assumption, they underline the particular behaviour of mercury in the tropical coastal environment. For example, Quevauviller et al. (1992) observed that most of the mercury is present as dimethyl-mercury in a mangrove sediment core, a form that is not observed in sediments of temperate areas. The problem is that this form is too volatile to be stored in the sediment. They suggest that organic matter, sulphide concentrations and bacterial development do affect mercury behaviour, but could not explain in detail the process.

In this work we have contributed to the knowledge of mercury's behaviour in a tropical coastal environment (Guanabara Bay/São João de Meriti river system) by establishing its distribution as related to physico-chemical gradients and depicting the geochemistry of its carriers.

Study Area

Guanabara Bay (Fig. 1) is a tropical ecosystem adjacent to Rio de Janeiro city. It is one of the biggest bays in Brazil, covering an area of 381 km^2. Its watershed area is more than 4,000 km^2, where 7 million people live. Its mean depth is 7.7 m, with a maximum depth of 50 m at the entrance of the bay. Some 80% of its area has a depth below 10 m. The residence time estimated for Guanabara Bay waters is 20 days (Coelho and Fonseca, 1981; Rebello et al, 1986).

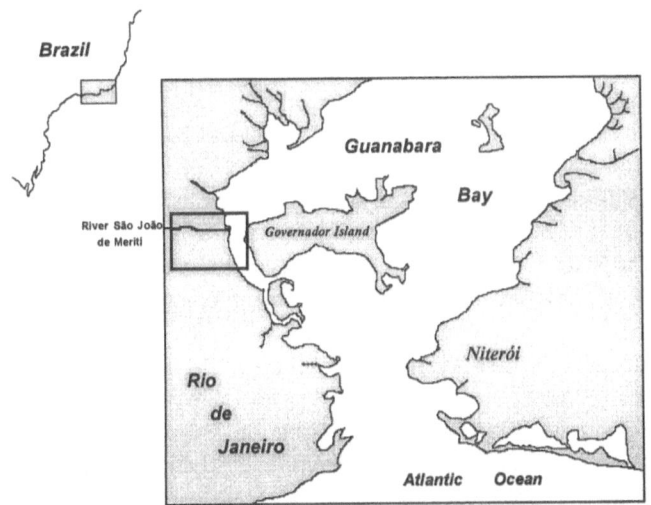

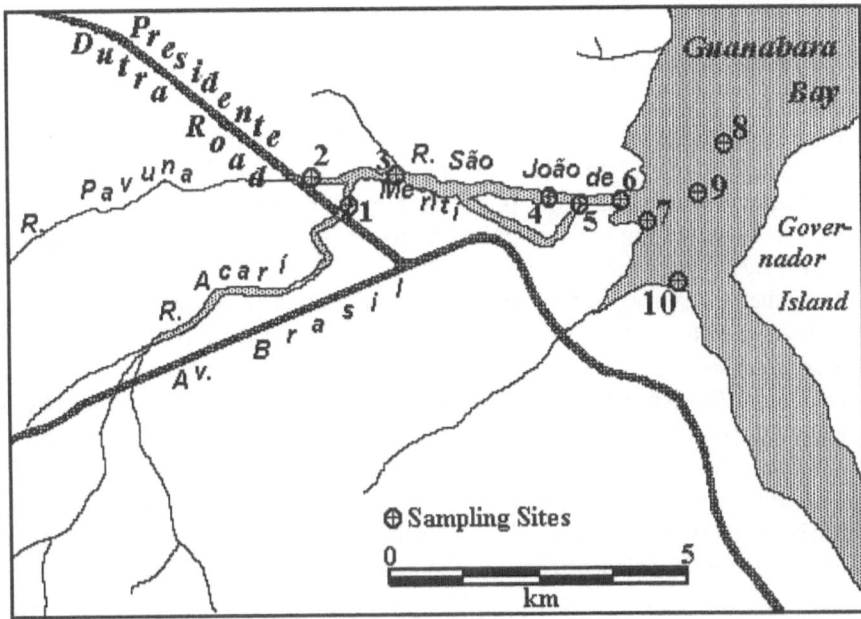

Fig. 1: Location of the study area and sampling sites

A number of rivers and channels flow into the bay, forming many sub-basins. The water quality data of the majority of these rivers show a critical situation, specially those located on the north-west coast. Although the continental water volume is small when compared to the inputs of marine waters (mean salinity is between 21 and 30), the mean load of material carried by the rivers is 2×10^5 ton year^{-1} and of variable nature (FEEMA, 1985; Rebello et al, 1986; FEEMA, 1990).

About 7,000 industries are installed in Guanabara Bay watershed. They are responsible for 25% of organic pollution and almost all toxic substances and heavy metals released in this area. Daily 22 kg of cyanides, 4,200 kg of phenols, 1,800 kg of sulphides and 4,800 kg of heavy metals are dumped by industrial effluents (FEEMA, 1985; 1990).

São João de Meriti River is one of the most polluted, and the main source of mercury for the bay. The highest concentrations of mercury in Guanabara Bay watershed were observed in this river estuary (Mósca, 1980; FEEMA, 1986; Rebello et al, 1986; Rego et al., 1993). These high concentrations originate from a chlor-alkali plant installed in its sub-basin. According to estimates from FEEMA (1985), this plant released 1.46 tons of mercury during the year of 1975. Later, in 1979, the installation of an effluent treatment system reduced the annual mercury loads to 20 kg (FEEMA, 1985). Nevertheless, in a recent survey, Rego et al. (1993) estimated that São João de Meriti River releases 160 kg Hg year^{-1} in Guanabara bay.

Material and Methods

Bottom sediments were collected with a titanium grab in the sites indicated in Fig. 1. Sampling sites were chosen with regard to a gradient from freshwater (upstream) to estuarine water (Guanabara Bay). Samples were stored in plastic bags and kept frozen until analysis. Physico-chemical parameters were measured *in situ* during samplings.

At the laboratory, samples were gently oven dried (not more than 50°C until constant weight). Total mercury levels were determined by cold vapour atomic absorption spectrophotometry (CVAAS) after acid extraction (adapted from Malm et al., 1989). Accuracy and precision of the analysis were verified by an intercalibration exercise with another laboratory (IPEN-SP) that uses neutron activation analysis (Figueiredo et al., 1994), showing errors never greater than 7%.

A sequential extraction procedure, adapted from Di Giulio and Ryan (1987) and Revis et al. (1989), was applied in order to evaluate the geochemical partitioning of mercury in the sediments. The procedure used is briefly described in Fig. 2. The difference between the sequential extraction and total mercury analysis was always smaller than 20%.

Other variables which have influence in mercury geochemistry in aquatic environments were also analysed. Silt-clay fraction was measured by wet sieving;

Other variables which have influence in mercury geochemistry in aquatic environments were also analysed. Silt-clay fraction was measured by wet sieving; Fe content of samples was measured by atomic absorption spectrophotometry (Loring and Rantala, 1977); total sulphur was analysed in an elemental analyser (ASTM, 1988); total and inorganic carbon were analysed following Verardo et al. (1990) in an elemental analyser; total hydrolyzable amino acids (THAA) and specific amino acids were measured in an amino-acid analyser by liquid chromatography (Müller et al., 1986).

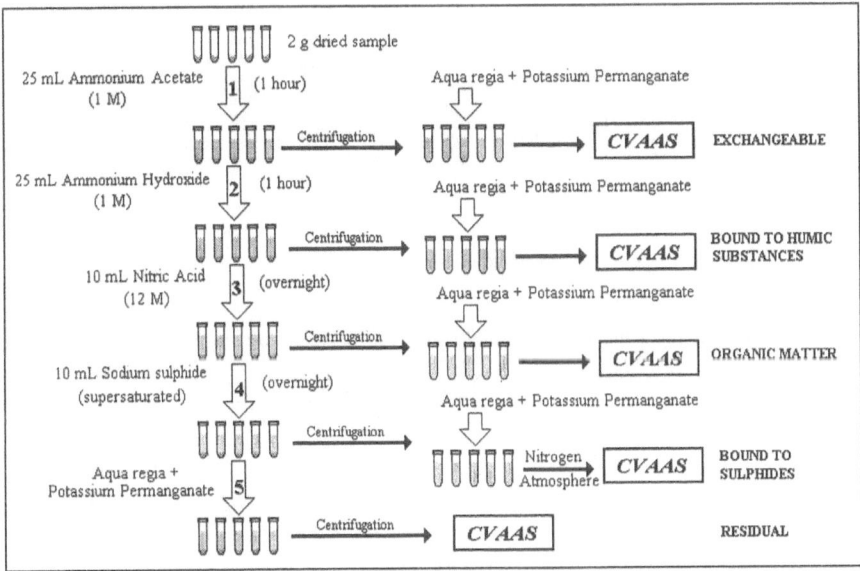

Fig. 2: Scheme of the sequential extraction (adapted from Di Giulio and Ryan, 1987 and Revis et al., 1989)

Results and Discussions

Total concentrations of mercury in the sediments ranged from 0.32 to 3.38 mg kg^{-1} (Fig. 3), showing a mild state of contamination when compared to other heavily contaminated areas in the world (Table 1). On the other hand, these values can be considered high when compared to uncontaminated environments (0.01 - 0.1 mg kg^{-1}; Vucetic et al, 1974) and above background values for Guanabara Bay, which were established as 0.05 mg kg^{-1} (FEEMA, 1986; Pinto, 1995).

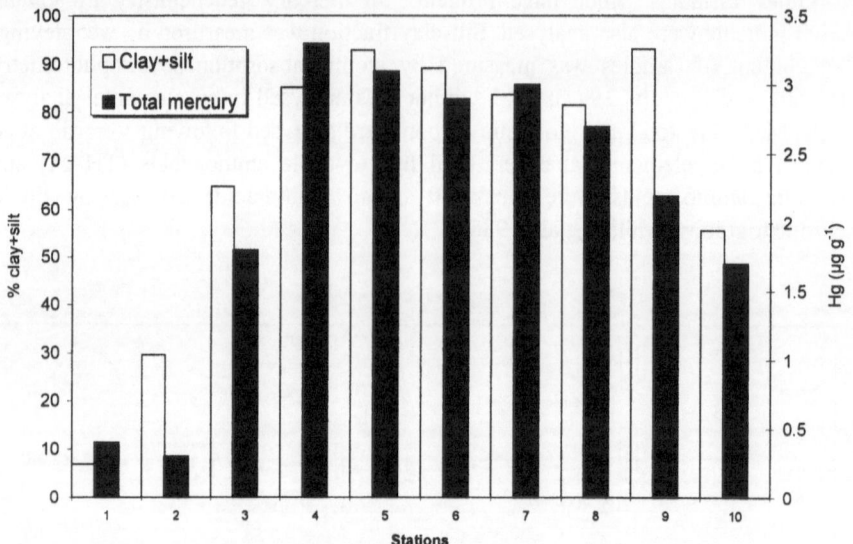

Fig. 3: Total mercury concentrations and clay+silt content

Table 1: Mercury loads in rivers and estuaries as reported in the literature (n.d.= non-detectable)

Study area	Range (mg kg^{-1})	Author
Sena river (France)	9.80 - 15.80	Batti et al. (1975)
Clyde river (England)	<0.05 - 3.68	Craig and Moreton (1986)
Dart estuary (England)	0.33 - 4.46	Craig and Moreton (1986)
Mersey estuary (England)	0.01 - 14.30	Craig and Morton (1976)
Thames estuary (England)	0.02 - 0.49	Smith et al. (1973)
Everglade estuary (USA)	0.22 - 1.86	Lindberg and Harriss (1974)
Chesapeake bay (USA)	n.d. - 1.12	Brinckman et al. (1973)
Minamata bay (Japan)	630 (mean)	Matida and Kumada (1969)
Tainheiros bay (Brazil)	n.d. - 37.48	CEPED (1975)
São João de Meriti river	0.32 - 3.38	This study

Figure 3 shows the behaviour of total mercury concentrations along the river, down to the bay. Increasing concentrations are observed until the estuarine area, where levels slightly fall. Grain size control seems to be occurring, significant correlation ($p<0.05$) between total mercury and clay+silt contents was observed (Fig. 3). This is well established in the literature, for mercury and other heavy metals (Förstner and Wittmann, 1983). This granulometric range is reported to be the main component of suspended matter (Förstner and Salomons, 1980). As mercury shows a preferential association with clay+silt, it can be suggested that in the São João de Meriti system, the main way of mercury transportation is associated with suspended matter. If we further consider that the silt+clay fraction is the main geochemical carrier of mercury and we suppose, as done elsewhere (deGroot et al. 1971), that very little mercury is present in the sand fraction, it is possible to correct concentrations of mercury to the fraction $\leq 63\mu m$ (Cameron, 1974).

Figure 3 shows that corrected concentrations are higher in the Acari river (station 1), where the main source of mercury is located, the chlor-alkali plant. Station 2 (Pavuna river), where there is no reported source of mercury, remains the less contaminated. Finally, the São João de Meriti river and estuarine sediments, downstream, seem to be a mixture of the two former river sediments.

Since the correlations between silt+clay fraction and mercury, as stated above, are very high, it is to be considered that the sand fractions of the sediment are quite non-reactive and frequently have a minor role on the geochemistry of elements. On the other hand, it is not possible to attribute the behaviour of mercury only to the silt+clay fraction, since other elements and complexes like organic matter, sulphur nitrogen and phosphorus are also trapped in this fraction. Furthermore, the relationship of mercury and the silt+clay fraction is consistent in our study area, but this behaviour is not common for every environment. In this volume, Miller and Lechler show an example where the silt+clay fraction does not control mercury concentrations in river sediments.

The relationship between organic matter and mercury is widely reported in the literature (Jenne, 1970; Moore and Ramamoorthy, 1984; Trent et al., 1989). Mercury forms stable complexes with humic acids (Kendorff and Schnitzer, 1980), amino acids (Rashid, 1972) and hydroxi-carboxylic acids (Moore and Ramamoorthy, 1984). In this work, it was observed that the variables related to organic matter, i.e., organic carbon and total nitrogen, were strongly correlated with mercury ($p < 0.05$), corroborating the above literature statements (Fig. 4).

The high correlations between these organic variables and the silt+clay fraction also suggest that mercury is associated with surface adsorbed nitrogen compounds, like amino-acids and amino-sugars, observed elsewhere (Carter and Mitterer, 1978; Rosenfeld, 1979). Furthermore, these organic complexes tend to be subjected to a similar process of scavenging in the estuarine environment (flocculation and coagulation) and accumulate in the sediments (Mantoura and Woodward, 1983; Chester, 1990).

The study of THAA and especially the specific amino-acids (aspartic acid and arginine) and their decomposition products (respectively β-alanine and ornithine) revealed the quality of organic matter within the system (Rosenfeld, 1979; Maita et al., 1982; Cowie and Hedges, 1992). The rates aspartic acid/β-alanine and arginine/ornithine increased in the estuarine samples and in the bay sediments (Table 2) showing that important sources are present in this area (probably the extensive mangrove stands in the border of the bay and estuary). It is interesting to note that the amino-acid concentrations observed in this area are similar to those

observed in the Paraiba do Sul river mouth, a large river located some 400 km to
the north of the study area, where there is a large mangrove area (Jennerjahn,
unpublished data).

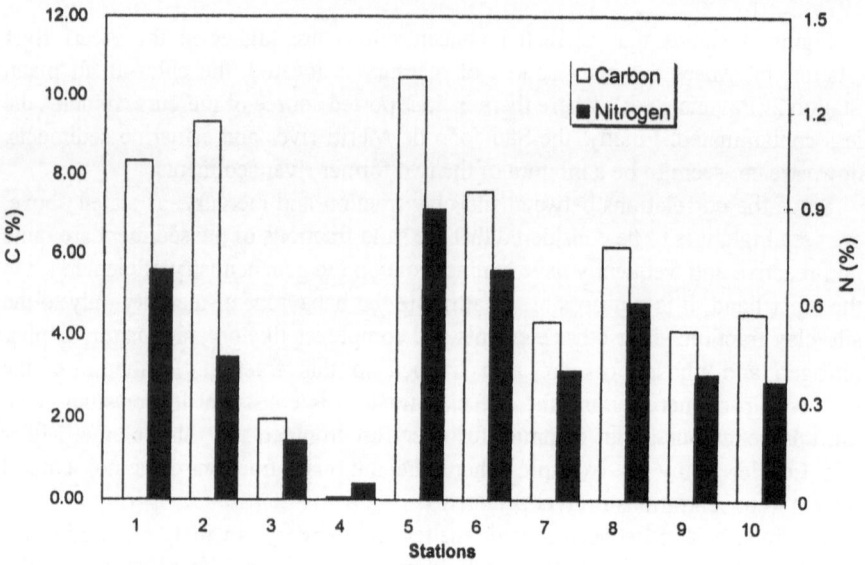

Fig. 4: Total nitrogen and organic carbon contents

Table 2: Total hydrolysable amino acids and specific amino acid concentrations

Sampling site	Aspartic acid (mg kg⁻¹)	β-Alanine (mg kg⁻¹)	Asp/β-Ala	Arginine (mg kg⁻¹)	Ornithine (mg kg⁻¹)	Arg/Orn	THAA (mg g⁻¹)
1	224.6	12.5	18.0	76.5	9.4	8.1	1.82
2	716.9	66.1	10.9	224.7	33.6	6.7	5.70
3	2106.3	117.1	18.0	644.2	92.1	7.0	16.77
4	3103.4	140.1	22.2	976.8	141.4	6.9	25.70
5	4813.2	117.5	41.0	1983.9	134.4	14.8	37.59
6	3675.0	96.0	38.3	1443.3	108.7	13.3	27.52
7	1822.5	67.8	26.9	681.8	14.3	47.7	12.92
8	3370.9	89.2	37.8	1204.3	52.8	22.8	23.82
9	1922.4	81.1	23.7	716.0	25.2	28.4	13.55
10	2264.1	59.0	38.4	939.5	37.2	25.3	16.23

Although thoroughly described in the literature (Jenne, 1970; Reimers et al., 1975; Belliveau and Trevorst, 1989), no significant correlation ($p<0.05$) between mercury and total sulphur was observed in the study area (Fig. 5). This lack of correlation can be explained by the difference in the magnitudes of concentrations of these two elements (sulphur expressed in % and mercury expressed in mg kg^{-1}), thus, mercury sulphide would represent only a small portion of the sulphur compounds present in the sample, masking the statistical correlations. However, results of a core sampled in this same estuary (Pinto, 1995) showed a correlation between mercury and sulphur. This correlation is probably related to mineralization of organic matter, the main geochemical support of mercury, in the sediment column. Therefore, released mercury can be complexed by the sulphur dissolved in the interstitial water, in deeper sediments.

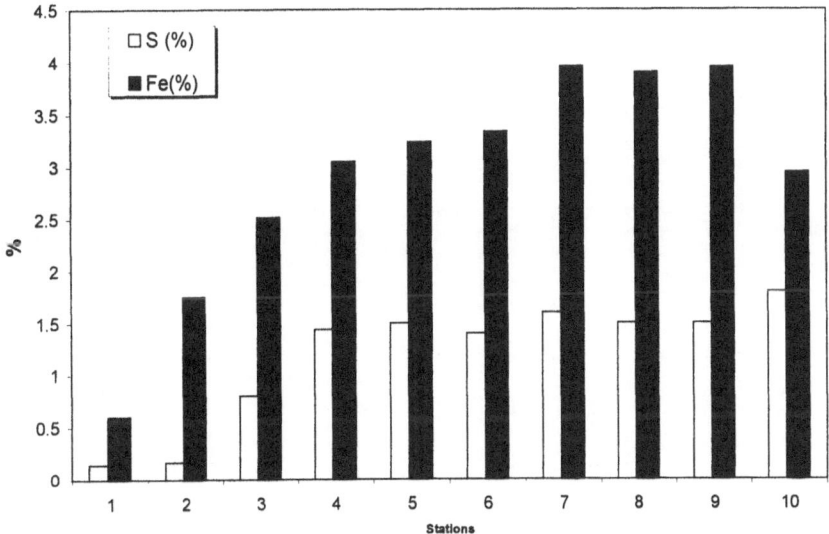

Fig. 5: Total sulphur and iron levels

Iron did not show such a significant correlation with mercury ($p<0.05$), corroborating the preceding discussion that mercury is preferentially linked to organic complexes. Furthermore, due to the very reducing conditions of the system, iron is present in its reduced form, which is not an important geochemical carrier for metals as hydroxides are. The good correlation ($p<0.05$) between iron and sulphur (Fig. 5) in such a reducing environment also suggests formation of pyrite and a competition between iron and other metals for the sulphide sites.

Only the sequential extraction results will clear up the problem of lack of correlation between sulphur and mercury, showing the importance of organic matter as a geochemical carrier. Figure 6 shows that more than 95% of the mercury is bound in phase III (residual organic compounds). This phase seems very stable since it is neither affected by an increase in salinity within the estuarine gradient, nor by the first two steps of the sequential extraction. This results corroborate the strong correlation between the mercury and organic matter presented above.

Step I of sequential extraction corresponds to the exchangeable or very weakly bound phase, and although very small concentrations of mercury appear in this phase upstream in the river, it completely disappears in the estuarine and bay sediments, confirming that mercury is not available in this system. Step II represents mercury bound to humic substances, which, as stated in Fig. 6, are not important carriers of mercury. Notwithstanding these results, the proceeding for extraction of humic substances in this step was not thoroughly tested and it can be stated that the method lacks efficiency, and part of the humic substances would only be extracted in phase III.

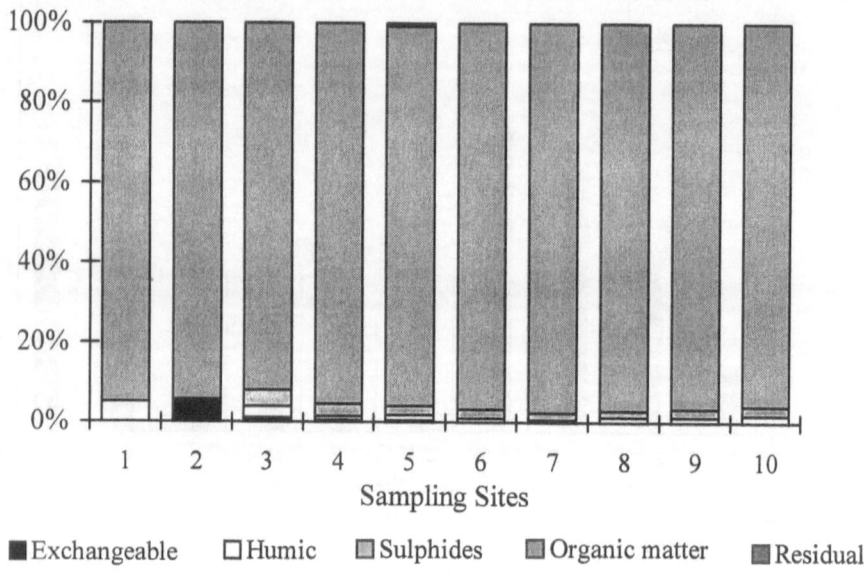

Fig. 6: Sequential extraction results

In the fourth step, mercury bound to sulphide, a slight increase was observed in concentrations within the estuarine area, suggesting that some of the mercury from phase I (weakly bound) is readily complexed as sulphides. This is shown to be a not very important process concerning mercury behaviour, since it affects only 2 % of the mercury present in the sediment. The residual phase (step V) did not show detectable concentrations and in the literature, it is not reported to be an important phase, except in areas of cinnabar mining.

Conclusions

The results of this work show evident mercury contamination of the sub-basin of the São João de Meriti river, confirming results presented in the literature for the area. However, the observed levels are not very elevated when compared to other contaminated environments.

The importance of organic matter in the geochemistry of mercury in the studied environment was established. Furthermore, the complexes formed seem to be very stable, and are not affected by common environmental changes (e.g. salinity changes). Thus, it seems that mercury is not bioavailable in the present environmental conditions. However, our results show the high toxicity potential of the sediments in this area, and considerable harm can be predicted if dredging activities occur in this area.

Despite the literature statement that sulphides are important mercury carriers, in the Guanabara Bay/São João de Meriti system it seems to be a minor carrier. This would reinforce the statements that mercury shows a distinctive behaviour in tropical coastal areas.

It should be also considered that the studied area is an impacted area, subjected to high levels of anthropogenic contamination of various types or sources and it is quite a difficult task to define exactly what kind of organic matter we have. Every kind of synthetic organics may be present, and many of them can be strong adsorbents, raising important questions concerning the methodology for the study of such complex systems.

References

ASTM - American Standard Methods (1988*). Standard test method for sulphur in petroleum products (High Temperature Method).* ASTM (Ed.) D 1552:606-611.

Barrocas, P.R.G. and Wasserman, J.C. (1993). The mercury in Guanabara Bay: a historical summary. *In: Proceedings of the International Symposium on Perspectives for Environmental Geochemistry in Tropical Countries.* J.J. Abrão, J.C. Wasserman and E.V. Silva-Filho (Eds.), pp.:454-460.

Barrocas, P.R.G., Wasserman, J.C., Jennerjahn, T. and Pivetta, F. (1993). Geochemistry of mercury in sediments of São João de Meriti estuary in Guanabara bay system. *In: Proceedings of the International Symposium on Perspectives for Environmental Geochemistry in Tropical Countries.* J.J. Abrão, J.C. Wasserman and E.V. Silva-Filho (Eds.), pp.:143-148.

Batti, R., Magnaval, R., Lanzola, E. (1975). Methylmercury in river sediments. *Chemosphere*, 1:13-14.

Belliveau, B.H. and Trevorst, J.T. (1989). Mercury resistance and detoxification in bacteria. *Applied Organometallic Chemistry*, 3:283-294.

Brinckman, F.E., Jewett, K.L., Blair, W.R. (1975). Mercury distribution in Chesapeake Bay. *In: KRENKEL, P.A. (Ed.) Heavy Metals in the Aquatic Environment.* Oxford: Pergamon Press,. pp:251-252.

Cameron, E.M. (1974). Geochemical methods of exploration for massive sulphide mineralization in the Canadian shield. *In: Geochem. Explor.* - Proc. 5th Inter. Geochem. Explor. Symp. Vancouver. Elliott, I.L. Fletcher, W.K. (eds.). pp:21-49.

Carter, P.W. and Mitterer, R.M. (1978). Amino acid composition of organic matter associated with carbonate and non-carbonate sediments. *Geochim. Cosmochim. Acta*, 42:1231-1238,.

CETESB. (1986a). Avaliação da contaminação por metais pesados e pesticidas organoclorados na água, ictiofauna e outros organismos aquáticos do complexo estuarino lagunar de Iguape-Cananéia. *Relatório Técnico CETESB -SP*.

CETESB. (1986b). Metais pesados no estuario e Baía de Santos. *Relatório Técnico CETESB - SP*, 75p.

Chester, R. (1990). *Marine Geochemistry.* London: Unwin Hyman Ltd. 698p.

Coelho, V.M.B. and Fonseca, M.R.M.B. 1981). Problemas de eutroficação no Estado do Rio de Janeiro. *Cadernos FEEMA*, 51p.

Cowie, G.L. and Hedges, J.I. (1992). Sources and reactivities of amino acids in a coastal marine environment. *Limnol. Oceanogr.*, 37(4):703-724.

Craig, P.J. and Morton, S.F. (1976). Mercury in Mersey estuary sediments, and the analytical procedure for total mercury. *Nature*, 261:125-126.

Craig, P.J. and Moreton, P.A. (1986). Total mercury, methylmercury, and sulphide levels in British estuarine sediments - III. *Wat. Res.*, 20(9):1111-1118.

deGroot, A.J., J.J.M.Goeij, C. Zegers (1971). Contents and behaviour of mercury as compared with other heavy metals in sediments from the Rhine and Ems. *Geol. Mijnbouw*, 50:393-398.

deGroot, A.J., Salomons, W., Allersma, E. (1976). Processes affecting heavy metals in estuarine sediments. *In: Estuarine Chemistry*. J.D. Burton and P.S. Liss (Eds.), pp.:131-157.

Elsokkary, I.H. (1989). Mercury biogeochemstry in the aquatic systems of Alexandria distric: an assessment of contamination and especiation. *In: Proceedings of the VII International Conference on Heavy Metals in the Environment*, Edinburgh, CEP Consultants, v. 2, pp:416-419.

Eysink, G.G.J., Pádua, H.B., Martins, M.C. (1988). Presença do mercúrio no ambiente. *Ambiente*, 2(1):43-50.

Farid, L.H. (1992). *Diagnóstico Preliminar dos Impactos Ambentais Gerados por Garimpos de Ouro em Alta Floresta/MT: Estudo de Caso.* Série Tecnologia Ambiental, 2, CETEM, 190p.

FEEMA. (1986). *Levantamento de Metais Pesados no Estado do Rio de Janeiro*: Relatório Final. FEEMA/RJ. 2v.

FEEMA. (1990). *Projeto de Recuperação Gradual do Ecosistema da Baía de Guanabara - Indicadores Ambientais de Degradação, Obras e Projetos de Recuperação.* Relatório Final. FEEMA/RJ. 2v.

FEEMA. (1985). *Levantamento de Metais Pesados no Estado do Rio de Janeiro.* Primeiro Relatório Trimestral.

Fernandez, M.A.S. (1994). *Geoquímica de Metais Pesados na Região dos Lagos, RJ: Proposta para um Estudo Integrado.* M.Sc. thesis. Programa de Pós-Graduação em Geoquímica - Universidade Federal Fluminense. 143 p.

Figueiredo, A.M.G, Fávaro, D.T.I. e Wasserman, J.C. (1994). Aplicação do método de análise por ativação com nêutrons à determinação de metais pesados em sedimentos. *In: Anais do V Congresso Geral de Energia Nuclear*, São Paulo, ABEN, V. III:789-790.

Förstner, U. and Salomons, W. (1980). Trace metal analysis on polluted sediments: Assessment of sources and intensities. *Environmental Technology Letters*, v. 1, p. 494-505.

Förstner, U. and Wittmann, G.T.M. (1983). *Metal Pollution in the Aquatic Environment.* 2 ed. Berlin, Springer-Verlag. 486 p.

Fujiki, M. (1972). The transitional condition of Minamata bay and neighboring sea polluted by factory waste matter containing mercury. *In: 6th Int. Conf. Water Pollut. Res.*, paper 12.

Di Giulio, R.T. and Ryan, E.A. (1987). Mercury in soils, sediments, and clams from North Carolina peatland. *Wat. Air Soil Pollut.*, 33:205-219,.

Hacon, S., Lacerda, L.D., Pfeiffer, W.C., Carvalho, D. (1990). *Riscos e Consequências do Uso do Mercúrio*. FINEP, IBAMA, Ministério da Saúde e CNPq. 314 p.

Jenne, E.A. (1970). Atmospheric and fluvial transport of mercury. *In: U.S. Geological Survey: Mercury in Environment*. Washington. pp:40-45 (Professional Paper, 713).

Kendorff, H. and Schnitzer, M. (1980). Sorption of metals on humic acid. *Geochim. Cosmochim. Acta*, 44:1701-1708.

Lacerda, L.D. and Salomons, W. (1992). *Mercúrio na Amazônia: Uma bomba relógio química?* Série Geoquímica Ambiental do Departamento de Geoquímica da UFF. CETEM, 78p.

Lindberg, S.E. and Harriss, R.C. (1974). Mercury-organic mattter associations in estuarine sediments and their associated interstitial water. *Environ. Sci. Technol.*, 8(5):459-462,.

Loring, D.H. and Rantala, R.T.T. (1977). *Geochemical Analyses of Marine Sediments and Suspended Matter*. Canadian Technical Report of Fisheries and Aquatic Sciences, n. 700, 58p.

Maita, Y., Montani, S., Ishii, J. (1982). Early diagenesis of amino acids in Okhotsk Sea sediments. *Deep Sea Res.*, 29:485-498.

Malm, O., Pfeiffer, W.C., Bastos, W.R., Souza, C.M.M. (1989). Utilização do acessório de geração de vapor a frio para análise de mercúrio em investigações ambientais por espectrofotometria de absorção atômica. *Ciência e Cultura*, 41(1):88-92.

Mantoura, R.F.C. and Woodward, E.M.S. (1983): Conservative behaviour of riverine dissolved organic carbon in the Severn estuary: chemical and geochemical implications. *Geochim. Cosmochim. Acta*, 47:1293-1309.

Matida, Y. and Kumada, H. (1969). Distribution of mercury in water, botttom sediments and aquatic organisms of Minamata Bay, the river Agano and other water bodies in Japan. *Bul. Freshwater Fish. Res. Lab.*, 19, 73p.

Moore, J.W. and Ramamoorthy, S. (1984). *Heavy metals in natural waters: Applied monitoring and impact assessment*. Berlin, Springer-Verlag, Chap. 7: Mercury, pp.:125-160.

Mósca, N.P. (1980). *Concentração de mercúrio nas águas e sedimentos da Baía de Guanabara, Rio de Janeiro, RJ Brasil*. Dissertação de Mestrado em Geociências - Universidade Federal Fluminense. Niterói, 101p.

Müller, P.J., SUESS, E., UNGERER, C.A. (1986). Amino acids and amino sugars of surface particulate and sediment trap material from waters of the Scotia Sea. *Deep Sea Res.*, 33:819-838.

Pinto, A.P.F. (1995). *Geoquímica do Mercúrio em Perfis Sedimentares de Manguezais da Baía de Guanabara - Rio de Janeiro - Brasil*. Dissertação de Mestrado do Programa de Pós-Graduação em Geoquímica - UFF, Niterói, 99 p.

Quevauviller, P., Donard, O.F.X., Wasserman, J.C., Martin, F.M., Schneider, J. (1992). Occurence of methylated tin and dimethyl mercury compounds in a mangrove core from Sepetiba bay, Brazil. *Appl. Organometallic Chem.*, 6:221-228.

Rashid, M.A. (1972). Amino acids associated with marine sediments and humic compounds and their role in solubility and complexing of metals. In: *Proceeding of the 24th International Geological Congress*, Montreal: Edited by Boyle, R.W. and Shaw, D.M., pp.:346-353.

Rebello, A.D.L., Haekel, W., Moreira, I., Santelli, R., Schroeder, F. (1986). The fate of heavy metals in an estuarine tropical system. *Marine Chemistry*, 18:215-225.

184   Mercury in Guanabara Bay

Rego, V.S., Pfeiffer, W.C., Barcellos, C.C., Rezende, C.E., Malm, O. and Souza, C.M.M. (1993). Heavy metal transport in the Acari-São João de Meriti river system, Brazil. *Environ. Technol.*, 14:167-174.

Reimers, R.S., Krenkel, P.A., Eagle, M. (1975). Sorption phenomenon in the organics of bottom sediments. In: *Heavy Metals in the Aquatic Environment*. Krenkel, P. A. (Ed.) Oxford: Pergamon Press. pp.:117-129.

Revis, N.W., Osborne, T.R., Sedgley, D., King, A. (1989). Quantitative method for determining the concentration of mercury (II) sulphide in soils and sediments. *Analyst*, 114:823-825.

Rosenfeld, J. K. (1979). Amino acid diagenesis and adsorption in nearshore sediments. *Limnol. Oceanogr.*, 24(6):1014-1021,.

Smith, J.D., Nicholson, R.A., Moore, P.J. Mercury in sediments from Thames estuary. *Environ. Pollut.*, 4:153-157, 1973.

Takeuchi, T. (1972). Distribution of mercury in the environment of Minamata bay and the inland Ariake Sea. In: *Environmental Mercury Contamination*. Hartung, R.,Dinman, B.D. (Eds.). Ann Arbor Sci. Pub. Inc., Ann Arbor, pp.79-81.

Trent, M. A. , Trent, J.D., Robertson, J.M. and Laguros, J.G. (1989). Interactions and effects of oxygen, pH, and mercury concentration on partioning of mercury into solution and Lahontan sediment. In: *Proceedings of the VII International Conference on Heavy Metals in the Environment*. CEP Consultants, Edinburgh, v. 2, pp.:485-484.

Vargas-Boldrini, C. and Navas-Pereira, D. (1987). Metais pesados na Baía de Santos e estuários de Santos e São Vicente. Bioacumulação. *Ambiente*, 1(3), pp.:118-127,.

Vargas-Boldrini, C. (1990). Mercúrio na Baixada Santista. In: *Riscos e Consequências do Uso do Mercúrio*. Hacon, S., Lacerda, L.D., Pfeiffer, W.C. and Carvalho, D. (Eds.), FINEP, IBAMA, Ministério da Saúde e CNPq, pp.:161-195.

Veiga, M.M. and Fernandes, F.R.C. (1991). Aspectos gerais do projeto Poconé. In: *Poconé: Um Campo de Estudo do Impacto Ambiental do Garimpo*. Veiga, M.M. and Fernandes, F.R.C. Série Tecnologia Ambiental 2, CETEM, pp.:1-25,.

Verardo, D.J., FROELICH, P.N., McINTRE, A. (1990). Determination of organic carbon and nitrogen in marine sediments using the Carlo Erba NA-1500 Analyser. *Deep Sea Res.*, 37:157-165.

Vucetic, T., Wernberg, W.B., Anderson, G. (1974). Long term fluctuation of mercury in the zooplankton of the East Central Adriatic. *Rev. Inter. Océanogr. Méd.*, 28:73-81.

Wasserman, J.C., Dumon, J.C. et Latouche, C.(1991). Importance des zostères (*Zostera noltii* Hornemann) dans le bilan des métaux lourds du Bassin d'Arcachon. *Vie et Milieu*, 41(2/3):81-86.

Wasserman, J.C., Dumon, J.C. et Latouche, C. (1992). Le bilan de 18 éléments traces et 7 éléments majeurs dans un environnement peuplé par des Zostères (*Zostera noltii* Hornemann). *Vie et Milieu*, 42(1):15-20.

Wasserman, J.C., Silva-Filho, E.V., Abrão, J.J., Patchineelam, S.R. e Bidarra, M. (1991). Carreadores geoquímicos de Cu, Fe, Mn e Zn na Baía de Sepetiba (RJ): trocas entre o material em suspensão e o sedimento. In: *III Congresso Brasileiro de Geoquímica*, São Paulo, pp:374-376.

14 Diagnosis of Environmental Problems Related to Vein Gold Mining in Colombia

Gloria R. Prieto and Myriam L. Gonzalez
INGEOMINAS Diagonal 53 # 34- 53, Bogotá, Colombia

Abstract

Since 1985 the annual gold production in Colombia has been fluctuating between 30 and 35 tons (1-1.4 million ounces troy). Exploitation plants can be found in vein and placer gold deposits.

During 1992 a preliminary study was undertaken, resulting in a diagnosis of problems in small scale mining in 6 gold areas (vein type) of Colombia. In order to evaluate the general impact caused to the environment due to mining activities, six gold districts located in Nariño, Antioquia, Bolivar, Valle and Caldas Departments were visited. Geochemical analysis (ES, AAS, HGAAS, GFAAS) of orebodies, tailings and waters were carried out, with results that showed high levels of heavy metals in the environmental compartments studied.

The intense use of mercury for amalgamation and cyanides for leaching during mining and processing of gold orebodies was notable. The study verified that exploitation of gold deposits in Colombia is mainly done by small scale mining, which is characterized by low technology and recovery rates (about 50%), causing harm to water resources (municipal, industrial and groundwater), sediments, soils, air, vegetation, animals and humans. The study concludes that environmental problems are intimately related to the lack of appropriate technologies for mining and processing gold. This is a key factor causing low economic revenues, and consequently there is no compliance among the miners to invest in the care of the environment.

Introduction

Gold mining in Colombia has been done since Spanial colony times, around the sixteenth century. In Colombia, different gold districts have been exploited and there are many possibilities of many other important deposits. Since 1985 the total gold production has fluctuated between 30 and 35 tons per year, which is mostly produced in placers (almost 75%), and also from vein deposits (Mines and Energy Ministry, 1993). The main gold districts are: South of Bolivar (veins and placers), Antioquia (veins), Santander (veins), Caldas (veins), Nariño (veins), Taraira (veins), Bajo Cauca (placers), Chocó-Atrato (placers), Chocó-San Juan (placers - gold/Platinum), Pacific Coast (placers), Guainia (placers).

The production in these districts is classified as small scale mining, characterized by low technology and low gold recovery (hardly 50%), and due to these features mining activities may cause a great deal of effects in the water resources (municipal, industrial and groundwater), sediments, soils, air, vegetation, animals and humans, even though quantitative and systematic evaluation of damage to the environment has not yet been done.

Many research groups in Colombia are starting studies on environmental effects caused by gold mining, especially on topics such as emission of mercury and cyanide effluents (Polensky 1980; Mines and Energy Ministry, 1991), though systematic researches must be done in order to quantify the extent of the damage and the fate of the contaminating agents.

Methods

During 1992, a diagnosis of the general conditions of small scale mining in Colombia related to gold deposits was carried out. In this study, six districts were documented, visited and evaluated (Fig. 1): Frontino (Antioquia), Marmato (Caldas), Ginebra (Valle), Sotomayor (Nariño), San Martín de Loba (south of Bolívar) and Vetas y California (Santander).

Mining and recovering processes were examined. Data on mercury and cyanide were considered and general possibilities of water contamination due to mining effluents were examined.

In every district visited, samples of orebodies, tailings and waters were collected, using standard sampling methods (INGEOMINAS, 1992). Geochemical analyses were carried out at the geochemistry laboratory of INGEOMINAS, following standardized methods: Emission Spectrometry (ES; APHA, AWWA, WPCF, 1989), Atomic Absorption Spectrophotometry (AAS; APHA, AWWA, WPCF, 1987), Hydride Generation Atomic Absorption Spectrophotometry (HGAAS; APHA, AWWA, WPCF, 1989) Graphite Furnace Atomic Absorption Spectrophotometry (GFAAS, INGEOMINAS, 1992).

Results and Discussion

The gold deposits studied are of hydrothermal origin and their mineralization is generally associated with base metal sulfides (Cu, Pb, Zn; INGEOMINAS, 1983; 1987). The mineralization shows Cu-Pb-Zn association and a high arsenic content (arsenopyrite). This is a characteristic feature of most deposits in the western Cordillera, which are related to magmatic events of the Cenozoic period (INGEOMINAS, 1983; 1987).

The mineralogy in these deposits (vein type) is mainly composed of quartz, gold, silver, pyrite, arsenopyrite, chalcopyrite, pyrrhotite, markasite, molybdenite, galene, sphalerite, calcite, (Escobar 1942; Lopez 1986), and according to analytical results in sediments and rocks some of them contain traces of antimony (Sb), cadmium (Cd), tellurium (Te), mercury (Hg), and bismuth (Bi).

Fig. 1: Mining areas studied

The contents of gold in rock samples and crushing heads are between 5 - 10 g ton^{-1} with gold recovery between 50 - 80 %. In every visited district mercury is used for gold amalgamation, and leaching with cyanide is used for gold recovery. A diagram of the process employed for gold recovery is shown in Fig. 2.

Ginebra is a typical gold district which is located in the Valle department in the western flank of the Central Cordillera (Fig. 1) into the contact zone of the intrusive Buga Batholith. There are several small mines in the zone, each one with low technology and very few processing facilities, which usually include piston crushers with plates covered with $AgNO_3$, and crushing barrels in which mercury is added to amalgamate free gold and cyanidation tanks.

During the process, the mercury of the amalgam is burned off to the atmosphere and gold is recovered. There is considerable emission of mercury to the atmosphere and contamination of running water. The miners handle mercury directly without any care. The tailings are dumped directly into the rivers and accumulate everywhere without treatment or precautions.

Several measurements of mercury were made in two rivers supplying water for several small cities in the region. Mercury in the waters was below detection limits (D.L.=10 μg L^{-1}) but in sediments it reaches values up to 1,130 μg kg^{-1}.

Problems between the miners and the community in relation to water use are affecting the economy of that area. In this department agriculture is the main economic support.

Similarly to other places in the world, it was found that the main environmental problems related to mining and processing of gold ores in the studied districts are: Acid drainage due to sulfide ore mining, acid precipitation (processing of sulfide ores), acidic tailings (mining and processing of sulfide ores), particulate matter (mining and processing of vein ores), oils (processing equipment, mills), phosphates (processing-surfactants), cyanide (processing), mercury (mining and processing), lead (mining and processing), arsenic (mining and processing), silver (mining and processing), etc. (United Nations Institute, Mexico 1978; Van Loon, 1990).

Few geochemical studies exist concerning metals associated with gold orebodies (especially heavy metals including mercury) Furthermore, no monitoring programs have ever been proposed to evaluate environmental impact caused by mining activities.

In addition it was found that besides Hg and Cn, toxic elements such as Cd, V, Tl, Pb, Be, As, Te, Cr^{+6}, Ru, Os, Pt must be evaluated, particularly since at a high level most elements can become toxic (Petrowsky, 1980; Van Loon, 1990).

Understanding and predicting the geochemistry of mine drainage will be a very important subject for treatment of mining effluents in gold vein deposits in Colombia.

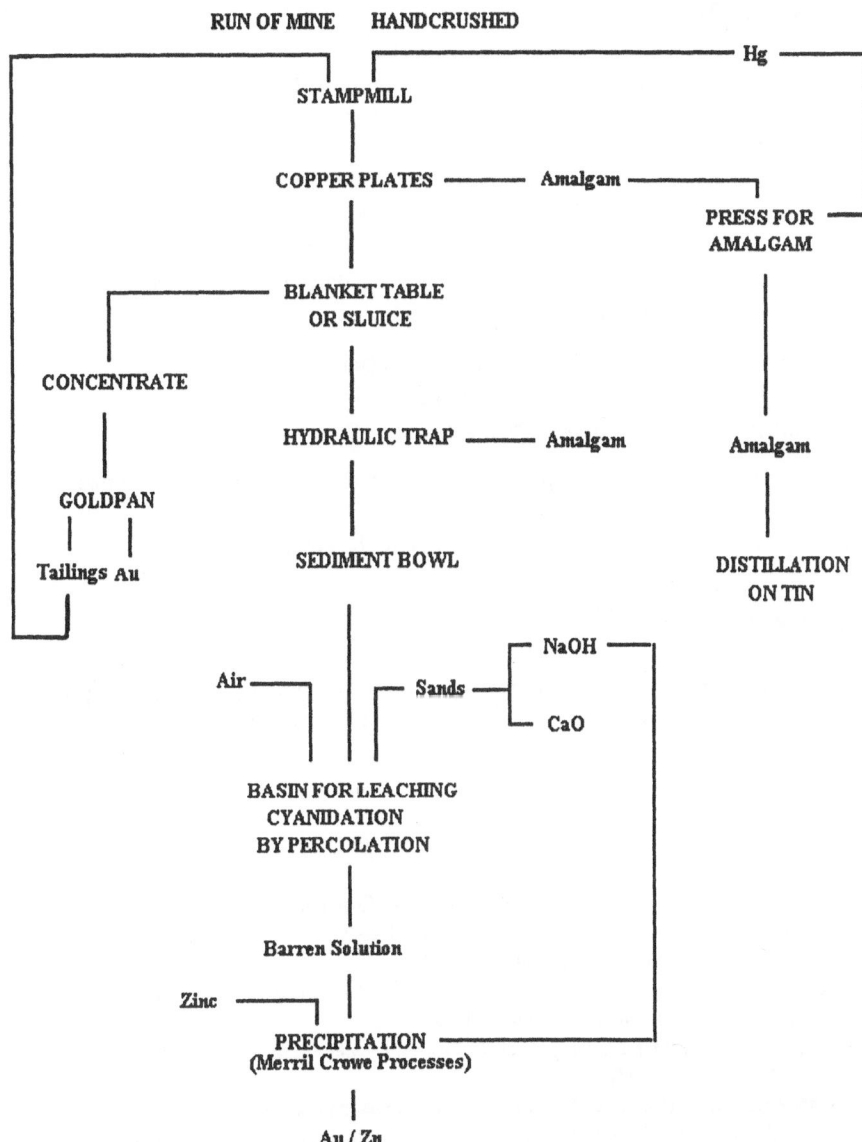

Fig. 2: General orebody processing

Conclusions

The overview of gold mining in Colombia permitted the conclusion that almost 90% of total production is generated in small scale mines which are characterized by low technology and recovery (about 50%), therefore causing negative effects in water resources (municipal, industrial and groundwater), sediments, soils, air, vegetation, animals and humans. The environmental problems are intimately related to the lack of appropriate technologies for mining and processing gold.

To summarize, like in other countries two main types of dangers exist in relation to gold mining and processing (Van Loon, 1990):

1. Exposure of humans to high levels of toxic substances
2. Discharge of toxic elements to the environment
   And the main questions to be solved are:

1. What are the effects of long term low level exposure of plants, animals and humans?
2. By what pathways do these substances cycle in the environment?

Acknowledgements

We thank INGEOMINAS for financial assistance, the Ministerio de Minas y Energia de Colombia for assistance and informations, and the mine managers and mine workers for their kind information.

References

APHA, AWWA, WPCF (1989). *Standard Methods for Examination of Water and Wastewater*. 17th ed. American Public Health Association, American Water Works Association, and Water Pollution Control Federation. Washington DC 1268p.

Escobar, J. and Echeverry, A. (1942). *Comentarios Relacionados a Minería de Filón y Cianuración*, Medellín, Fundición Escobar. 145p.

INGEOMINAS (1983). *Mapa Geológico de Colombia*, escala 1:1.500.000, Memoria Explicativa. INGEOMINAS, Bogotá. 71p.

INGEOMINAS (1987). *Recursos Minerales de Colombia*. Bogotá, INGEOMINAS.v.1. 563p.

INGEOMINAS (1992). *Manual de Técnicas Analíticas*. Laboratorio de Geoquímica. Bogotá, INGEOMINAS. 80p.

Lopez, J.H. (1986). *Estudios Relacionados con la Minería del Oro en Colombia*. Medellín, Fundición Gutierrez. 65p.

Mines and Energy Ministry of Colombia (1991). *I Mining and Environmental Symposium*, informe técnico. Bogotá, Ministerio de Minas y Energía. 55p.

Mines and Energy Ministry of Colombia (1993). *Colombia: Mineral Potential and Investment Opportunities*. Bogotá, INGEOMINAS - Ministerio de Minas y Energía. 20p.

Piotrowsky, J.K. and Coleman, D.O. (1980). *Environmental Hazards of Heavy Metals: Summary Evaluation of Lead, Cadmium and Mercury*. University of London. 30p.

Polensky, E. (1990). *Biological treatment of golden minerals, geochemistry study of mining in Vetas and California Santander.* Bogotá, Col. Cienc., 100p.

Priester, M., Hentchel, T., Benthin, B. (1992). *Heavy Metal Contents of Stream Sediments in a Gold Mining Area near Los Andes, Southern Colombia: Technical and Ecological Perspectives.* Pasto, CORPONARIÑO. 20p.

United Nations Institute for Training and Research (1978). *I International Conference on Small Scale Mining.* Mexico. 40p.

Van Loon J.C. (1990). *Environmental Analysis related to the Mining and Processing of Geological Materials.* Ontario, University of Toronto. 20p.

15 Assessment of the Heavy Metal Pollution in a Gold "Garimpo"

Saulo Rodrigues-Filho[1] and John Edmund L. Maddock[2]

[1] CETEM, Ilha do Fundão, Rio de Janeiro, RJ, Brazil
[2] Departamento de Geoquímica Universidade Federal Fluminense, Outeiro de São João Batista s/n°, Centro, 24020-150, Niterói, RJ, Brazil

Abstract

Gold has been exploited in the Poconé regions, in Mato Grosso state in Brazil, during the past 13 years using garimpo methods. In this study, background levels of metals were determined by analysing sediments and soils unaffected by mining activities, located upstream of the anthropogenic inputs. The study done in Poconé focuses on Hg, Cu, Pb, Zn, Fe and Mn. In addition, the study aims to evaluate the level of contamination in sediments, soils and water, taking into account drainage waters directly affected by gold mining. As the physical-chemical parameters of the waters are subjected to seasonal variations, they were monitored throughout the year to interpret the behaviour of the metals in the studied environments. In general, the concentrations of Cu, Pb, Zn, Fe and Mn presented values very close to the background in sediments, while the concentrations of Hg in sediments showed anomalous geo-accumulation indexes.

Introduction

Mining inevitably involves upsetting the balance between the various environmental compartments lying in the area of influence. This environmental impact can, however, be minimised by using environmentally compatible mining and ore dressing techniques, and by developing new technologies for treating ores, effluents and contaminated solid tailings.

Informal mining, in the form of garimpos, as a rule uses rudimentary mining and processing methods, which reflect the unreliable environmental control practiced by the "garimpeiros".

The garimpo region of Poconé-MT was originally founded on the discovery of gold-bearing occurrences in 1777, by the "Bandeirantes" (explorers). After the main occurrences had been depleted, the founders of the town then took to cattle-raising, which has been an important activity in the region until today (CETEM/CNPq, 1991).

With the rise in the price of gold in the international market in the early 1980s, a second gold cycle began in the region, when the development of the hitherto uneconomic mineral deposits became promising. The traces left by the Bandeirantes, such as the horizons of reworked rocky fragments, were used in many cases as a reference when prospecting ore-bearing areas.

This study was developed under the Environmental Technology Development (DTA), Programme, set up at CETEM/CNPq - Centro de Tecnologia Mineral - in 1989, which focuses on harmonising mining activities with environmental protection.

Researches on environmental geochemistry presuppose the establishment of a reference level which expresses the natural concentration of a given element in the environment, commonly known as the background level. For such a regional standard to be established, analyses of traces are necessary in samples that are free from anthropogenic inputs.

While the concentrations of trace metals in aquatic systems are extremely variable, as much on a global as on a local scale, besides analytical conclusions, geological reconnaissance of the watershed is necessary, paying special attention to mineralogy, so that it is possible to estimate the availability of each element with regard to the action of weathering (Salomons and Förstner, 1984).

Many possibilities have been proposed for establishing background values for trace metals in sediments. Förstner and Wittman (1979) have described the following alternatives:

1. Average concentration in schists as a standard value at global level.
2. Average concentration in sedimentary rocks, typifying the deposition environment and taking into consideration autochthonous and allochthonous natural mechanisms and factors.
3 Average concentration in recent unpolluted sediments.
4. Concentration and dating in sediment layers, making it possible to identify the historical record of the natural events that occurred in a certain drainage water.

The purpose of this study was to determine the distribution of the natural background concentrations of mercury (total Hg) and other heavy metals (Cu, Pb, Zn, Fe and Mn), found in soils, sediments and water of the garimpo region of Poconé-MT, using that distribution as a reference for assessing the degree of contamination of the aquatic system by heavy metals.

For recognising the background levels, drainage waters were chosen which had not been directly affected by gold mining, so as to avoid the occurrence of anomalies resulting from mineral dressing (concentration of minerals rich in heavy metals, addition of mercury, etc.).

Additionally, the study sought to assess the degree of contamination in soils, sediments and water, using as a reference drainage waters that have been clearly affected by the garimpo activity.

Areas Studied

The municipality of Poconé-MT is in the region of the upper stretch of the Paraguay River, along the northern edge of the Pantanal Matogrossense, 100 km southwest of Cuiabá, the state capital.

The subject of the study was the sub-basin of the Bento Gomes River, whose main tributaries are the streams Guanandi, Formiga and Piraputanga, lying in an area of approximately 1700 km^2 of the Poconé Sheet (SE.21-X-A-1/DSG), with latitudes between 16°00' and 16°30'S and longitudes between 56°30' and 57°00'W.

The Poconé region is represented by the detritic meta-sedimentary sequence of the Cuiabá Group, of older Proterozoic age (800 to 600 m.y.), composed of sericitic, graphitic and pyritic phyllites, micaceous ferruginous quartzites and meta-conglomerates showing flattened pebbles sub-oriented parallel to the schistosity, imposed by metamorphism in the green schist facies.

In that region, the "garimpeiros" have mined gold from both lateritic overburden and from quartz veins rich in gold which, as they are hydrothermal in origin, concentrated the gold found in the surrounding rocks of the Cuiabá Group. It is, therefore, a primary gold deposit with supergenic enrichment.

The mining is usually done jointly by the owners of the heavy machinery (loader, drag-line, hammer mill, centrifuge, etc.) and the manual workers. The former are responsible for working the surface layers of alteration of rocks ("bench") and for dressing the ore, whereas the latter are in charge of opening trenches along the ore-bearing veins. This joint work in some cases can go down as deep as 50 m, and is predominantly marked by the following characteristics:

1. Exploration of "dry" areas, disassociated from the drainage waters;
2. The use of grinding in the dressing process;
3. Deposition of tailings in containment dams;
4. Closed-circuit amalgamation (using water tanks); and
5. Retorts are not used.

Materials and Methods

The sampling campaign at Poconé was planned to cover areas where there were different variables that would affect the concentration of heavy metals in sediments, soils and water, that is, differentiations with regard to the rocky substrate, characteristics of the drainage waters, vegetation, pedological horizon (soils) and nearness to the garimpos. Hence, the points where sampling of sediments and water (PG notation) and soils (GB notation) was to be done were selected by identifying places with distinct characteristics in relation to such factors, amounting to a total of 21 samples of current sediments, 13 samples of water and 27 soil samples.

The current sediment samples were collected in compound form, that is, at the same collection station, an average of 5 aliquot parts of samples were removed in a radius of approximately 10 meters. The soil and sediment samples were cooled (0°C) after collection, and packed in plastic bags, whereas the filtered water samples were acidified with concentrated HNO_3, in the proportion of 2 ml of acid for each litre of water, and stored in polyethylene bottles.

The subject of the study was the grain size fraction <74 μm of sediments (silt and clay), because this fraction most represents the solid-liquid interaction processes (Salomons and Förstner, 1984). Furthermore, that fraction had been used by other authors in studies for assessing contamination by heavy metals in water, thereby represent a good correlation parameter (Ackermann, 1980; Malm et al., 1989; Lacerda et al., 1991).

At all the sampling points (PG notation) in situ measurements of temperature, pH, Eh and drainage conductivity were taken. The physical-chemical parameters were determined using special DIGIMED models DMPH-PV and CD-2P pH, redox potential and conductivity meters.

The volume of filtered water at each point, averaging 0.5 litre, was used to assess metals in the dissolved phase.

The analytical methodologies used in this study were divided according to the chemical element involved and the type of sample to be assessed. All the analytical determinations were made in duplicate.

The assessments of total mercury in soils and sediments were done using an atomic absorption spectrophotometer with cold vapour generation, a CG 7000 MAX 8 spectrophotometer belonging to CETEM/CNPq's analytical chemistry laboratory. The methodology used followed that optimised by Malm et al. (1989), where the samples are oxidised with potassium permanganate 5% for 30 minutes at 60° C and neutralised with hydroxylamine hydrochloride 12% (EPA, 1983).

Additionally, inter-laboratory calibration was done between CETEM and the Sedimentology Institute of Heidelberg University, where the mercury analyses are done with an atomic absorption spectrometer used only for mercury analysis - "Mercury Analyser Hg-254 A" Seefelder Messtechnik. This analytical method was originally developed by Poluektov and Vitkun 1963 (in Welz, 1985) and is based on mercury vapour generated by a nebuliser immersed in an aqueous medium, through reduction of the ionic mercury by a stannous chloride solution. The mercury vapour is then purged into the absorption cell by the air flow produced by a peristaltic pump.

The sediment samples digestion procedure used by the Sedimentology Institute uses *aqua regia* for 3 hours at 150° C, while each reaction tube has a condenser to avoid losses during digestion.

The water samples were analysed for mercury, using the Hg-254 A mercury analyser, and for Pb, Cu and Ni, using a Perkin-Elmer 3030 B atomic absorption spectrophotometer and a HGA-600 graphite furnace.

The methodology used for analysing the mercury in water is based on oxidation of the metal by adding potassium permanganate, nitric acid, sulphuric acid and potassium persulphate. It is left in a double boiler for 2 hours at 95° C. Then neutralisation by hydroxylamine hydrochloride follows and reduction by stannous chloride, after which a reading is taken (methodology suggested by the manufacturer).

In the graphite furnace analysis of the metals in water does not require any preliminary treatment, because the sample is subjected to a series of temperatures rises before atomisation, which frees it from interfering substances.

The methodology used for analysing Cu, Pb, Zn, Fe and Mn, in sediments, soils and rocks, uses the atomic absorption spectrometry technique, while the samples are digested in a triacid solution of HNO_3, HCl and HF (2:2:1) at 120° C in Teflon crucibles and recovered with HCl (Welcher, 1975).

Results and Discussion

Physical-Chemical Description of the Poconé Drainage Waters

As they are subject to seasonal variations, the physical-chemical variables were determined at different times of the year, so that their variation intervals could be known. In this study, the field stages considered the dry season from September to November, and rainy season from January to May.

The pH values recorded in Poconé during the dry season (September, 1992) showed slightly alkaline neutrality conditions, varying from 6.1 to 8.0. The highest pH values (7.8 and 8.0) were recorded in drainage waters which cross the savannah, particularly in the Formiga stream and at the point PG-04 of the Bento Gomes River.

During the rainy season (January, 1992), the pH values were slightly acid, varying from 6.1 to 6.9, while the biggest drops in pH were noted in the drainage waters of the lowlands of the Pantanal. Hence, it is right to expect that these drainage waters would be more subject to the influence of rain water containing dissolved CO_2, and leached humic substances.

The Eh values revealed a more oxidising potential in the dry season, both in savannah and Pantanal drainage waters, varying from 163 to 210 mV.

Data obtained on the electrical conductivity as a rule showed low values, oscillating between 48 and 124 μS cm^{-1}. There was an exception at some sampling points in a savannah environment: PG-04 in the Bento Gomes River, Formiga stream and Japão stream. In those places the concentrations of dissolved salts were considerably higher, with conductivity values of up to 1000 μS cm^{-1}. In September, 1990, the waters of the Formiga stream showed 770 and 1000 μS cm^{-1}. The decline of these values in the rainy seasons (400 and 250 μS cm^{-1}) showed a strong seasonal dependence which in this case is represented by the effect of dilution caused by the rains. The experimental error involved in the electrical conductivity measurements may reach variations of up to 2 μS cm^{-1} around each determination.

The Eh-pH stability diagram of the mercury, according to Hem (1970), reveals the stability of the metallic mercury, aqueous Hg^0, for the conditions found in the Poconé drainage waters, imparting low solubility (<25 μg L^{-1}) to the metal which is poured directly into the rivers and streams as liquid effluent from the gold concentration amalgamation process. Even with the minor Eh and pH variations resulting from seasonality, all the measurements lay within the stability field of the Hg^0. However, it is possible to note a tendency of the measurements made in the dry season as they approached the high solubility region, where the stable forms of the metal are Hg_2^{2+} and $Hg(OH)_2$.

Very little is yet known about the oxidation of Hg^0 to Hg^{2+} and the subsequent methylation to CH_3Hg^+ and $(CH_3)_2Hg$. It is known that microbial activity is related to the two reactions, but the bacteria which are able to oxidise the Hg^0, in general, are not the same as cause the methylation of the Hg^{2+} (Silver, 1984). Therefore, the stability diagram of the mercury must be analysed as a means of partial interpretation, because it does not consider the biotic component (microbial) which can intermediate the reaction of metal oxidation.

According to Salomons and Förstner (1984), the mono-methylation of mercury phenomenon is accelerated under mildly acid and low electrical conductivity conditions, while under alkaline and high conductivity conditions, the main product of the methylation is di-methylmercury, which is highly volatile and much less stable than the mono-methylated form.

Besides mercury, lead and copper were also considered for interpreting the thermodynamic equilibrium of these elements, under the physical-chemical conditions found in the drainage waters of the Poconé region.

In the lead stability diagram, according Rose et al. (1979), it can be noted that the values obtained in the dry season are included in the stability field of cerussite ($PbCO_3$) with low solubility ($<10^{-6}$ M). During the rainy season, however, the values obtained favoured the stability of the Pb^{2+} ion, making the metal more mobile, although in the presence of the Cl^- ion, the formation of $PbCl_2$ would occur, which shows low stability.

In the Eh-pH stability diagram of copper we noted behaviour similar to that of lead. During the dry season there is a predominance of measurements in the stability fields of tenorite (CuO) and cuprite (Cu_2O), where the molarity of the dissolved copper is lower than 10^{-7}. During the rainy season a shift to the stability field of the Cu^+ ion, where the molarity of the dissolved copper increases, was noticed.

Distribution of Heavy Metals in Sub-drainage Basins of the Poconé Region

Sub-basin of the Formiga Stream

The concentrations of heavy metals found in samples of bottom sediment - fraction <74 μm - of the Formiga and Japão streams showed values compatible with the background levels known at world level, except for mercury which showed considerably higher values, with an average of 0.91 µg g^{-1}.

Assessments of the concentration of heavy metals done by Förstner and Müller (1974) in sub-recent sediments of the Rhine river in Germany, which were free from anthropogenic inputs, showed levels compatible with the values found in the Formiga stream for copper, lead, zinc, iron and manganese. The average concentration of mercury, however, is almost five times higher in the sediments of the Formiga stream.

In this study, a comparison will be made of the geo-accumulation indexes (Igeo) of metals in sediments, as a methodology for assessing the pollution level of heavy metals in an aquatic environment (Müller, 1979). The Igeo is defined by the following expression:

$$Igeo = \frac{Log_2 C_n}{1.5 \times B_n}$$

where C_n is the measured concentration of the element in the fraction <2 µm (clay) and B_n is the background value of the element found in sub-recent clayey sediments and schists. Hence, the Igeo in class 0 indicates absence of contamination, and the Igeo in class 6 represents the upper limit of maximum contamination.

The grain size fraction <74 µm (silt and clay) used in this study can also be applied to the Igeo, according to Salomons and Förstner (1984), provided there is a definition of the background values existing in that grain size fraction.

Sub-basin of the Guanandi Stream

The Guanandi stream, which flows into the Bento Gomes River, is the drainage water most directly affected by the physical and chemical disturbances resulting from garimpo activity in Poconé. The principal mining and ore dressing fronts of the garimpos are scattered through areas near the spring, and principally in the last 10 km of its course, downstream of the MT-111 highway.

It is interesting to note the concentrations of mercury in the PG-01 and PG-03 stations. At these sites, the samples were essentially formed of sandy sediments and rocky fragments (tailings), due to the nearness of the dressing fronts of the garimpos. The concentration troughs ("cobra fumando"), which are particularly prevalent in these areas, besides concentrating the ore, cause the remobilisation of fine particles which end up being deposited in stretches downstream of the garimpos. It was found that the concentrations of mercury in these samples were as low as the detection limit, of 0.04 µg g^{-1}, while the samples originating in areas further away from the ore dressing fronts, and rich in silt-clayey fraction, showed mercury concentrations corresponding to the Igeo in classes 1 and 2 (Table 1).

The fact that the mercury concentrations are so low in areas near the sources of mercury emission seems to be due to the coarse grain size of those sediments which, because they do not participate significantly in the solid-liquid interaction processes, of the physical and/or chemical adsorption type, give the mercury emitted by the amalgamation process an accentuated heterogeneous distribution. Such evidence strengthens that presented by other authors, who claim that the transportation of mercury in an aquatic medium is controlled by the suspended sediments (Murdoch and Clair, 1986; Lacerda et al., 1991; Silva et al., 1991).

With regard to copper and lead, entirely distinct behaviour was noted. The greatest concentrations of these metals were found mostly in the PG-01 and PG-03 stations, where there is a predominance of tailings from the garimpos. In these cases, the source of emission was said to be partly anthropogenic and partly lithogenic, because the gravity concentration process used by the "garimpeiros" retains not only the gold, but also all the high density minerals, i.e. oxides, sulphides and ferro-magnesium silicates. These minerals may contain anomalous concentrations of copper, lead and iron, among other metals (Loring, 1978). After the ore concentrate amalgamation stage, these accessory minerals then form part of the amalgamation tailings, which often pass through the trough again, and are then mixed with the gravity concentration tailings.

Table 1: Concentration of heavy metals in bottom sediments – <74 μm – of the sub-basin of the Guanandi Stream, Poconé

Guanandi Stream	Hg		Cu		Pb		Zn		Fe		Mn	
	$\mu g\ g^{-1}$	Igeo	$\mu g\ g^{-1}$	Igeo	$\mu g\ g^{-1}$	Igeo	$\mu g\ g^{-1}$	Igeo	%	Igeo	$\mu g\ g^{-1}$	Igeo
PG-01	<0.04	0	324	5	45	0	35	0	4.7	1	790	2
PG-02	4	0	8	0	28	0	46	0	3.2	0	130	0
PG-03	0.08	0	32	2	52	1	76	1	2.6	0	820	2
PG-11	0.04	2	41	2	28	0	60	0	2.9	0	110	2
PG-12	0.50	2	40	2	36	0	120	1	4.3	1	0	3
PG-14	0.60	1	12	0	32	0	106	1	1.7	0	178	1
PG-15	0.20	1	15	0	20	0	95	1	2.4	0	0	0
PG-16	0.15	0	24	1	36	0	70	1	2.5	0	420	1
	0.04										330	
											410	
Average	0.21	1	62	3	35	0	76	1	3.0	0	722	1
Standard Deviation	0.21		98		11		31		1.0		442	
Local Background	0.10		10		32		43		2.6		260	

Igeo = Geo-accumulation Index

Although the concentrations of these metals in sediments may be somewhat high, the sediments cannot be considered absolutely polluted, due to a probably low bioavailability of the metals they contain. These elements would have reached the drainage waters not only due to weathering of their minerals, which would increase mobility and possibly their bioavailability, but by the mechanical action of mining and dressing in the garimpos.

The low mobility of trace metals in the sub-basin of the Guanandi stream is indicated by the low concentrations obtained in filtered water samples (dissolved phase). Only the PG-12 sample showed an anomalous concentration of copper (Table 2).

As to the soils of that sub-basin, no significant differences were noted in the behaviour of heavy metals, compared with soils of the Formiga stream sub-basin. Nonetheless, for all the metals, the average concentrations found at horizon A of the Guanandi sub-basin was slightly higher, whereas at horizon B, the average concentrations of mercury and copper were slightly lower than those noted in the Formiga stream (Table 3).

It is also possible to identify a direct correlation between the mercury and copper concentrations in the soils of the Guanandi sub-basin (Fig. 1). This correlation seems to indicate a common lithological origin for those elements, possibly related to weathering of sulphide rocks. The coefficient of correlation between the mercury and copper concentrations in the soils, calculated at 0.765, confirms the existence of a marked correlation between those two elements in the samples studied.

Table 2: Concentrations of heavy metals in drainage waters of the Guanandi stream sub-basin compared with uncontaminated rivers

Drainage waters		Hg (μg L^{-1})	Cu (μg L^{-1})	Pb (μg L^{-1})	Author
Guanandi					
	PG-01	nd	nd	1.0	
	PG-02	nd	nd	nd	
	PG-03	nd	nd	nd	
	PG-11	nd	nd	nd	
	PG-12	nd	9.2	0.9	
	PG-13	nd	0.4	nd	
	PG-15	nd	nd	nd	This study
Uncontaminated rivers		0.01	1.0	0.2	Drever (1982); Salomons
at world level		0.07	7.0	1.0	and Förstner (1984)
Uncontaminated rivers of the Amazon		< 0.04	-	-	Pfeiffer et al. (1989)

nd < 0.1 μg L^{-1} (not detectable)

Sub-basin of the Bento Gomes River

As the main drainage water of the Poconé region, the Bento Gomes River receives the flows of all the drainage waters described above, except for the Formiga stream (Fig. 4). Hence, it is the final depository of the matter carried by the tributary drainage waters and, therefore, is the main focus of this study.

At the lake where station PG-24 is situated, there is an accentuated drop in the carrying capacity of the Bento Gomes River, where the land slopes gradually to then form part of the wetlands of the Pantanal Matogrossense.

The reduced carrying capacity results in higher sedimentation rates, particularly of suspended sediments. This means that the waters in such places have very low turbidity. This is characteristic of the Pantanal Matogrossense and consists essentially of suspended organic particles.

Table 3: Concentrations of heavy metals in soils of the Guanandi stream sub-basin compared with background concentrations found in the Formiga stream sub-basin

Soils	Type of soil	Hg (μg g^{-1})	Cu (μg g^{-1})	Pb (μg g^{-1})	Zn (μg g^{-1})
Guanandi	(Horizon)				
GB-1B	Organic (A)	0.07	10	25	13
GB-1C	Lateritic (B)	0.05	16	69	28
GB-1D	Saprolitic (C)	0.06	23	50	17
GB-3B	Organic (A)	nd	15	37	58
GB-4A	Organic (A)	nd	35	50	60
GB-4B	Lateritic (B)	nd	40	53	42
GB-9A	Organic (A)	0.07	76	112	64
GB-9B	Lateritic (B)	0.08	110	71	81
GB-9C	Lateritic (B)	0.07	62	32	50
GB-12B	Organic (A)	nd	35	26	20
GB-20A	Organic (A)	0.17	40	4	26
GB-20B	Lateritic (B)	0.14	96	6	19
GB-20C	Saprolitic (C)	0.24	54	32	21
GB-21B	Organic (A)	0.12	74	22	32
GB-22B	Sandy (B)	0.32	136	5	84
Average	Horizon A	0.08	41	39	38
Average	Horizon B	0.12	77	39	50
Background	Horizon A	0.05	28	34	29
Background	Horizon B	0.15	89	21	5

nd < 0.04 μg g^{-1}

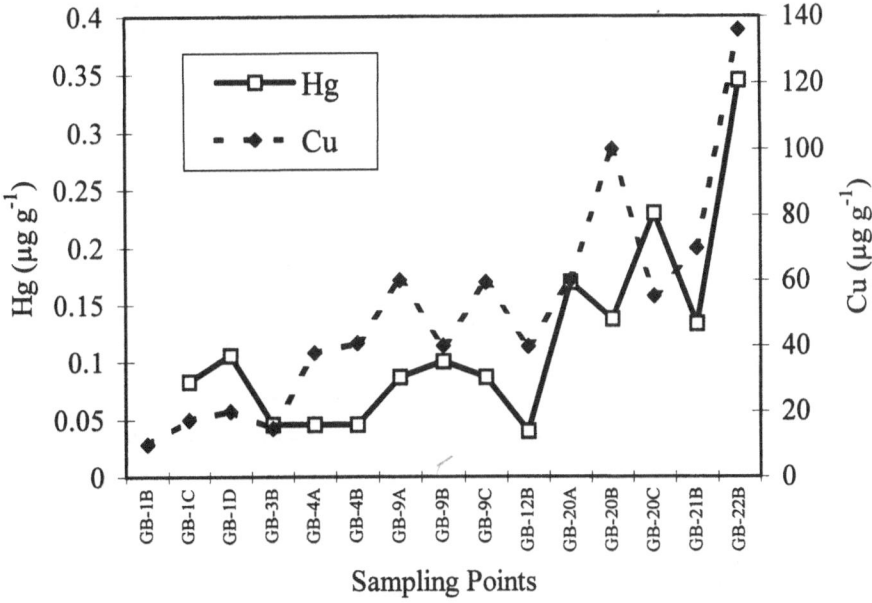

Fig. 1: Correlation between the Hg and Cu concentrations in the soils of the Guanandi stream sub-basin

Approximately 40 km upstream of station PG-24 is station PG-04, where the Bento Gomes River drains a typical savannah environment without impacts caused by garimpo activity. The concentrations of metals in that station are very similar to the concentrations found at station PG-02, in the Traíras stream, where also there are no activities related to the "garimpo".

The average concentrations noted using stations PG-04 and PG-02 as a basis, must therefore reflect the natural levels of occurrence of heavy metals in sediments (background). The mercury background, equivalent to 0.10 $\mu g\ g^{-1}$, found in current sediments in the fraction <74 μm, was higher than the value found in lacustrine sediments in remote areas of the Pantanal Matogrossense, which is equivalent to 0.02 $\mu g\ g^{-1}$ of mercury (Lacerda et al., 1991). Probably, this lower value in lacustrine sediments is due to the great distance from the main lithogenic sources of mercury, allied to the extremely low carrying capacity of the waters throughout the Pantanal lowlands. On the other hand, the local mercury background corresponds to half the average concentration noted in uncontaminated clayey sediments (fraction <2 μm) of the Rhine, in Germany, of 0.20 $\mu g\ g^{-1}$ (Förstner and Müller, 1974).

The relatively high local background concentration of mercury seems to be related mainly to the dissemination of pyrites in the meta-sedimentary rocks of the

Cuiabá Group, which make up the region's rocky substrate, and which, as verified, can contain anomalous mercury concentrations (the weathered pyrite crystals contain an average 0.13 µg g^{-1}).

The study done by Andrade et al. (1988), in the garimpo region of Pilar de Goiás, shows the occurrence of a very high (0.77 µg g^{-1}) mercury background level in sediments, in the fraction <106 µm. That region is included in the Crixás Greenstone Belt, formed of basic and ultra-basic meta-sedimentary and metamorphic rocks, among others. The gold mineralisation is associated with the occurrence of arsenopyrites and chalcopyrites.

Table 4: Concentrations of heavy metals in bottom sediments of the Bento Gomes river sub-basin and geo-accumulation indexes

Bento Gomes	Hg µg g^{-1}	Igeo	Cu µg g^{-1}	Igeo	Pb µg g^{-1}	Igeo	Zn µg g^{-1}	Igeo	Fe %	Igeo	Mn µg g^{-1}	Igeo
PG-04	0.12	0	12	0	36	0	40	0	2.0	0	390	0
PG-06	0.30	1	25	1	28	0	46	0	1.9	0	180	0
PG-08	0.10	0	23	1	29	0	32	0	1.7	0	640	1
PG-17	1.10	3	40	2	44	0	90	1	3.1	0	820	2
PG-18	0.25	1	24	1	36	0	60	0	1.9	0	470	1
PG-19	1.85	4	36	2	58	1	76	1	4.2	1	840	2
PG-24	0.70	3	92	3	76	1	120	1	2.8	0	1340	2
Average	0.63	3	36	2	44	0	66	1	2.5	0	668	1
Standard Deviation	0.60		24		16		29		0.8		350	
Local Background	0.10		10		32		43		2.6		260	
World Background[1] C.S.[2]	0.20		45		20		95		4.7		600	

[1] Turekian and Wedepohl (1961)   [2] Clayey Sediments

The local background levels of the other metals studied in the Poconé region revealed concentrations that were still lower than those reported by Turekian and Wedepohl (1961) for uncontaminated clayey sediments at world level, except for lead (Table 4).

It is noted that starting from station PG-19, near the garimpo called Fazenda Salinas, as far as the vicinity of the confluence of the Guanadi stream, stations PG-08, PG-17 and PG-18, the concentrations of metals in the sediments of the Bento Gomes River become higher, showing the ingress of heavy metals polluted sediments (Fig. 4).

At station PG-24, where the river forms a large fan, high concentrations of all the metals were observed, denoting a sink for heavy metals, because downstream of the lake the concentrations in sediments are considerably reduced, coming close to the background levels (Fig. 2). This is further evidence that the transportation in an aquatic medium not only of mercury, but also of the other metals studied, is mainly related to the suspended sediments.

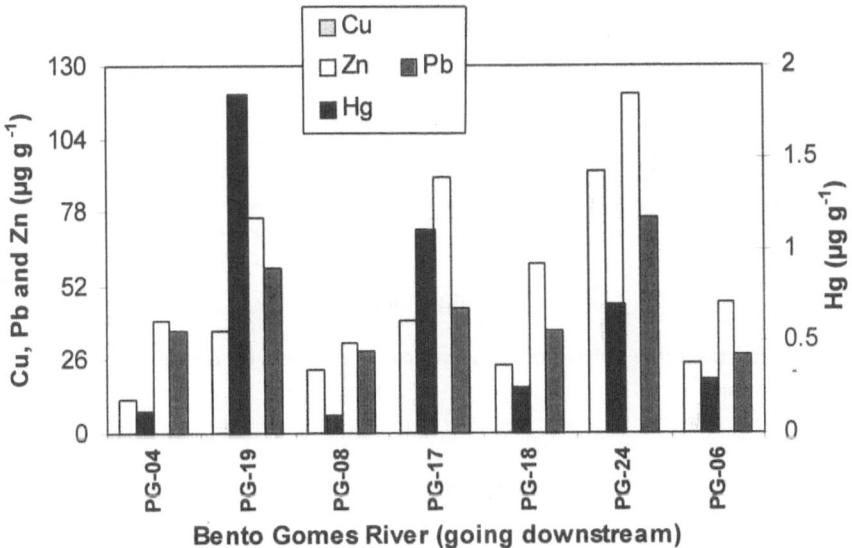

Fig. 2: Distribution of heavy metal concentrations in bottom sediments along the Bento Gomes River

When the correlation is established between the concentrations of mercury and iron found in current sediments, grouping together all the affected drainage waters, it is noted that there is a clear affinity in the behaviour of the two elements, with a coefficient of correlation of 0.717. This shows that the iron oxides and hydroxides play an important role in retaining/transporting the mercury present in such drainage waters (Fig. 3).

The geo-accumulation indexes noted at that site were the highest of these drainage waters, but even so they indicate an only moderate degree of pollution (Table 4 and Fig. 4).

If the origin of the mercury found in these samples is to be investigated, that is, to find out the extent to which they contribute to the anthropogenic and lithogenic sources, it would be necessary to make analytical determinations using selective extraction methods. However, the reliability of the results of these methods is still subject to verification, with regard to mercury (Veiga and Fernandes, 1990).

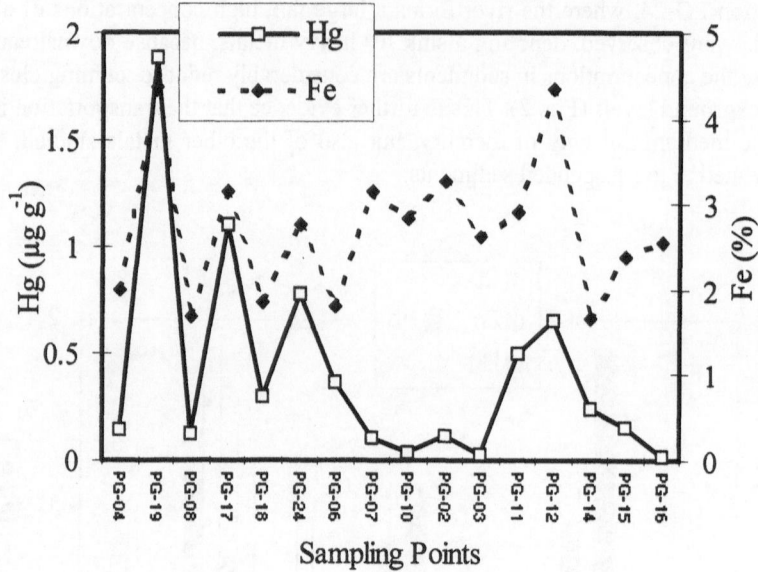

Fig. 3: Correlation between the mercury and iron concentrations in bottom sediments of the affected drainages

Table 5: Concentrations of heavy metals in river waters of the Bento Gomes River sub-basin compared with uncontaminated rivers

Drainage Waters	Hg ($\mu g\ L^{-1}$)	Cu ($\mu g\ L^{-1}$)	Pb ($\mu g\ L^{-1}$)	Author
Bento Gomes				
PG-04	nd	nd	nd	
PG-06	nd	nd	nd	
PG-08	2.0	1.5	nd	
PG-24	nd	3.9	nd	This study
Uncontaminated rivers at world level	0.01	1.0	0.2	Salomons and Förstner (1984)
Uncontaminated rivers of the Amazon	< 0.04	-	-	Pfeiffer et al. (1989)

nd < 0.1 $\mu g\ L^{-1}$

With regard to the concentrations of metals dissolved in samples of water from the Gomes River, low values were noted for all the metals, except for the PG-08 samples which showed 2.0 $\mu g\ L^{-1}$ of mercury and 1.5 $\mu g\ L^{-1}$ of copper, and PG-24 with 3.9 $\mu g\ L^{-1}$ of copper. Due to the nearness of that sampling station to the garimpos and to the gold buying shops, one could attribute this mercury anomaly to the atmospheric deposition of the metal which is emitted as vapour from the amalgam burning places (Table 5).

Conclusions

It was noted that the concentrations of mercury in bottom sediments, water and soils of the drainage sub-basins studied in the Poconé regions showed anomalous indexes, compared with the local background values for sediments and soils, and compared with the known mercury concentrations for waters of uncontaminated rivers. In general, the metals copper, lead, zinc, iron and manganese showed concentrations in sediments very close to the background values. The same occurred in the soils for copper, lead and zinc, and in waters for copper and lead.

In the sub-basin of the Formiga stream, the anomalous concentrations of mercury in sediments, with an average Igeo in class 3, seem to be related to the marked occurrence of pyrite mineralisations in the phyllites, which were probably releasing the metal from their crystalline structure through weathering processes.

The electrical conductivity values found in the sub-basin of the Formiga stream were high, reaching up to 1000 μS cm^{-1}, as well as in the other stations in the savannah environment. The pH values also were higher in those places, compared with the Pantanal environments, particularly during the dry season.

Considering all the physical-chemical measurements carried out in the Poconé region, it was noted that in the rainy season (January, 1992) there is more acidity, pH of 6.1 to 6.9, and a less oxidising potential in the runoff waters, 163 to 210 mV. Although the Eh and pH values may indicate greater copper and lead solubility in the rainy season, anomalous values were not found in the region's drainage waters, using as a reference the known values for uncontaminated river waters at world level. Only two water samples from the Bento Gomes River and one from the Guanandi stream showed slightly higher concentrations of copper, 1.5, 3.9 and 9.2 μg L^{-1}, respectively.

In the sub-basin of the Guanandi stream, where the presence of garimpo activities is strong, fairly high concentrations of mercury were noted, of 0.2 to 0.6 μg g^{-1}, associated with sediments farther away from the gold ore dressing fronts, with an Igeo in class 1 and 2. The samples collected at points near the garimpo tailings showed minimum concentrations, of around 0.04 μg g^{-1}, which may be explained by the predominance of quartz in those tailings whose grain size is coarse, and by the allochthonous origin of those materials.

With regard to copper and lead, the opposite behaviour was noted, that is, the higher concentrations were associated with sediments near the gravity concentration tailings. The accumulation of copper and lead in those places could therefore be the consequence of the concentration of accessory minerals (more dense) from the gold ore dressing. The indication of presence of these minerals points to a low bioavailability of copper and lead in that drainage water.

A direct correlation was found between the mercury and copper concentrations in the soils of this sub-basin, indicating a common mineralogical origin for the two elements, possibly associated to the iron sulphides.

The geo-accumulation indexes of mercury in sediments of the Bento Gomes River indicate a relatively high degree of contamination at some points, even reaching class 4 (1.85 µg g^{-1}) at the PG-19 station, near the garimpo of the Salinas Ranch. However, when they reach the Pantanal Matogrossense, the mercury concentrations drop considerably, reaching 0.30 µg g^{-1}. This is due to the accumulation of metals observed in the sediments of the lake of the Ipiranga Ranch (PG-24), which retains a large part of the sediments transported by the Bento Gomes River. The preferential accumulation noted in that lake also occurred for the metals copper, lead, zinc, iron and manganese.

A great affinity was noted between the mercury and iron concentrations in sediments of this sub-basin, indicating that the iron oxides and hydroxides play an important role in transporting mercury along the Bento Gomes River.

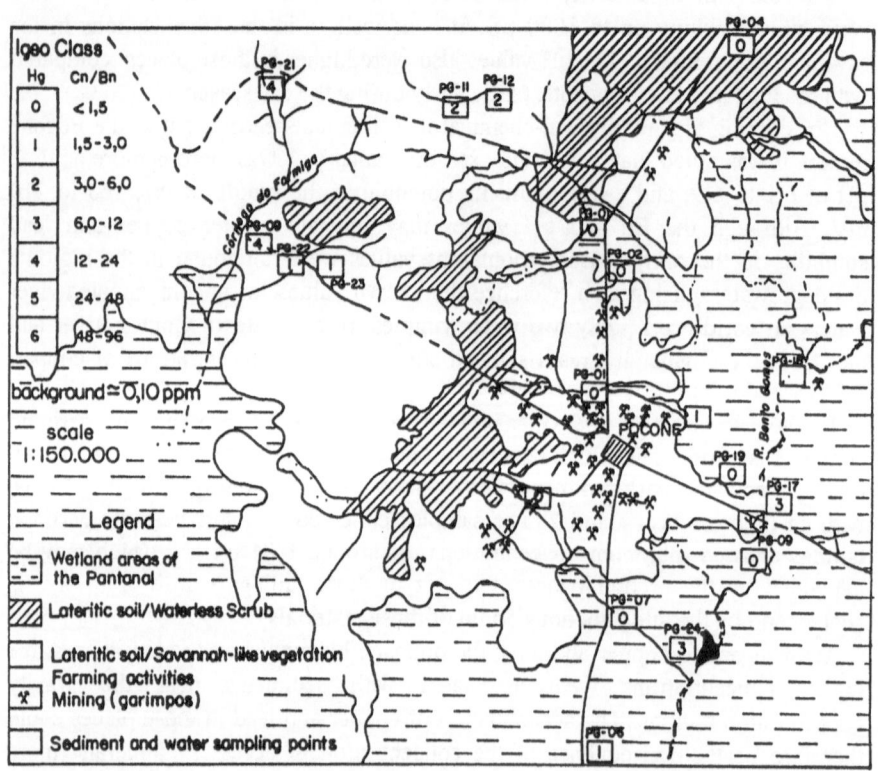

Fig. 4: Map of distribution of mercury geo-accumulation indexes in sediments of the Poconé region

References

Ackermann, F. (1980). A procedure for correcting the grain size effect in heavy metal analysis of estuarine and coastal sediments. *Environmental Technology Letters*, 1:518-527.

Andrade, J.C.; M.I.M.S., Bueno; P.V. Soares and A. Choudhuri (1988). The fate of mercury released from prospecting areas (garimpos) near Guarinos and Pilar de Goiás, GO (Brazil). *An. Acad. Bras. Cienc.* 60:293-303.

CETEM/CNPq (1991). *Poconé: um campo de estudos do impacto ambiental do garimpo* (seg. ed. rev.). Org. por M.M.Veiga e F.R.C. Fernandes (Tecnologia Ambiental 1). Rio de Janeiro. 113p.

Drever, J.I. (1982). *The geochemistry of natural waters.* New York, Prentice-Hall. 387p.

EPA-Environmental Protection Agency (1983). *Methods for chemical analysis of water and wastes.* Cincinnati. (EPA 600/4-79/020).

Förstner, U. and G. Müller (1974). *Schwermetalle in Flüssen und Seen als Ausdruck der Umweltverschmutzung.* Springer, Berlin, 225p.

Förstner, U. and G.T.W., Wittmann (1979). *Metal Pollution in the Aquatic Environment.* Springer, Berlin. 486p.

Hem, J.D. (1970). Chemical behaviour of mercury in aqueous media. In: U.S. Geol. Survey. *Mercury in the Environment* (Professional Paper, 173). Washington. 67p. p.40-46.

Lacerda, L.D.; W.C.Pfeiffer; R.V.Marins; S.Rodrigues; C.M.M.Souza e W.R. Bastos (1991). Mercury dispersal in water, sediments and aquatic biota of a gold mining tailing deposit drainage in Poconé, Brazil. *Water, Air and Soil Pollution*, 55:283-294.

Loring, D.H. (1978). Geochemistry of zinc, copper and lead in the sediments of the estuary and gulf of St. Lawrence. *Can. J. Earth Science.*, 15:757-772.

Malm, O.; Pfeffer, W.C.; W.R.Bastos e C.M.M.Souza (1989). Utilização do acessório de geração de vapor frio para análise de mercúrio em investigações ambientais por espectrofotometria de absorção atômica. *Ciência e Cultura*, 41(1):88-92.

Mudroch, A. and T.A. Clair (1986). Transport of arsenic and mercury form gold mining activities through an aquatic system. *Science of Total Environment.*, 57:205-216.

Müller, G. (1979). Schwermetalle in Sedimenten des Rheins - Veränderungen seit 1971. *Umschau.,* 79:778-783.

Pfeiffer, W.C.; L.D.Lacerda; O.Malm; W.R.Bastos; C.M.M.Souza and E.G.Silveira (1989). Mercury contamination in inland waters of Rondônia, Amazon, Brazil. *Science of Total Environment,* 87/88:223-240.

Rose, A.W.; H.E. Hawkes and J.S.Webb (1979). *Geochemistry in mineral exploration.* Academic Press, London. 658p.

Salomons, W. and U.Förstner (1984). *Metals in the Hydrocycle.* Springer-Verlag, Berlin. 349p.

Silva, A.P. (1991). *Estudos biogeoquímicos sobre o mercúrio em ambientes aquáticos de Poconé.* Rio de Janeiro, CETEM/CNPq (Série Tecnologia Ambiental, 1), p.61-83.

Silver, S. (1984). Bacterial transformations of and resistance to heavy metals. In: Nriagu, J.O. (Ed.) *Changing metal cycles and human health.* Springer-Verlag, Berlin. p.199-224.

Turekian, K.K. and K.H.Wedepohl (1961). Distribution of the elements in some major units of the earth's crust. *Bull. Geol. Soc. Am.,* 72:175-192.

Veiga, M.M. and F.R.C.Fernandes (1990). Poconé: an opportunity for studing the environmental impact of the gold fields. In: *Intern. Symp. Environm. Stud. Tropic. Rain Forests*, 1, 1990, Manaus-Brazil, Proceedings, v.1 p.185-194.

Welcher, F.J., ed. (1975). *Standard methods of chemical analysis.* 6 ed. Robert E. Krieger Publ. v.IIA, IIB, IIIA.

Welz, B. (1985). *Atomic Absorption Spectrometry.* Weinheim, VHC Publishers. 506p.

16 Mercury Partitioning Within Alluvial Sediments of the Carson River Valley, Nevada: Implications for Sampling Strategies in Tropical Environments

Jerry R. Miller[1] and Paul J. Lechler[2]

[1] Indiana University, Purdue University, School of Sciences, Department of Geology, 723W Michigan Street, Indianapolis, Indiana, USA

[2] Nevada Bureau of Mines and Geology, University of Nevada, Reno, Nevada 89557, USA

Abstract

The Carson River system of west-central Nevada was subjected to a massive influx of mercury-enriched tailings derived from mining and milling of the Comstock Lode near the end of the 1800s. Detailed investigations have shown that these contaminated tailings were deposited along the Carson River Valley during a period of channel and floodplain aggradation. Following cessation of mining, incision re-exposed the historical sediment in banks of the modern channel. Concentrations of mercury in these deposits are commonly two to three orders of magnitude above background levels observed in other sediments and soils of the region. Most of the mercury is carried on fine-grained particles as would be expected from previous studies of trace metal partitioning in alluvial deposits. However, erosion and reworking of the historical sediment during lateral channel migration leads to a loss of fine-grained particles and a concentration of mercury-gold/silver amalgam grains in the modern channel bed. As a result, mercury is distributed in both fine- and coarse-grained sediment fractions of the channel floor.

The enrichment of trace metals in fine-grained sediment of most aquatic systems has caused some investigators to argue that geographical and temporal trends in trace metal concentrations should be determined only after applying one of the several methodologies developed to correct for the effects of varying grain size. While this concept seems to apply well to the historical deposits of the Carson River, it would lead to erroneous conclusions if applied to the modern channel bed sediments. In addition, mercury, gold and silver are partitioned along the channel in to specific depositional sites, possibly in the form of amalgam grains that are concentrated as placers. It is therefore imperative that downstream trends in mercury concentrations are assessed by examining data collected from similar depositional environments.

It is debatable at this time whether the partitioning of mercury observed along the Carson River occurs in other aquatic systems such as those found in the humid tropics. Nevertheless, the data collected herein illustrate that the partitioning of mercury as a function of both grain-size and depositional environment should be carefully evaluated before developing detailed sampling strategies

Introduction

In an excellent review of the historical development of mercury amalgamation mining, Nriagu (1994) notes that the Roman smiths of ancient Italy were well aware of the use of mercury in the processing and refining of gold and silver. It was not, however, until the 1554 development of the versatile patio process in Mexico that mercury amalgamation was used on an industrial scale (Nriagu, 1994). The patio process, and its many variations, was so efficient at refining large volumes of low grade ore, that its utilization continued nearly unchallenged through the end of the 19th century. Unfortunately, it also led to the release of unprecedented amounts of mercury to the environment, particularly in North, Central and South America. In the United States, for example, Nriagu (1994) estimated that approximately 68,380 tons of mercury were dispersed to the environment between 1820 and 1900. This figure is supported by our recent documentation of more than 1600 sites in the US where mercury is suspected to have been used in association with historical mining activities. The loss of mercury during mining operations in Central and South America between 1570 and 1900 may have been even more extensive, totaling as much as 196,000 tons (Nriagu, 1994).

While the advent of the cyanidation process in 1890 has largely replaced mercury amalgamation in most temperate environments, it continues to be widely used by non-organized prospectors of the tropics. The number of individuals engaged in mercury amalgamation is debatable, but the SCOPE (Scientific Committee on Problems of the Environment) mercury project estimated that in 1993 there were 4 ± 1 million miners currently involved in its use (Jernelov and Ramel, 1994). The most significant areas of mining activity include Brazil (650,000 miners), Tanzania (250,000), Indonesia (250,000) and Vietnam (150,000) (Jernelov and Ramel, 1994); other countries where mercury amalgamation is a significant problem include Bolivia, Colombia (*cf.* Chap. 14), Peru, the Philippines and Venezuela.

The tremendous number of poor and unskilled people involved in alluvial gold mining in virgin rainforest, and the potential impacts of mercury on native populations, have provoked public concern with regard to mercury amalgamation on a world-wide scale (Reuther, 1994). As a result, numerous investigations aimed at assessing the ecological impacts of both historical and modern mining activities, and the behavior of mercury in the various environmental compartments of tropical ecosystems (e.g., air, water, sediments and biota) have been conducted during the past five to ten years, particularly in Brazil (Martinelli et al., 1988; Malm et al., 1990; Pfeiffer et al., 1991; Lacerda et al., 1991a, 1991b; Lacerda and Salomons,

1992; Reuther, 1994). As pointed out in the SCOPE mercury-project report (Jernelov and Ramel, 1994), these investigations suggest that "tropical ecosystems with their higher degree of complexity and specialized predators might respond differently to mercury contamination than do temperate and boreal ecosystems". It can, therefore, be argued that our understanding of mercury cycling in tropical environments is still in its infancy.

The success of future investigations in assessing the ecological impacts of mercury in tropical climatic regimes will undoubtedly rest on the adequacy of the sampling strategies utilized and the analytical methods invoked. It is, perhaps, with regard to these two topics that studies conducted in temperate and boreal localities can prove most useful to future investigations in the humid tropics.

In temperate and boreal climates, most studies of sediment contamination within lakes, estuaries and low-gradient rivers have shown that significantly higher trace metal concentrations occur in the finer-grained particle fractions (Whitney, 1975; Helmke et al., 1977; Ackermann, 1980; Jenne et al., 1980; Förstner, 1982; Salomons and Förstner, 1984). This trend is primarily attributed to a number of chemical and physical factors which increase the capacity of fine-grained particles to retain trace metals, including surface area, cation exchange capacity, surface charge, concentration of iron and manganese oxides and hydroxides, organic matter content and the concentration of clay minerals (Horowitz and Elrick, 1988). In general, fine sediment tends to be chemically reactive, while particle sizes ranging from coarse silt to sand, are composed of quartz, feldspar and carbonate minerals, that are largely inert (Salomons and Förstner, 1984). All of this is important, of course, only for metals which are at least partially soluble in surface waters and are, therefore, available in solution to be sorbed to fine particles.

The effects of grain-size on trace metal chemistry are significant in that even small spatial or temporal changes in the percentage of fine particles in the bed sediment of aquatic systems can (1) influence the spatial and temporal concentration patterns observed in nature, and (2) hinder the determination of localized inputs of trace metals from either natural sources (e.g., ore bodies) or anthropogenic sources (e.g., industrial sludge) (Horowitz and Elrick, 1988). This fact has prompted some investigators to argue that a standard analysis of the less than 63 µm fraction should be utilized in studies of trace metal transport and accumulation (see, for example, Förstner and Salomons, 1980). Standardizing the analysis allows the results of different investigators working in diverse environment to be compared. Analyzing the less than 63 µm fraction has the advantage over other size ranges in that (1) it reduces the percentage of the sample that is chemically inert, thereby, eliminating the effects of grain-size on trace metal concentrations, (2) it can be extracted from the bulk sample relatively quickly via sieving, a process which does not alter trace metal chemistry, and (3) it is the particle size most commonly carried in suspension by aquatic systems and may, therefore, be most readily distributed over the environment. Other investigators have suggested, however, that the effects of grain-size can be overcome by mathematically normalizing metal concentrations with respect to the percentage of fine material of a given range (e.g., the < 63 µm fraction) as determined from independent analysis of the bulk sample (Förstner and Wittmann, 1979; Jenne et al., 1980; Ackermann et al., 1983). Use of the latter method provides the investigator with the concentrations actually found in the deposits, but poorly documents the concentrations that would be measured in the finer-size fractions.

Regardless of the methods invoked, most recent studies use some correction procedure to eliminate the effects of grain-size on the temporal and spatial distribution of heavy metal concentrations in sediments and soils. However, Moore et al. (1989) have convincingly shown that the utilization of grain-size correction methods would lead to erroneous conclusions if applied to the coarse-grained floodplain of the Clark Fork River, Montana, a river contaminated with mine and milling wastes. Their argument is based on the fact that most or all size fractions contribute to the high metal concentrations observed in the floodplain sediment, presumably because the coarse-grained fractions are not dominated by quartz and feldspar, as would be expected had they been produced by natural weathering processes. Rather, the sediment consists of crushed mica, primary sulfides (ore), and smelter slag glasses, containing elevated levels of trace metals. Clearly, the results from Moore et al. (1989) question whether grain-size correction procedures should be applied to high-gradient rivers that have been contaminated by mining and milling wastes.

Mercury in the form of cinnabar can, and often is associated with mining and milling debris where sulfide ore bodies are being worked. With respect to operations utilizing the amalgamation process, metallic mercury has the potential to be transported as liquid droplets (at least close to the sites of utilization), and as amalgamated mercury/gold or mercury/silver particles. All three of the above forms of mercury are likely to be associated with the coarse-grained fraction of alluvial deposits. In addition, a significant portion of the mercury will presumably be sorbed onto fine-grained particles. Thus, the question arises as to whether the bulk sample should be analyzed during studies of mercury transport and accumulation, or a split of the sample representing a particular size fraction.

In this chapter, we will utilize data from the Carson River valley of west-central Nevada to assess (1) the partitioning of mercury derived from the amalgamation milling process within various grain-size fractions of alluvial sediments, (2) the chemical form of mercury within the alluvial deposits and (3) whether grain-size correction techniques are appropriate for investigations dealing with sites contaminated with mining derived mercury. Undoubtedly, care must be taken when applying the results from this semi-arid environment to tropical regimes. Nonetheless, the conclusions presented below may provide useful insights into the development of sampling strategies and the study of mercury transport mechanics.

Study Area

The Carson River heads on the eastern flank of the Sierra Nevada, a mountain range located along the eastern border of California (Fig. 1). From its headwaters, which receive between 64 and 127 cm of precipitation per year, it flows in a east-northeasterly direction through a series of alluvial valleys and canyons until reaching the Carson Sink, a large, hydrologically closed playa system, located in the Carson Desert. The downstream regions often receive less than 13 cm of annual precipitation (Fig. 1).

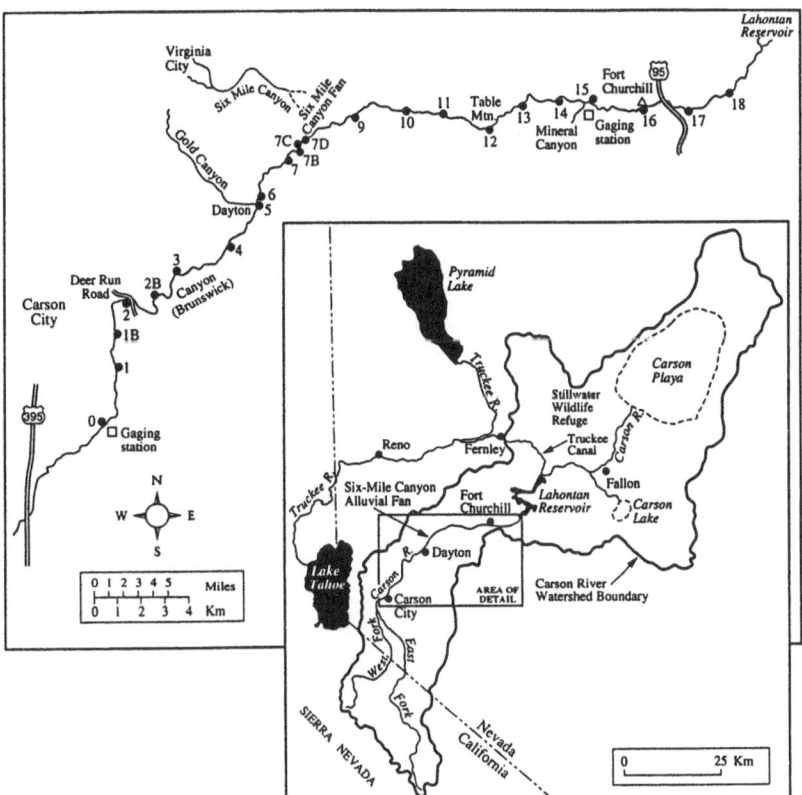

Fig. 1: Location map showing sampling sites along the Carson River

In 1850, gold was discovered near Dayton, Nevada and for nearly a decade, mining consisted of working placer deposits within Gold and Sixmile Canyons (Fig. 1). These mining operations led upstream to the discovery of the Comstock Lode in 1859, near Virginia City. For the next thirty years (1860-1890), substantial quantities of precious metals were obtained from what become known as one of the richest gold and silver bearing ore bodies in mining history.

In general, precious metal extraction consisted of shipping ore obtained from deep-shaft mines to one or more mills located within Gold Canyon, Sixmile Canyon and along the Carson River, from Carson City to Dayton (Fig. 2). At the mills, the ore was pulverized by stamps and arrastras, after which a slurry containing various salts was created and charged with mercury. The resulting mercury/gold and mercury/silver amalgam particles were separated from the slurry and retorted to produce pure forms of the precious metals. In the process, substantial quantities of mercury, gold and silver were lost with the discarded mill tailing fines. Although the estimates vary, the most commonly cited figure is that between 7,000 and 7,500 tons of mercury were released to the environment (Smith, 1943).

THE BRUNSWICK MILL–CARSON RIVER.

Fig. 2: The Brunswick Mill located along the Carson River downstream of sampling site 2 (Fig. 1). Light colored sediments along the channel are mercury-enriched tailings (courtesy of The Bancroft Library)

Reconnaissance level investigations since the early 1970s have demonstrated that contaminated tailings have been eroded from the milling sites and redistributed along the Carson River valley, from Carson City to the Carson Sink (Fig. 1) (Van Denburgh, 1973; Cooper et al., 1985). Moreover, concentrations of mercury in many fish species were found to exceed 1 μg g^{-1}, the US Food and

Drug Administration's action level (Richins, 1973; Richins and Risser, 1975; Cooper et al., 1985). As a result, a reach of the Carson River valley extending downstream of the milling operations was placed in 1990 on the US National Superfund Priorities List. Since that time, investigations funded by the National Institutes of Environmental Health Sciences have shown that mercury concentrations well in excess of background levels occur in air (Gustin et al., 1994; 1995), water (Bonzongo et al., in press) and sediments (Lechler et al., 1995; Miller et al., 1995a; 1995b). These highly elevated levels make the Carson River valley an excellent setting to examine the form and distribution of mercury in alluvial deposits.

Methods

In order to document the distribution of mercury in the alluvial deposits of the Carson valley, 20 sampling sites were randomly selected at nearly uniform intervals along the river between Carson City and Lahontan Reservoir (Fig. 1). At each of the selected sampling locations, sediments were collected from the upper 10 cm of pool, riffle and point bar deposits (where present). At several of these sites, samples were obtained during the summers of both 1993 and 1994. Temporal changes in bed concentrations were generally less than 15%. Thus, the analytical results derived from both sampling periods were averaged prior to their utilization in statistical analysis. Integrated samples were also collected from each of the stratigraphic units delineated in the channel banks on the basis of stratigraphic position, grain size distribution, color, induration and degree of sediment weathering. All samples were shipped to the Nevada Bureau of Mines and Geology and frozen until being analyzed for Hg, Au and Ag.

Mercury, gold and silver were determined by weighing 10 g of wet sediment sample into a 250 mL borosilicate beaker, adding 15 mL of *aqua regia* (1 part nitric acid and 3 parts hydrochloric acid), and heating at a low boil for one hour. The samples were then brought to a 50 mL volume with 3 N hydrochloric acid and silver was determined by flame atomic absorption spectrometry. An aliquot of this solution was used to determine mercury by cold vapor atomic absorption spectrometry. Gold was determined by graphite furnace atomic absorption spectrometry on a 5 mL aliquot of the above solution after extraction into 2 mL methyl isobutyl ketone and washing with 10 % (v/v) hydrochloric acid. Moistures were determined at 105°C on separate aliquots of the sediment samples and all data were corrected to a dry basis. Lower limits of detection by this method are 0.2, 0.01 and 0.005 $\mu g \, g^{-1}$ for silver, mercury and gold, respectively.

Total organic matter content was also determined for each of the collected samples. The procedure involves sample oxidation, followed by total carbon determination using a spectrophotometer as described in Metson (1961). Particle size distribution in terms of the percentages of sand, silt and clay was determined using wet-sieving and pipette analysis as modified from Day (1965) and Jackson (1969).

All statistical analysis were conducted using SYSTAT® for Windows® (version 5.04). Examination of the data using stem-and-leaf diagrams and normal probability plots revealed that the data were not normally distributed and transformation to log values did little to improve the homogeneity of the data. Thus, the statistical analysis relies heavily on non-parametric methods.

Results and Discussion

Relations Between Trace Metal Concentrations and Particle Size

Field observations have revealed that the banks of the Carson River downstream of mining and milling operations are composed of two primary units (Figs. 3 and 4). Generally observed near the base of the banks are massive, yellowish-brown to dark gray deposits, dominated by fine-sands, silts and clays. At some localities, the upper surface of this unit can be identified by a soil profile exhibiting a 10 to 20 cm thick, organic-rich A-horizon. More commonly, however, the soils have been truncated and the upper boundary is denoted by an erosional surface. In either case, this unit is typically buried by light yellowish-brown to brown deposits, characterized by prominent stratification and/or channel fills of yellow tailings-like sediment. These upper and, thus, younger sediments have been interpreted on the basis of geochemical data and the occurrence of buried artifacts as a valley fill of historic age, consisting in large part of mercury contaminated materials. The thickness of these historical sediments is highly variable, but at most sites is on the order of 1.5 to 2.0 m. In upstream areas (i.e., between sampling sites 5 and 12; Fig. 1), historical deposits form most or all of the channel banks. Farther downstream, the thickness of the deposits decreases and they commonly occur as (1) large channel fills, that are cut into the older pre-mining sediments, or (2) fill terraces inset into older valley deposits. The deposition of historical sediment is thought to be related to a phase of channel and floodplain aggradation, initiated by a massive influx of mine and mill tailings during the late 1800s, while their current exposure is the result of an episode of channel incision that followed the cessation of mining and milling operations (Miller et al., in preparation).

The range of mercury, gold and silver concentrations measured in the historic valley fill deposits are shown in Table 1. Total mercury concentrations are often two orders of magnitude above background levels determined to be in the range of 0.01 μg g^{-1} to 0.1 μg g^{-1} for sediment and soils in this region (Miller et al., in preparation).

Fig. 3: Bank exposure near sampling site 2 showing contaminated historical sediments (2) overlying pre-mining valley fill (1). Dark colored soil A-horizon separates the two units (photo by J. Miller)

Fig. 4: Stratified tailings material exposed in channel bank near sampling site 7 (photo by J. Miller)

The statistical relationships between selected trace metals and the percent sand, silt, clay, mud (defined as particles < 63 μm in size), and organic carbon are shown in Table 2A for the historical deposits. No statistically significant ($p = 0.05$) relationships were identified between trace metal concentrations and the percentage of organic carbon in the deposits. This may be due to the relatively low levels of carbon within bank materials of this semi-arid environment. The percentages of organic carbon ranged from 0.02% to 1.68% and averaged 0.63%.

Table 1: Summary of mercury, gold and silver concentrations in delineated deposits of the Carson River. Modern channel sediments include channel, pool, riffle and point bar deposits. Background concentrations are on the order of 0.01 to 0.1 μg g^{-1}

Site	Minimum	Maximum	Mean	Std. Dev.
Channel Deposits* (n=3)				
Mercury (μg g^{-1})	0.051	2.920	1.219	1.507
Gold (μg kg^{-1})	< 5.0	69.0	24.7	38.4
Silver (μg g^{-1})	< 0.2	1.7	0.6	1.0
Pool Deposits (n=26)				
Mercury (μg g^{-1})	0.018	4.820	1.768	1.484
Gold (μg kg^{-1})	< 5.0	175.0	19.5	38.1
Silver (μg g^{-1})	< 0.2	2.6	0.9	0.8
Riffle Deposits (n=12)				
Mercury (μg g^{-1})	0.029	2.690	1.067	1.020
Gold (μg kg^{-1})	< 5.0	222.0	22.5	66.2
Silver (μg g^{-1})	0.2	2.5	0.8	0.6
Point Bar Deposits (n=24)				
Mercury (μg g^{-1})	0.025	11.800	2.516	2.516
Gold (μg kg^{-1})	< 5.0	806.0	84.5	178.3
Silver (μg g^{-1})	< 0.2	4.9	1.7	1.3
Historical Valley Fill Deposits (n=42)				
Mercury (μg g^{-1})	0.039	887.000	99.866	176.581
Gold (μg kg^{-1})	< 5.0	426.0	168.6	110.8
Silver (μg g^{-1})	< 0.2	117.8	16.3	20.4

* Channel Deposits refer to reaches where pools, riffles and point bars could not be delineated.
Note: At several localities, multiple samples were collected in summer 1994, from selected pools, riffles and point bars to determine within unit variability at a particular site. These data are not included above.

The strongest positive correlations occur, in decreasing order, between mercury and mud, mercury and silt, silver and silt, and silver and mud (Table 2A). Weaker, but statistically significant correlations ($p = 0.05$) also occur between gold and silt, and gold and mud. All three of the trace metals are inversely related to the percentage of sand in the samples (Table 2 A). The strong significant relationships suggest that within the historical deposits, the sediment-bound trace metals are primarily associated with fine grained particles, presumably in association with grain coatings or sorbed onto particle surfaces as reported in previous investigations. In contrast, trace metal-grain size relationships do not exist for sediments extracted from the modern channel bed (Table 2B). Undoubtedly, this failure to identify statistically significant relationships is governed by the nature of the material in the channel floor.

Table 2: Spearman's rank correlation coefficients for total metal concentration, percent size fractions and organic matter. Data utilized represent average values at sites where samples were collected in summers of both 1993 and 1994

(A) Historical Valley Fill Deposits (n=42)

	Hg	Ag	Au	Sand	Silt	Clay	Mud	Carbon
Hg	1.000							
Ag	0.891*	1.000						
Au	0.749*	0.755*	1.000					
Sand	-0.612*	-0.561*	-0.305*	1.000				
Silt	0.656*	0.616*	0.366*	-0.992*	1.000			
Clay	0.376*	0.256	0.224	-0.603*	0.550*	1.000		
Mud	0.673*	0.610*	0.355*	---------	0.973*	0.667*	1.000	
Carbon	0.289	0.292	0.037	0.250	0.252	0.098	0.268	1.000

(B) Channel Bed Deposits (n=39)

	Hg	Ag	Au	Sand	Silt	Clay	Mud	Carbon
Hg	1.000							
Ag	0.718*	1.000						
Au	0.639*	0.667*	1.000					
Sand	0.052	0.068	-0.178	1.000				
Silt	0.003	-0.022	0.191	-0.946*	1.000			
Clay	-0.116	-0.066	0.156	-0.904*	0.823*	1.000		
Mud	-0.056	-0.033	0.191	---------	0.945*	0.953*	1.000	
Carbon	0.076	-0.113	0.119	0.104	0.104	0.057	0.050	1.000

* Significant ($p \leq 0.05$)

With regards to particle-size, we find that the most important difference between the historic and channel bed sediment is the percentage of particles less than 63 μm in intermediate diameter (Table 3). In contrast to the historical sediment, the channel bed materials generally possess less than 7% mud (Table 3).

A comparison of aerial photographs taken in 1965 and 1991 illustrates that the river has laterally migrated across the valley floor at extremely high rates (a mean of 1.03 m yr^{-1}; Fig. 5) during this 26 year period (measurements were made at three locations along 28 reaches). The reworking of the contaminated historical deposits during bank erosion should supply the modern channel with large quantity of sediment and trace metals. The lack of fine-grained particles in the channel bed suggests, however, that the silt- and clay-sized fractions observed in the historical sediments are not incorporated to any significant degree into the channel bed (Table 3). Apparently, these particles are transported as wash load downstream until they reach Lahontan Reservoir (Fig. 1). In support of this conclusion, Miller et al. (1995a) have shown that deep-water and deltaic environments of the reservoir contain more than $8.5 \times 10^6 \ m^3$ of sediment that is dominated by silt- and clay-sized materials.

Table 3: Summary of the percentages of sand, silt, clay and mud in delineated deposits of the Carson River

Site	Minimum	Maximum	Mean	Std. Dev.
Channel Deposits* (n=3)				
Sand (%)	1.29	96.43	63.57	53.58
Silt (%)	1.75	5.01	2.93	1.80
Clay (%)	1.53	3.43	2.48	1.34
Mud (%)	1.75	8.44	4.59	3.46
Pool Deposits (n=26)				
Sand (%)	91.07	99.05	96.85	2.18
Silt (%)	0.06	4.90	1.78	1.10
Clay (%)	0.30	8.56	2.00	2.02
Mud (%)	0.95	13.46	3.78	3.01
Riffle Deposits (n=12)				
Sand (%)	97.40	98.96	98.39	0.64
Silt (%)	0.46	2.34	1.03	0.57
Clay (%)	0.21	1.33	0.59	0.35
Mud (%)	1.04	2.70	1.62	0.66
Point Bar Deposits (n=24)				
Sand (%)	71.52	98.96	93.57	7.66
Silt (%)	0.48	22.69	4.92	6.37
Clay (%)	0.20	5.78	2.10	1.87
Mud (%)	1.04	28.47	7.02	8.15
Historical Valley Fill Deposits (n=42)				
Sand (%)	22.05	98.28	72.48	20.09
Silt (%)	0.88	72.29	21.51	17.79
Clay (%)	0.67	17.56	6.61	4.49
Mud (%)	1.72	77.95	28.13	19.87

* Channel Deposits refer to reaches where pools, riffles and point bars could not be delineated.

Geochemical data reveal that trace metal concentrations measured in the channel bed are well below the values observed in the historical sediments that are locally exposed in the channel banks (Table 1). The reduction in concentration is likely the result of (1) a loss of fine-grained particles enriched in mercury, gold and silver, (2) dilution with sediments derived from reaches upstream of mining and milling operations and from tributaries devoid of polluted materials, and (3) dilution with "clean" pre-mining valley fill deposits that locally form the channel banks. The channel bed sediment is, nonetheless, enriched in mercury, reaching concentrations two orders of magnitude above the regional background levels (Table 1). Both the low percentages of mud in the deposits (Table 3) and the statistically insignificant correlations between trace metal concentrations and the percent sand, silt, clay and mud in the channel bed (Table 2B) suggest that the existing mercury is distributed throughout the various grain-size fractions.

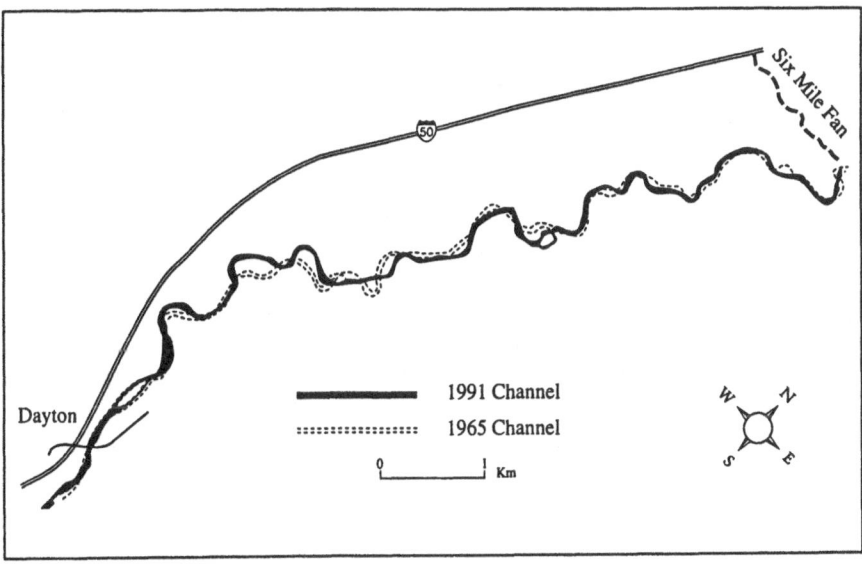

Fig. 5: Map showing the position of the Carson River in 1965 and 1991. The channel has migrated laterally over much of the river's course allowing mercury-enriched valley fill deposits to be reworked in the process

To further test this hypothesis, geochemical analyses were carried out on the less than and greater than 63 µm fraction of selected samples (Table 4). Mercury concentrations are between 8 and 33 times higher in the finer fraction. This was not unexpected, given that mercury is enriched in the fine-grained sediment of the historical deposits. It is significant, however, that concentrations are elevated well above background levels in coarse-grained materials, particularly at sampling sites 11 and 16. Moreover, calculations of the quantity of mercury in each particle-size fraction of the bulk sample show that a majority of the mercury is actually associated with sediment greater than 63 µm in size (Table 4).

Table 4: Mercury concentrations and quantities found in the less than and greater than 63 μm fraction of selected samples. See Fig. 1 for sample site locations

| Sample Site | Hg (μg g⁻¹) | <63μm/>63μm | μg of Hg /100 g of sediment | |
			<63μm	>63μm
CR-1, >63μm	0.035	14.8	0.40	3.47
<63μm	0.517			
CR-1B, >63μm	0.029	8.9	0.74	2.82
<63μm	0.257			
CR-2, >63μm	0.055	33.1	2.39	5.43
<63μm	1.820			
CR-11, >63μm	2.230	18.7	29.93	221.40
<63μm	41.700			
CR-16, >63μm	3.320	9.18	5.20	331.43
<63μm	30.470			

Perhaps the most obvious question that arises at this point in our analysis pertains to the form of mercury in the coarse-grained material. Callahan et al. (1994) noted that amalgam particles well in excess of 100 μm in size were present in rivers impacted by gold mining operations in North Carolina which occurred in the 1800s and early 1900s. During mining of the Comstock Lode, substantial quantities of gold, silver and mercury were lost from the mills, apparently in the form of amalgam. These losses did not go unnoticed as the mill tailings were extensively reprocessed. In fact, some extraordinary techniques were invoked to capture the heavy particles as they flowed away from the mills. Smith (1943), for example, notes one attempt utilizing blanket concentration. He states that:

> "The tailings from these mills (at Virginia City and Gold Hill), estimated at not less than 600 tons a day, are allowed to run into Gold Cañon and Six Mile Cañon where they are passed over a great length of blankets from five to six miles in length in each cañon."

It was unclear whether the above methodology proved successful. There is little doubt, however, that much of the tailings material flowed out of the canyons and into the Carson River. In fact, dredging operations were initiated in 1895 near Dayton to "recover the millions of wasted amalgam" particles which had been lost from the mills during the previous twenty years (Appeal, 18 February; cited in Dangberg, 1975). Given the above, it is reasonable to assume that much of the mercury found in the coarse-grained deposits of the Carson River is associated with grains of gold and silver.

The existence of amalgam grains in the channel bed of the Carson River is supported by strong significant relationships between mercury and silver, and to a lesser extent, mercury and gold, within both the historical valley fill deposits and the channel bed sediments (Table 2). In addition, chemical speciation data reveal that greater than 90% of the mercury in the channel bed is in the elemental form, as could be expected if the amalgam particles were present (Lechler et al., 1995).

The above observations led us to separate heavy concentrates from selected samples of high energy reaches of the channel floor in order to examine them with a scanning electron microscope equipped with an energy dispersive spectrometer. All five of the samples examined contained either mercury-silver or mercury-gold amalgam particles. Thus, particles of the amalgam do exist, supporting our previous statement that they are responsible for at least part of the mercury observed in the sand-sized fraction of the channel bed.

As mentioned earlier, the reworking of the historical deposits during lateral channel migration may represent a primary source of trace metals found in the floor of the modern channel. If this is the case, coarse-grained amalgam should exist within the historical sediments, given that it is found within the channel bed, a conclusion that seems to run contrary to the inverse relationships observed between trace metal concentrations and the percentage of sand in the historical deposits (Table 2A). One possible explanation is that the trace metals associated with the grains of amalgam make up only a very small percentage of the total found in the historic valley fill, the majority of the metals being associated with fine-grained particles (i.e., those < 63 μm in size). The increase in the significance of amalgam related trace metals in the channel bed, presumably results from the winnowing of fine particles from historical valley fill as the banks are eroded. This speculation is significant in that it suggests that mercury found in the banks, and later in suspension following bank erosion, is chemically and physically different from that found in the channel floor and moved primarily as bedload.

The traditional view, as pointed out in the introduction to this chapter, has been that mercury is primarily associated with the highly reactive fine-grained sediment fraction composed of clays, organic carbon and other particles coated by various oxides and hydroxides. The coarser fraction, containing relatively low levels of mercury, acts as a dilutant which lowers the elemental concentrations observed in the bulk sample. Geographical and temporal trends of mercury dispersal away from a particular source should, therefore, be determined by applying one of the methodologies developed to correct for the effects of grain size. While this concept seems to apply well to the historical valley fill deposits of the Carson River, the coarse-grained sediments of the channel bed exhibit mercury concentrations one to two orders of magnitude above background levels and contain most of the mercury. This enrichment presumably results, at least in part, from the existence of amalgam grains which are concentrated by sediment reworking in the channel bed. Because the mercury concentrations in the channel bed materials do not follow the expected partitioning with respect to particle size, the application of grain-size correction factors would lead to erroneous conclusions about geographic and temporal mercury patterns within the drainage system.

It is debatable at this time whether the partitioning of mercury observed in the various grain-size fractions of the Carson River occurs in the other aquatic environments. It will likely depend, in large part, on the nature of the channel. Coarse amalgam particles may become a significant component of the bed sediment in high gradient rivers, because these systems allow for the reworking and concentration of the heavier grains found in point and non-point sources of mining debris. It will also depend on the form in which mercury is initially released to the system, and the magnitude of the inputs relative to the size of the river.

Mining operations by non-organized prospectors in many tropical regions are small and inefficient compared to the techniques used along the Carson Valley. Lacerda and Salomons (1992) note, for example, that two primary methodologies of amalgamation mining occur in Brazil. The first process, used widely in central Brazil and northeastern sectors of the Amazon, involves the extraction of gold from alluvial soils and rock. The procedure consists of digging large amounts of gold-bearing material from localized areas, and passing it through grinding mills and centrifuges, in order to produce a gold-rich gravity concentrate. The concentrate is subsequently moved to small amalgamation ponds, a few square meters in size, or to amalgamation drums, where it is mixed with liquid mercury. The amalgam is then separated and roasted.

The second process is carried out along most amazonian rivers. In this case, gold is extracted from bottom sediments which have been dredged from the river beds. The gold-bearing materials are passed through iron nets to remove the larger particles before the material is passed over a set of carpeted riffles which retain the heavier grains. This operation lasts for 20 to 30 hours after which the dredging stops and the heavy fraction is collected in barrels for amalgamation. The amalgam is then separated and roasted.

Both of the processes described by Lacerda and Salomons (1992) result in losses of mercury to the atmosphere during the roasting process. In addition, it is likely that these relatively crude methods (by today's standards) result in the release of mercury to the river as liquid droplets and as particles of amalgam. Both forms may prove to be a significant component of the mercury found in the channel bed sediments of these tropical river systems.

Reuther (1994) determined, for example, that mercury concentrations in both the Madeira River and one of its tributaries (the Mutum-Paraná River) are elevated in the greater than 1 mm size fraction. He attributes the enrichment in this case to the presence of macroscopic, liquid mercury droplets, released to the rivers during mining activities. To our knowledge, however, Reuther (1994) did not look for the amalgam grains within the sediments, and thus it is possible that part of this enrichment is caused by the presence of amalgam. In either case, it suggests that the application of methods to correct grain-size effects should be carefully evaluated before they are utilized in investigations of the impacts of mercury amalgamation mining in tropical systems.

Trace Metal Partitioning Along the Channel and Its Implications for Sampling Strategies

Many heavy minerals including, to mention a few, gold, silver, cassiterite, chromite, magnetite and zircon are separated from similar sized silts and clays and deposited as placers with coarser, but hydraulically lighter sands and gravels. This hydraulic sorting is primarily a function of their specific gravities, shape, and entrainment velocities (Guilbert and Park, 1986), and the resulting bands of concentrated minerals may vary in scale from units measured in kilometers to individual laminae a few millimeters thick (Smith and Minter, 1980).

Although placer minerals, particularly gold, have been mined for hundreds of years, relatively little is known about the physical processes that are responsible for their concentration (Smith and Beukes, 1983; Guilbert and Park, 1986). Nevertheless, important placer deposits are often found along the alluvial-bedrock contact (Adams et al., 1978) where they are associated with channel lag sediments that presumably come to rest during floods. Placers have also been observed in zones of flow convergence downstream of tributary confluences (Mosely and Schumm, 1976) and at the surface of modern channels where flows are greatly constricted (*cf.* Smith and Beukes, 1983). The concentration of heavy minerals in these latter two locales is generally attributed to the winnowing of finer and lighter sediments as a result of substantial scour and reworking of the bed sediments. In meandering systems, placers are also known to form where heavy minerals are dropped in zones of decreasing velocity, including the lower margins of point bars (Bateman, 1950; Boyle, 1979). For that reason, point bars are often targeted for placer exploration (Guilbert and Park, 1986).

Considering that the densities of mercury, gold and silver are 13.53 g cm^{-3}, 19.28 g cm^{-3} and 10.50 g cm^{-3}, respectively, it should be recognized that amalgam particles have the potential to be hydraulically separated from similar sized, lighter materials and deposited as placer-like concentrations along the river's course. Such a partitioning of trace metals into specific depositional sites could lead to the development of erroneous downstream trends in concentration if care is not taken to sample and compare similar depositional environments.

Within the Carson River, three primary depositional units are observed along the channel floor and pools, riffles and point bars (Fig. 6). Point bars and pools are the most distinctive units, occurring adjacent to one another at meander bends. In contrast, well-defined riffles were often difficult to delineate either in the field or on aerial photographs, but where they could be defined, they occur as shallow water zones, between successive meander bends. All of the deposits are typically dominated by gravel sized materials in upstream areas (between Carson City and Dayton) and sand-sized materials further downvalley.

Fig. 6: Pool (1), riffle (2) and point bar (3) deposits located near sampling site 3 (*above*) and sampling site 16 (*below*). These depositional environments become finer textured downstream (photos by J. Miller)

The concentrations of mercury, gold and silver measured in these deposits are presented in Table 5. Unfortunately, riffle samples were not collected at sites 12 through 14 during the summer of 1993 and difficulties in obtaining site access prohibited later collection of these sediments. As a result, data associated with riffles are somewhat limited.

The data that were obtained have been tested for statistical population similarity, using the Wilcoxon Rank Sum Tests (Table 6). This analysis show that there is a tendency for metal concentrations to differ between point bar and pool/riffle environments, and trace metal concentrations are elevated in point bar deposits (Tables 5 and 6). We speculated that these differences are related to the development of placer-like concentrations composed, in part, of amalgam grains, along the lower margins of the point bars. The lack of a statistical difference in trace metal concentrations between pools and riffles for all three elements reinforces this hypothesis (Table 6).

Table 5: Trace metal concentrations in pool, riffles, and point bars of selected sampling sites. See Fig. 1 for site locations. Data represent average values at sites where samples were collected both in the summers of 1993 and 1994

Site	Mercury (μg g^{-1})			Gold (μg kg^{-1})			Silver (μg g^{-1})		
	Pool	Riffle	Point Bars	Pool	Riffle	Point Bars	Pool	Riffle	Point Bars
CR-1	0.048	0.099	0.032	< 5.0	< 5.0	< 5.0	0.2	0.5	0.3
CR-2	0.023	0.029	0.450	< 5.0	< 5.0	< 5.0	< 0.2	0.5	0.8
CR-3	0.521	0.116	0.684	< 5.0	< 5.0	< 5.0	0.4	0.4	2.0
CR-5	0.789	-------	2.600	< 5.0	-------	15.0	0.4	-------	0.8
CR-6	1.631	0.435	1.320	< 5.0	< 5.0	2.5	1.1	0.7	0.9
CR-7	1.070	1.910	2.660	< 5.0	< 5.0	64.0	0.5	0.7	1.9
CR-9	2.270	1.320	2.510	11.3	< 5.0	323.0	1.5	1.2	2.9
CR-10	2.100	-------	2.370	< 5.0	-------	< 5.0	1.4	-------	2.1
CR-11	3.995	1.880	3.600	71.4	< 5.0	67.0	2.2	0.8	2.2
CR-12	3.190	-------	11.800	39.0	-------	86.0	1.2	-------	4.5
CR-13	4.820	-------	2.760	20.0	-------	34.0	2.2	-------	1.7
CR-14	2.250	-------	2.680	< 5.0	-------	< 5.0	1.1	-------	1.5
CR-16	1.730	1.183	1.858	6.8	< 5.0	11.0	0.4	0.3	0.5
CR-17	-------	1.510	1.170	-------	222.0	17.0	-------	0.7	0.5
CR-18	1.600	-------	1.730	32.0	-------	12.0	0.6	-------	0.4

Table 6: Tests for differences in trace metal concentrations between pool, riffle and point bar deposits

Metal	Depositional environments compared	Wilcoxon rank sum test
Mercury	Pool / Riffle	NS
	Pool / Point Bar	NS
	Riffle / Point Bar	$p \leq 0.05$
Silver	Pool / Riffle	NS
	Pool / Point Bar	$p \leq 0.05$
	Riffle / Point Bar	$p \leq 0.05$
Gold	Pool / Riffle	NS
	Pool / Point Bar	$p \leq 0.01$
	Riffle / Point Bar	NS

NS - Concentrations are not significantly different ($p < 0.1$)

Our sampling capabilities did not allow us to collect sediments at significant depths below the channel bed. Thus, the development of amalgam placers deposited as channel lags during flood has yet to be tested. Nonetheless, it is suspected on the basis of previous investigations of placer formation that they exist.

Earlier, we suggested that the application of methods to correct for grain-size effects would lead to erroneous conclusions about spatial and temporal trends of mercury within the channel bed of the Carson River. It is also important, however, to reduce or eliminate the effects of trace metal partitioning into specific depositional environments of the channel floor. These differences observed between point bar and pool/riffle deposits could lead to erroneous interpretations of downstream trace metal trends, if the data from the three depositional environments were mixed.

In light of the above, we argue here that it is imperative that spatial (downstream) trends in trace metal concentrations be examined by comparing data collected from similar depositional sites. Ideally, sampling should be carried out on each of the depositional units that can be delineated along the channel bed, such as point bars, alternate bars, longitudinal bars, pools, riffles, etc. Unfortunately, such a detailed sampling program can be prohibitively expensive. An alternative is to sample deposits which consistently occur along the channel. In the case of the Carson River, we would have selected channel pools in that they were identified in most sampling localities, and in those reaches where pools, riffles and point bars could not be delineated, the channel bed most closely resembles channel pools in terms of bed sedimentology and morphology.

Conclusions

Within the highly contaminated historical deposits, mercury is primarily associated with the fine-grained sediment fraction as would be expected from previous studies of trace metal partitioning in alluvial environments. Apparently, the historical deposits also contain small percentages of amalgam grains which are concentrated in the modern channel bed during bank erosion. As a result, mercury is distributed in both the fine- and coarse-sediment fractions of the modern channel floor.

The enrichment of trace metals in fine-grained sediments of most aquatic systems has led to the suggestion that geographical and temporal trends in metal concentration should be determined after applying one of the several procedures designed to correct for the effects of varying grain-size. Because mercury is distributed throughout the channel bed sediment of the Carson River, the application of these procedures to this system would lead to erroneous conclusions.

Mercury is also partitioned as a function of depositional environment along the Carson River, possibly because amalgam grains are concentrated as placers near the lower margins of point bars. It is, therefore, important to keep the depositional environments sampled in a river system consistent, when assessing spatial trends in mercury concentration.

Acknowledgements

This study was funded, in part, by the National Institutes of Environmental Health Sciences (contract # P42ESO5961). Their support is greatly appreciated. Thanks also go to Suzanne Orbock-Miller and Dale F. Ritter for reviewing earlier drafts of the manuscript. Susan Tingley and Kris Pizarro created the graphic illustrations in the paper and for that we thank them.

References

Ackermann, F. (1980). A procedure for correcting for grain size effect in heavy metal analyses of estuarine and coastal sediments: *Environ. Technol. Let.*, v.1, p.518-527.

Ackermann, F., Bergmann, H., and Schleichert, U. (1983). *Environ. Technol. Lett.*, v.4, p. 317-328.

Adams, J., Zumpfer, G.L., and McLane, C.F. (1978). Basin dynamics, channel processes, and placer formation: a model study: *Econ. Geol.*, v. 73, p. 416-426.

Bateman, A.M. (1950). *Economic Mineral Deposits*: New York, John Wiley and Sons.

Bonzongo, J.C., Heim, K.J., Warwick, J.J., and Lyons, W.B. (in press). Mercury levels in surface waters of the Carson River-Lahontan Reservoir system, Nevada: Influence of historic mining activities: *Environ. Pollut.*

Callahan, J.E., Miller, J.W., and Craig, J.R. (1994). Mercury pollution as a result of gold extraction in Northern Carolina, USA: *Appl. Geochem.*, v.9, p.234-241.

Cooper, J.J., Thomaz, R.O., and Reed, S.M. (1985). *Total Mercury in Sediment, Water, and Fishes in the Carson River Drainage, West-Central Nevada*: Nevada Division of Environmental Protection.

Dangberg, G. (1975). *Conflict on the Carson: Carson Valley Historical Society*, Minden, Nevada.

Day, P.R. (1965). Particle fractionation and particle-size analysis: In C.A. Blake, D.D. Evans, J.L. White, L.E. Ensminger, and F.E. Clark, (eds.), *Methods of Soil Analysis*, Part 1, no. 9, p. 545-567.

Guilbert, J.M. and Park, C.F., Jr. (1986). *The Geology of Ore Deposits*. New York, W.H. Freeman and Company.

Helmke, P.A., Koons, R.D., Schombers, P.J., and Iskandar, I.K. (1977). Determination of trace element contamination of sediments by multielement analysis of clay-size fraction: *Environ. Sci. Technol.*, v. 11, p. 984-989.

Förstner, U. (1982). Cumulative phases for heavy metals in limnic systems: *Hydrobiol.*, v.91, p. 299-313.

Förstner, U. and Salomons, W. (1980). Trace metal analysis on polluted sediments, part I: assessment of sources and intensities: *Environ. Technol. Lett.*, v. 1, p. 494-506.

Förstner, U. and Whitmann, G. (1979). *Metal Pollution in the Aquatic Environment*: New York, Springer-Verlag.

Gustin, M.S., Taylor, G.E., Jr., and Leonard, T.L. (1995). Atmospheric mercury concentrations above contaminated mill tailings Carson River Drainage Basin, Nevada: *Water, Air, Soil Pollut.*, v.80, p.217-220.

Gustin, M.S., Taylor, G.E., Jr., and Leonard, T.L. (1994). High levels of mercury contamination in multiple media of the Carson River drainage basin of Nevada: implications for risk assessment: *Environ. Health Perspec.*, v. 102, p. 772-778.

Horowitz, A.J. and Elrick, K.A. (1988). Interpretation of bed sediment trace metal data: methods for dealing with the grain size effect: In J.J. Lichtenberg, J.A. Winter, C.I., Weber,, and L.Fradkin, (eds.), *Chemical and Biological Characterization of Sludges, Sediments, Dredge Spoils, and Drilling Muds*, ASTM STP 976, American Society for Testing and Materials, Philadelphia, p. 114-128.

Jackson, M.L. (1969). *Soil Chemical Analysis*, Advance course, published by author (2nd edition), Madison, Wisconsin.

Jenne, E.A., Kennedy, V.C., Burchard, J.M., and Ball, J.W. (1980). Sediment collection and processing for selective extraction and total trace element analysis: In R.A. Baker, (ed.), *Contaminants and Sediments*: Ann Arbor, Michigan, Ann Arbor Science Publishers, v.2, p. 169-191.

Jernelov, A. and Ramel, C. (1994). *Mercury in the Environment, Synopsis of the Scientific Committee on Problems of the Environment* (SCOPE) meeting on mercury held at the Royal Academy of Sciences Stockholm 28-30 October 1993, *Ambio*, v. 23, p. 166.

Lacerda, L.D. and Salomons, W. (1992). *Mercúrio na Amazônia: Uma Bomba Relógio Química?* CETEM/CNPq, Série Tecnologia Ambiental, 3, Rio de Janeiro.

Lacerda, D., Pfeiffer, W.C., Marins, R.V., Rodrigues, S., Souza, C.M.M., and Bastos, W.R. (1991a). Mercury dispersal in water, sediments, and aquatic biota of a gold mining tailing deposit drainage in Poconé, Brazil: *Water, Air, Soil Pol.*, v.55, p.283-294.

Lacerda, L.D., Salomons, W., Pfeiffer, W.C., and Bastos, W.R. (1991b). Mercury distribution in sediment profiles from lakes of the high Pantanal, Mato Grosso State, Brazil: *Biogeochem.*, v. 14, p. 91-97.

Lechler, P.J., Miller, J.R., Hsu, L.C. and Desilets, M.O. (1995). Understanding mercury mobility at the Carson River superfund site, Nevada, USA: Interpretation of mercury speciation results from mill tailings, soils and sediments. *Proceedings if the 10th International Conference on Heavy Metals in the Environment*, Hamburg, Germany, V.1:315-318.

Lechler, P. and Desilets, M. (1991). *The NBMG Standard Reference Material Project*. Nevada Geology, # 10:1-2.

Malm, O., Pfeiffer, W.C., Souza, C.M.M. and Reuther, R. (1990). Mercury pollution due to gold mining in the Madeira River Basin, Brazil. *Ambio*, 19:11-15.

Martinelli, L.A., Ferreira, J.R., Forsberg, B.R. and Victoria, A. (1988). Mercury contamination in the Amazon: A gold rush consequence. *Ambio*, 17:252-254.

Metson, A.J. (1961). Methods of Chemical Analysis for Soil Survey Samples. New Zealand Department of Scientific and Industrial Research, *Soil Bureau Bul.*, 12.

Miller, J.R., Lechler, P.J., Rowland, J., Desilets, M. and Hsu, L.C. (1995a). An integrated approach to the determination of the quantity, distribution and dispersal of mercury in Lahontan Reservoir, Nevada, USA. *J. Geochem. Explor.*, 52:45-55.

Miller, J.R., Lechler, P.J., Warwick, J.J. and Heim, K. (1995b). Hydrologic and sedimentologic controls on total mercury concentration within the Carson River, Nevada, USA. *Proceedings of the 10th International Conference on Heavy Metals in the Environment*, Hamburg, Germany, V.2:1-4.

Moore, J.N., Brook, E.J. and Johns, C. (1989). Grain size partitioning of metals in contaminated coarse-grained river floodplain sediment: Clark Fork River, Montana, USA. *Environ. Geol. Water Sci.*, 14:107-115.

Mosley, M.P. and Schumm, S.A. (1976). Stream junctions - a probable location for bedrock placers. *Econ. Geol.*, 72:691-697.

Nriagu, J.O. (1994). Mercury pollution from the past mining of gold and silver in the Americas. *Sci. Tot. Environ.*,149:167-181.

Pfeiffer, W.C., Malm, O., Souza, C.M.M., Lacerda, L.D., Silveira, E.G. and Bastos, W.R. (1991). Mercury in the Madeira River ecosystem, Rondônia, Brazil. *Forest Ecol. Managem.*, 38:239-245.

Reuther, R. (1994). Mercury accumulation in sediment and fish from rivers affected by alluvial gold mining in the Brazilian river basin, Amazon. *Environ. Monitor. and Assess.*, 32:239-258.

Richins, R.T. (1973). Mercury content of aquatic organisms in the Carson River drainage. M.Sc. Thesis, University of Nevada, Reno.

Richins, R.T. and Risser, A.C. (1975). *Pesticide Monit. J.*, 9:44-54.

Salomons, W. and Förstner, U. (1984). *Metals in the Hydrocycle*. Springer Verlag, New York. 256 p.

Smith, G.H. (1943). *The History of the Comstock Lode*, 1850-1920. University of Nevada Bull., Geol. Mining Series, 37.

Smith, N.D., Beukes, J.K. (1983). Bar to bank flow convergence zones: A contribution to the origin of alluvial placers. *Econ. Geol.*, 78:1342-1349.

Smith, N.D. and Minter, W.E.L. (1980). Sedimentological controls and uranium in two Witwatersrand paleoplacers. *Econ. Geol.*, 75:1-14.

Van Denburgh, A.S. (1973). *Mercury in the Carson River and Truckee River basin of Nevada*, US. Geological Survey Open-File Report.

Whitney, P.R. (1975). Relationship of manganese-iron oxides and associated heavy metals to grain size in stream sediments. *J. Geochem. Explor.*, 4:251-263.

17   Hydrogeochemical Description of Groundwaters from a Coastal Region of Rio de Janeiro State, Brazil

Emmanoel V. da Silva-Filho[1], Décio Tubbs[2] and John E.L. Maddock[1]

[1] Departamento de Geoquímica, Universidade Federal Fluminense, Niterói, RJ, Brazil

[2] Departamento de Geociências, Universidade Federal Rural do Rio de Janeiro, Brazil

Abstract

Groundwater quality and aquifer hydrogeochemistry were investigated in a suburban area of the city of Niteroi where domestic water supply is mainly from this source. Samples of water from deep wells, shallow wells and springs were analysed for parameters indicative of aquifer geology and hydrology: conductivity, alkalinity, HCO_3^-, Cl^-, Na, Fe, silica, Ca, Mg, K and Mn, and those indicative of contamination: the first five of the preceding list, plus: sulphate, ammonia, nitrate and phosphate. Evidence of contamination by domestic sewage and salinization of waters were shown to occur in a localized manner. The occurrence of reduced iron in well waters was generalized. Indices of vulnera-bility of the aquifers of various sub-areas are presented.

Introduction

Niteroi has been expanding in recent years, mainly because of its proximity to Rio de Janeiro and the availability of green spaces near the sea. However, the lack of public water supply and sewerage in the area where this expansion has been occurring has obliged the population to be dependent on groundwater for supply. Exploitation of groundwater in the suburbs of Niteroi is known to have reached considerable proportions, with an estimated more than 150 deep bore holes in operation, as well as thousands of shallow wells and abstraction from springs.

Despite this situation, no municipal or other authority controls abstraction or has even compiled any statistics about the amount of water use or potential supply capacity. Any decisions about where to bore wells and how much to abstract are left to individual property owners.

Field observations and the results of usual hydrogeochemical analyses of waters from the region are presented here and used to identify contaminated areas and to study the natural vulnerability of aquifers. These results allow us to identify impacts on groundwaters arising from the lack of environmental management, the non-adherence to technical norms and the unavailability of information to the community, all of which, of course, are aggravated by the lack of urban infrastructure.

Study Area

This study covered an area of about 100 km^2 of the coastal region of the municipality of Niteroi, located to the east of the city of Rio de Janeiro (Fig. 1). The region has a humid tropical climate, mean temperature 22°C and rainfall slightly greater than 1100 mm year^{-1}. The rainy season is from December to March and it is drier in the winter months, from June to August, but without any completely dry period (Prefeitura da Cidade de Niteroi, 1992).

Two topographic forms dominate the area: the coastal range of hills and the coastal plain. The first consists of sharply pointed bare rock hills along the shoreline, which become small upland ridges further inland. The coastal plain consists of lowlands, lagoons and sand bars. Although the overall topography is of steep hills, the ridges rise to a maximum of only 410 m.

The surface drainage network is poor, consisting of small streams which have been reduced to open sewers, which run into the lagoons.

Geological Aspects

The region is dominated by highly polymetamorphic augen gnaisses which show zones of ductile strain, consisting mainly of mega-crystals of microcline (>15 cm) and biotite. Other types of gnaisses, pegmatites and charnockites also occur, as well as basic intrusions in the form of dikes which may be more than 100 m wide. Sandy, silty, clayey and organic sediments are found only in depressions and the low lying plain.

The main aquifer which is exploited is of the unconfined, fracture type, where water circulation occurs along fractures and structural discontinuities, magnified in shear zones. Initial bore hole flow rates can be as high as 29 m^3 day^{-1} but mean values oscillate around 3 to 5 m^3 day^{-1}, depending on the location.

Large areas of weathered gnaisses occur, forming residual soils which are quite porous because of the mega grain size. However, this alteration zone is not exploited directly as an aquifer. Bore holes have been drilled with the object of obtaining water only from fractures.

Domestic consumption accounts for most of the water abstracted, principally in condominiums (59%), commerce (16%), gas stations and car washes (11%), schools (8%), with clubs, industries and hospitals accounting for less.

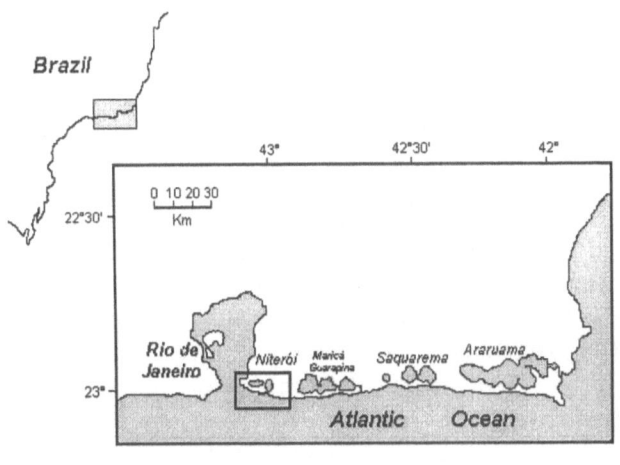

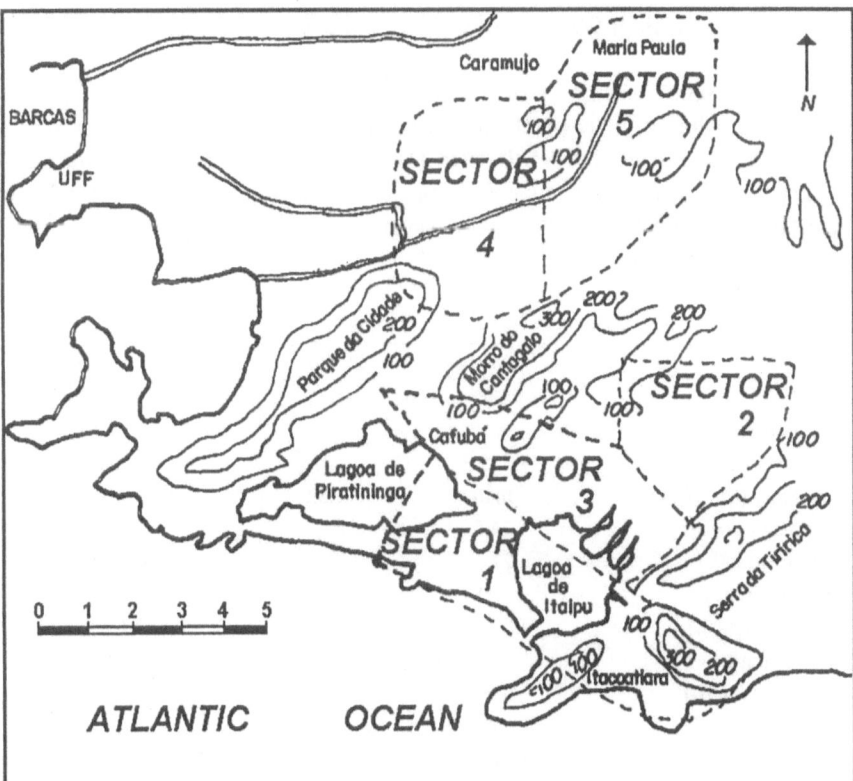

Fig. 1: Study area and location of the sampled wells

Methods

Water samples were collected from 51 points, including deep tubular wells, shallow wells and springs. Sampling techniques proposed by Hirata and Kimmelman (1986) and Wood (1973) were employed. The majority of the wells had been drilled since 1990, two between five and ten years ago and two were more than ten years old (Fig. 1). Temperature, pH, conductivity and alkalinity were determined directly after sampling. Chloride, nitrate, ammonia, phosphate and sulphate were determined by ion chromatography, within 48 hours of sampling, in subsamples which had been maintained under refrigeration at 4°C. Metals were determined by conventional flame AAS.

Results and Discussion

Hydrogeochemical Characterization of the Groundwaters

Table 1 presents a summary of the data obtained. In general the concentrations found and values of other parameters are compatible with those expected from the lithology, as described by Schoeller (1962) and White (1963). Compared to examples in Brazil from similar environments (Szikszay, 1993), the waters show a greater mineral content, increasing in deeper wells further from the coast.

Table 1: Mean values of parameters, as a function of well depth

Parameter	Springs	Shallow wells	Deep wells
pH	5.92	6.11	6.8
Depth (m)	-	13.2	76.7
Conductivity (µS/cm)	177.3	421.1	584.2
Alkalinity (Meq/L)	0.40	1.53	3.30
Bicarbonate (mg/L)	24.5	87.7	198.9
Chloride (mg/L)	27	59.9	73.5
Sulphate (mg/L)	9.15	21.1	30.2
Ammonia (mg/L)	0.20	0.17	0.67
Nitrate (mg/L)	157	255	185
Phosphate (mg/L)	0.68	0.90	0.73
Silica (mg/L)	5.45	6.02	6.58
Calcium (mg/L)	17.3	24.8	47.2
Magnesium (mg/L)	4.36	5.04	13.4
Sodium (mg/L)	43.6	56.0	71.3
Potassium (mg/l)	5.10	5.84	5.63
Iron (mg/L)	0.90	1.07	2.73
Manganese (mg/L)	0.06	0.10	0.19
Hardness	123.4	241.8	338.2

Some unusual features can be observed, such as low chloride and potassium concentrations, and surprisingly high hardness values. The elevated hardness values can be attributed to organic matter oxidation and nitrate reduction, two processes that are very common in polluted shallow wells. Mathess and Harvey (1992) explain that the interaction of the carbon dioxide with alkaline metals engenders the increase in the hardness.

The majority of the parameters presented here have values within the legally established limits for drinking water supply. However, some exceptions occur, in iron concentrations, for example.

The uniformity of composition of the country rock and its dissolution into the waters indicate that the hydrogeochemical variations observed can be attributed to the circulation of waters through basic rocks, coastal sediments and the entrance of water from surface fractures. Thus, high concentrations occur in cases of slow/deep circulation or contact with basic rocks. On the other hand, waters with a lower concentration of dissolved substances are associated with the sedimentary environment or wells which capture water from surface fractures. As observed by Tubbs (1994), these latter present characteristics very similar to those of shallow wells (depths <20 m) and springs.

Piper-Hill diagrams (Piper, 1944)(Fig. 2) show some dispersion in parameter values, mainly in wells from the coastal plain (quaternary, Fig. 2a), caused by a more complex environment. However, in general, a mixture of calcium bicarbonate and sodium chloride bearing waters can be identified, typical of circulation through crystalline rocks, altered crystalline material, basic rocks and sediments (Fig. 2b).

A few of the waters were slightly sulphate bearing. This could have had two origins, by oxidation of dispersed sulphides in the basic rocks, associated with passage through organic sediments.

Ionic concentration ratios indicated a bicarbonate facies extending throughout the area, including the coastal plain, where the presence of bicarbonates was slightly more pronounced than in the waters which had percolated through basic rocks. Sodium was prevalent among the cations, followed by calcium and magnesium. Theoretically, hydrogeochemical facies predominant in the region could be distinguished by lithologies: $HCO_3^- > Cl^- > SO_4^{-2}$, $Na > Ca > Mg > K$, the most commonly occurring type, corresponding to facoidal gnaisses and other granitic lithologies, $HCO_3^- >> Cl^-$, SO_4^{-2}, $Na \geq Ca \geq Mg > K$, corresponding to waters which had been in contact with basic rocks, and $Cl >> SO_4^{-2} \geq HCO_3^-$, $Na >> Mg > Ca$, indicating salt water input in waters of the coastal plain. In general, these classes were observed. Additionally, waters were sampled which exhibited high SO_4^{-2}/Cl^- ratios. These can be explained by oxidation of disseminated sulphide or of organic material in sediments.

High Cl^-/HCO_3^- ratios were found only in waters from wells in the coastal/lagunar region, where pumping rates were incompatible with the aquifer, inducing the incursion of salt water. Later measurements, made by Mariani et al.

(1995), showed even higher chloride concentrations in waters from the same wells, confirming the progressive process of salinization.

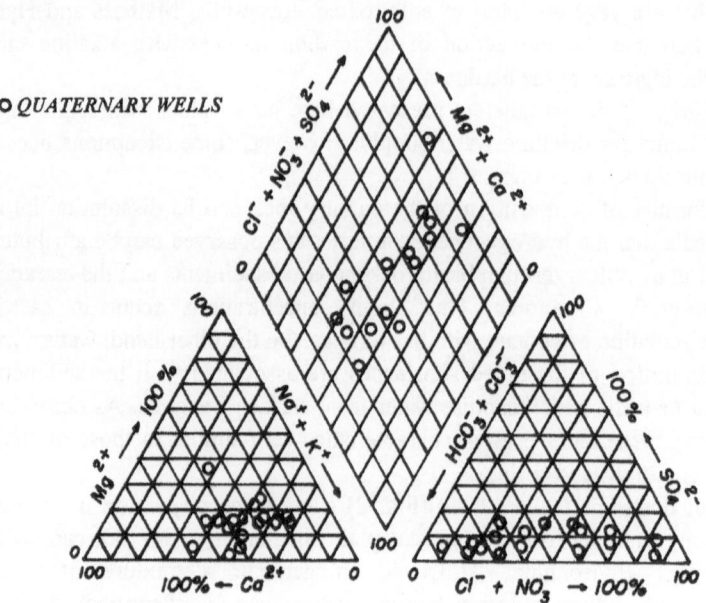

Fig. 2 (a): Piper-Hill diagram of wells in Quaternary grounds

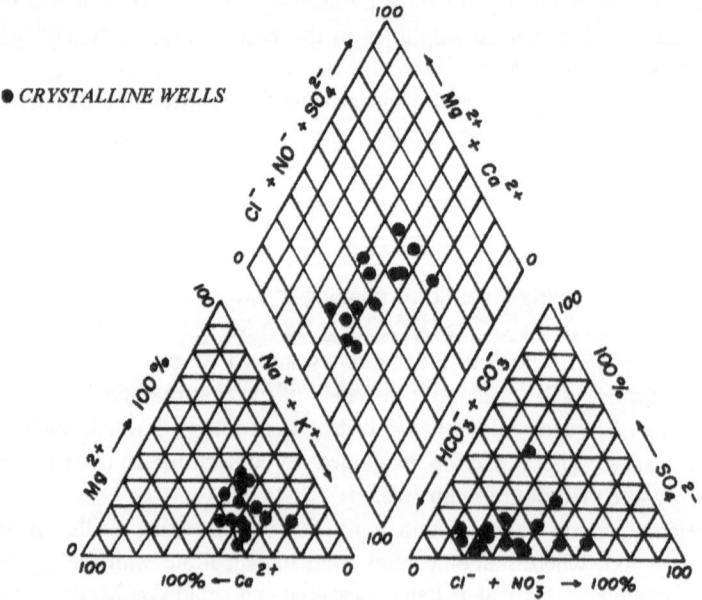

Fig. 2 (b): Piper-Hill diagram of wells in crystalline grounds

Measures of Water Quality

As well as infiltration of salt water, the possible contamination of groundwater by waste waters was investigated, using nutrient concentrations as indicators and the presence of high iron concentrations as an indicator of anaerobic conditions in the aquifer.

Nitrate

The concentrations of nitrate in the waters of shallow wells were acceptable by current legislation (Resolução CONAMA n° 20). However various wells exhibited concentrations greater than is normal for the types of lithology found in the area. 25% of samples contained more than 20 mg L^{-1} and 10% had concentrations in the range 40 to 45 mg L^{-1}. One deep well in the coastal plain showed a maximum of 247 mg L^{-1}. This well is located in a densely populated area where waste water disposal obeys no norms or technical criteria. The other high nitrate values are restricted to shallow wells which, in general, are constructed without due care and thus collect surface runoff, resulting in some nitrate in the waters.

It is likely that continued urban expansion in the area will increase the pollution load in the area and that nitrate levels will increase in future.

Ammonia

The presence of ammonia was first reported in 1988 when 27 samples from shallow wells were analysed, on the initiative of residents. Ammonia was found in 17 of these, as well as coliform bacteria (Prefeitura de Niteroi, 1996). In the present study, concentrations of up to 5 mg L^{-1} were found, mainly in shallow wells, yet again demonstrating the lack of basic sanitation. Five deep wells also showed ammonia contamination, with concentrations in the range 1.5 to 3 mg L^{-1}. The occurence of ammonia in waters at depth may be an indication of recent pollution and rapid deterioration of the aquifer. Three of the contaminated deep wells were situated next to a small stream contaminated with domestic sewage and the waters from the other two exhibited high alkalinity and conductivity, which makes it likely that they were contaminated with sewage which had percolated rapidly through rock fractures.

Phosphate

Mathess and Harvey (1982) stated that, due to absorbtion by organisms, phosphate concentrations in natural waters should be lower than 0.5 mg L^{-1}. Values above 1 mg L^{-1} indicate polution. Anthropogenic phosphorus arises from detergents, domestic effluents and insecticides.

Phosphate was detectable in nearly all samples and had a mean concentration of nearly 1 mg L^{-1}. However, some of this may be of natural origin, from the basic rocks. Contamination by domestic effluents was evident though, in some of the deep wells and most of the shallow wells, where concentrations up to 3.7 mg L^{-1} were encountered. Detection of anionic surfactants in waters of the region (Souza and Wasserman, 1995) is further evidence of this.

Iron

The evaluation of iron in groundwaters destined for human consumption is only undertaken to determine its *organoleptic properties*. Recent studies have indicated that *haemochromatosis* or iron intoxication, which is associated with genetic disturbances, is aggravated by iron ingestion, and that this disturbance is much commoner than had previously been supposed (Monmaney, 1992). Cavalcante and Vasconcelos (1990) suggested that values above 3 mg L^{-1} may be toxic.

The maximum limit established by CONAMA (Resolution no. 20) is 0.3 mg L^{-1}.

In the study area nearly 40% of samples contained more than the recommended concentration of iron, commonly twenty or thirty times more. Although many users have installed equipment for iron removal, these were observed not to work in many cases, possibly because of iron complexation by organic matter or because of the presence of hydrogen sulphide, especially in wells in the coastal plain sedimentary environment. The CONAMA limit for manganese is 0.1 mg L^{-1}. Thirty percent of samples had higher concentrations of this metal, as high as ten times greater.

Iron Bacteria

It is not surprising that iron bacteria occur in the waters of the region but their great prevalence in wells was not expected. However, it is known that they can occur in low iron waters (0.1 - 0.3 mg L^{-1}) as well as high (10 -30 mg L^{-1}). As well as precipitating great quantities of ferric hydroxide, various iron bacteria can excrete gelatinous extracelular polysaccharides and are considered agents for pipe encrustation (Hackett and Lehr, 1991). These authors suggested possible causes of occurrence of iron bacteria in groundwaters as direct or indirect contamination by surface water, bad quality of material used in well linings, lowering of the average water table and over-extraction. Most of these causes imply poor practices in well construction and water use, rather than problems with the aquifer water quality. However, in the study region, salinization of some deep wells suggests over-extraction and this may also be causing the appearance of iron bacteria. Thus the appearance of iron bacteria, as well as indicating poor well construction, can be taken here as a sign of general environmental degradation of the area, including of the aquifers.

Salinization of Deep Wells

Strong indication of salinization was observed in the coastal plain between the lagoon and the sea. Cl^- / HCO_3^- ratios ten times higher than those considered normal for fresh waters (0.5) were found. Marianni et al. (1995), during a study of the variation of water quality during three years, in this area, found values up to forty times this norm. A summary of these results, for four deep wells, is reproduced here, in Fig. 3. Well 1 (48 m) was bored in a basic dyke (olivine/diabase) and well 2 (60 m) in facoidal gnaiss, close to a basic dyke. Wells 3 (90 m) and 4 (155 m) were sunk in the coastal plain, between the lagoon and the sea, in medium sand sediments, interspersed with clay layers and, at greater depths, consisting of coarse sand with layers of gravel. The total thickness of this layer was between 8 and 18 m.

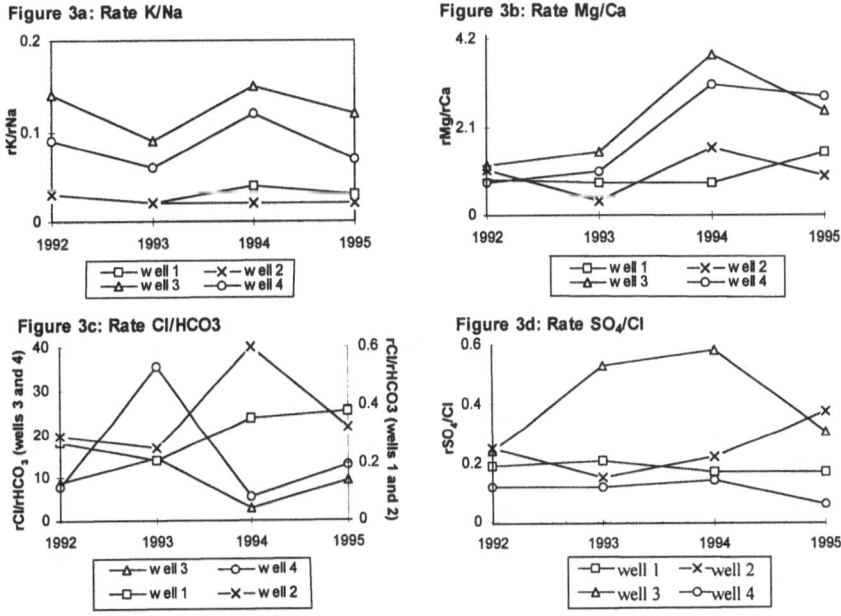

Fig. 3: Variations in element ratios characteristic of groundwaters in the coastal region of Niteroi, RJ

The ratio K/Na (Fig. 3a) is most characteristic of environmental variations within the area studied, which is typically dominated by granitic rocks (Mathess and Harvey, 1982). Wells 1 and 2 showed less fluctuation (0.02 to 0.04) while this ratio in wells 3 and 4 varied from 0.06 to 0.15. All four wells are subject to the same rainfall and higher values, and the greater variations in wells 3 and 4 than in wells 1 and 2, whose behaviour was very similar (Fig. 3a), has been attributed to different groundwater circulation, possibly due to the presence of clay layers in the area of 3 and 4 (Moreira-Nordemann, 1990).

Mg/Ca ratios (Fig. 3b) are higher than the values, 0.25 - 0.33, which Mathess and Harvey (1982) quote as typical for groundwaters from granitic rocks. The values, generally greater than 0.9, are typically associated, according to these authors, with basic rocks. On the other hand, in wells 3 and 4 these high ratios may be due to the influence of sea salts.

Cl^-/HCO_3^- ratios (Fig. 3c) for wells 1 and 2, with values around 0.5, are typical of aquifers with no saline contamination (Alvarado, 1990). This ratio for wells 3 and 4 is greater than 10, well above the value, 6.6, considered by Alvarado to be indicative of saline contamination. The wide variations which were observed on different sampling occasions at each of these latter two sites, in the Cl^-/HCO_3^- ratio, can be attributed to variations in saline intrusion caused by seasonal variation in both rainfall and domestic water consumption. Higher than normal consumption and drought combine to increase salinization.

SO_4^{-2}/Cl^- ratios (Fig. 3d) did not follow a clear pattern. This was probably due to variable oxi-reduction state of the waters, both producing sulphate by oxidation of mineral sulphides in the fracture aquifers and removing it as H_2S in reducing regions of the sedimentary aquifer. This behaviour has been discussed by Custódio and Lhamas (1976). Water samples from some of the wells in both environments emitted characteristic H_2S odours.

Aquifer Vulnerability

The vulnerability of an aquifer to pollution is a measure of its tendency to contamination by pollution at the surface. Albinet and Margat (1970) stated that aquifer vulnerability could be presented on thematic maps, allowing very simple visualization. Parisot and Rebouças (1982) showed that this vulnerability could be expressed in terms of hydrogeological and human factors. The vulnerability of aquifers in the study region was evaluated for different areas, shown in Fig. 1 (sectors), which were defined by geological, geomorphological and urban

Table 2: Vulnerability of the areas studied

Sector	Aquifer Condition	Lithology	Depth of Water Table	Vulnerability Index
1	0.6 - 1.0	0.6 - 0.8	0.8 -0.9	0.28 - 0.72
				low / high
2	0.6	0.6 - 0.8	0.8 - 0.9	0.28 - 0.57
				low / moderate
3	0.6	0.6 - 0.8	0.8 -0.9	0.28 - 0.57
				low / moderate
4	1.0	0.8	0.6 - 0.8	0.48 - 0.64
				moderate / high
5	1.0	0.8	0.6 - 0.8	0.48 - 0.64
				moderate / high

occupation factors. This evaluation was undertaken by the method proposed by Foster and Hirata (1988), which uses simple indices, derived from field data: aquifer condition, lithology and depth of the water table. Table 2 shows the results.

Conclusion

The only viable water supply at present in the region studied is groundwater. This study shows how the use of this resource, with no study or control by the authorities and no technical information available to users, is leading to degradation of the quality and quantity of water available. However, most of the contamination observed was localized or restricted to individual wells, rather than generalized for a whole aquifer.

References

Albinet, M. and Margat, J. (1970). Cartographie de la vulnerabilité à l'évolution des nappes d'eau souterraines. *Bulletim du B.R.G.M.*, Paris, n° 4, p.13-22.

Alvarado, J.R. (1990). Salinização de aquíferos costeros de Venezuela por intrusão de água do mar. *In: VI Congresso Brasileiro de águas subterrâneas.* Porto Alegre, RS, Brazil. p.277-292.

Cavalcante, I.N. and Vasconcelos, S.M.S. (1990) Qualidade das águas subterrâneas de Fortaleza - CE. *Revista de Geologia*, vol.3, pp.:89-97.

Custódio, E. and Lhamas, R. (1976) Hidrologia Subterrânea. Omega S/A. Barcelona, Vol.II. 357 p.

Foster, S. and Hirata, R. (1988). Determinações de riscos de contaminação da águas subterrâneas. *Bol. Inst. Geológico*, 10, São Paulo, Brasil. 92p.

Hackett, G. and Lehr, J.H. (1991). Ferro bactérias em poços artesianos: ocorrências, problemas e métodos de controle. *Cadernos Técnicos da ABAS* (Associação Brasileira de Águas Subterrâneas), 1. 65 p.

Hirata, R.C.A. and Kimmelmann, A.A. (1986). Águas Subterrâneas. In: Manual de Coleta de Amostras em Geociências. São Paulo, n° 2:5-10.

Mariani, R. L., Tubbs, D. and Silva-Filho, E.V.(1995) Variações temporais em águas subterrâneas da Região Litorânea de Niterói, RJ. *In: Anais do V Congresso Brasileiro de e III Congresso de Geoquímica dos Países de Língua Portuguesa*, Niterói, RJ, Brazil (CD-ROM).

Matthess, G. and Harvey, J. C. (1982). *The Properties of Groundwater*. John Wiley and Sons Ltd. London. 466 p.

Monmaney, T. (1992) Atração pelo ferro. *Rev. Nova Ciência*, ano 3(4):32-37.

Moreira-Nordemann, L. M. and Rebouças, A.C. (1986). Química e geologia de águas subterrâneas no SE do Brasil. *In: IV Congresso Brasileiro de Águas Subterrâneas*. Brasília. pp: 221-235.

Parisot, E.H. e Rebouças, A.C. (1982). Qualidade das águas subterrâneas da região centro-oeste do Município de São Paulo. *In: Anais do II Congresso Brasileiro de Águas Subterrâneas*, Salvador, BA, Brazil. pp:84-89

Piper, F. (1944). A new method for the study of groundwater chemistry. *Trans. Am. Geophys. Union*, 25:914-928.

Prefeitura da Cidade de Niterói (1992) Diagnóstico Ambiental de Niterói. Niterói, RJ, Brazil. 245 p.

Prefeitura da Cidade de Niterói (1996). Diagnóstico Ambiental de Niterói. Niterói, RJ, Brazil, 76 p.

Schoeller, M. (1963). *Les Eaux Souterraines*. Masson, Paris, France. 642 p.

Souza, N.M. and Wasserman, J.C. (1995). Diurnal variation of anionic surfactants and forms of phosphorus in a polluted stream (Rio de Janeiro, Brazil). *Toxicol. Environm. Chem.*, 55:173-181.

Szikszay, M. (1993) Geoquímica das águas. *Boletim IG. USP*, Série Didática n° 5, Universidade de São Paulo, 34 p.

Tubbs, D. (1994) *Caracterização Hidrogeoquímica da Águas Subterrâneas da Região Litorânea do Município de Niterói, Estado do Rio de Janeiro*. M.Sc. thesis, Departamento. de Geoquímica, Universidade Federal Fluminense, Niterói, RJ, Brazil. 163 p.

White, D. E. (1963). Chemical Composition of Subsurface Waters. *Geological Survey Professional Papers*, 440f. US Goverment Printing Office, Washington. 63 p.

Wood, W.W. (1973). Guidelines for collection and field analysis of groundwater samples for selected unstable constituents. *In: Technique of Water Resources Investigations of United States Geological Survey*. Washington, Book 1:1-24.

18 Long Term Atmospheric Aerosol Characterization in the Amazon Basin

Paulo Artaxo[1], Fábio Gerab[2] and Marcia A. Yamasoe[1]
[1]Instituto de Física, Universidade de São Paulo, Brazil
[2]Instituto de Pesquisas Energéticas e Nucleares, São Paulo, Brazil

Abstract

This chapter presents a characterization of atmospheric aerosols collected in different places in the Amazon Basin. Both the biogenic aerosol emission from the forest and the particulate material which is emitted to the atmosphere due to the large scale man-made burns during the dry season were studied. The samples were collected during a three year period at three different locations in the Amazon (Cuiabá, Alta Floresta and Serra do Navio), using stacked filter units. Aerosol samples were also collected directly over fires of cerrado vegetation and tropical primary forest burns The samples were analyzed using several techniques for a number of elements. Gravimetric analyses were used to determine the total atmospheric aerosol concentration. Multivariate statistical analysis was used in order to identify and characterize the sources of the atmospheric aerosol present in the sampled regions. Cerrado burning emissions were enriched compared to forest ones, specially for Cl, K and Zn. High atmospheric aerosol concentrations were observed in large amazonian areas due to emissions from man-made burns in the period from June to September. The emissions from burns dominate the fine fraction of the atmospheric aerosol with characteristic high contents of black carbon, S and K. Aerosols emitted in biomass burning process are correlated to the increase in the aerosol optical thickness of the atmosphere during the Amazonian dry season. The Serra do Navio aerosol is characterized by biogenic emissions with strong marine influence. The presence of trace elements characteristic of soil particulate associated with this marine contribution indicates the existence of aerosol transport from Africa to South America. Similar composition characteristics were observed in the biogenic emission aerosols from Serra do Navio and Alta Floresta.

Introduction

Tropical rainforest vegetation formations are characterized by intense release of biogenic gases and aerosols. The Amazon Basin has the world's largest rainforest, and is a region with intense convective activity (Gartang et al., 1988), resulting in a rapid vertical mixing of biogenic gases and aerosols to high altitudes where they can be transported over long distances and have an impact on the global atmospheric chemistry. The tropical rainforests of the world are in a delicate nutrient-limited environment (Salati and Vose, 1984; Vitousek and Sanford, 1986). Several airborne compounds, like phosphorus-bearing ones, are critical in these ecosystems and could limit their annual net primary production (Swap et al., 1992). Therefore, it is necessary to increase our knowledge of the chemical processes that determine the background composition of the atmosphere in these areas, and to understand the biosphere-atmosphere interactions. It is also important to obtain a better understanding of the effects on the atmospheric composition due to changes in land use in tropical forests.

Biogenic Aerosol Particle Emissions in the Amazon Region

Vegetation plays a fundamental role in the control and composition of the atmosphere composition in tropical forest regions. The vegetation is an important source of sulfur gases, such as COS, DMS (dimethyl sulfide), DMDS (dimethyl disulfide), SO_2, H_2S and CS_2. The tropical forest is also regarded as the major source of organic particles to the atmosphere (Crozat et al., 1978; Cachier et al., 1985). The biogenic aerosol emitted is composed of different kinds of particles, and an important fraction of them originate from gas to particle conversion processes.

The Amazon rain forest, at around 4 million km^2, is the Earth's greatest tropical forest, and it is responsible for a large amount of the world's biogenic aerosol production. Intense meteorological convective activities in that region (Greco et al., 1990) contribute to particles and gases emitted reaching high altitude levels and being transported great distances with an impact on the global atmospheric chemistry (Artaxo et al., 1994). The primary biogenic particulate matter is composed basically of pollen, bacteria, fungi, and other microorganisms, insect excrement and leaf fragments (Simoneit, 1989). The major part of the sub-micrometer biogenic particulate matter is produced in gas to particle conversion processes, generating aerosol enriched in organic compounds, sulfur and nitrogen.

Biogenic and biological aerosols have a role as cloud condensation nuclei (Dingle, 1966; Vali et al., 1976; Maki and Willoughby, 1978; Schnell, 1982). S, P, K, Ca and other elements are emitted during the respiration and guttation processes. Heavy metals are also emitted in eolian leaf abrasion. In a recent work Artaxo and Hansson (1995) studied the Amazon Basin biogenic aerosol size distribution, using cascade impactor samplers in three height levels from the ground level to above the Amazon forest canopy. Elements like S, Zn, and Sr are present in the fine fraction aerosol (average aerodynamic diameter ~ 0.5μm), that reaches its maximum concentration over the forest canopy. The coarse fraction biogenic aerosol (average aerodynamic diameter ~ 3μm) is characterized by the presence of P, K, Cl and Zr. This particle-size fraction has a maximum concentration at ground level. Artaxo and Hansson also detected the peak concentration of soil dust aerosol at the upper levels of the canopy. The mineral dust contribution is almost negligible at ground level, indicating long range transportation of the mineral dust. Several works refer to Amazon mineral particle importation. These particles are supposed to be Saharan dust transported over the Atlantic Ocean (Parkin et al., 1972; Savoie and Prospero, 1977; Prospero et al., 1987; Talbot, 1990). This transport usually occurs at altitudes lower than 6,000 m. Swap and colleagues (1996) estimated the minimun total annual Amazon particulate matter importation by this pathway to be 13×10^9 kg. For P, they estimate a deposition rate of 1 - 4 kg ha^{-1} y^{-1}. This can be a significant amount for the Amazon nutrient biogeochemical cycles.

Biomass Burning Aerosol Emissions in the Amazon Region

The rapid deforestation now occurring in tropical regions has the potential to change the composition of the atmosphere with a regional climatic impact that could affect a large portion of the equatorial region. Biomass burning is one of the most important anthropogenic sources of particulate matter and gaseous emission into the atmosphere (Crutzen and Andreae, 1990). More than 80% of biomass burning emissions originate in the tropics. The high rate of tropical biomass burning emissions during the last decade is mainly a result of burning of savannah in Africa and deforestation in the Amazon Basin. (Setzer and Pereira, 1991). Estimates of total biomass consumed on a global basis range from 2 to 10×10^{15} g year^{-1} (Crutzen and Andreae, 1990) and in terms of total particulate matter emissions are around 104×10^{15} g year^{-1}. Biomass burning emits a significant amount of fine fraction aerosol particles (Artaxo et al., 1994). The composition and size distribution of aerosol particles were measured in forest fires in Rondonia, southwest of the Brazilian Amazon, with large emissions of K, S, Si, Zn and organic matter (Artaxo et al., 1993). Black carbon associated with potassium was identified as a tracer of biomass burning plumes in remote oceanic areas (Andreae, 1983).

Experimental Methods

Sampling Location

Three aerosol sampling stations have been continously operated since the end of 1991 in different vegetation formations of the Amazon Basin. Figure 1 shows a map of South America with the locations of the three sampling stations. The first station is situated in Cuiabá, at the "cerrado" (the Brazilian Savannah). The site is heavily affected by regional cerrado biomass burning. The second monitoring station is installed at Serra do Navio, in the northern part of the Amazon Basin. This station is located 190 km north of the equator, in a primary tropical rain-forest. The Serra do Navio sampling site is relatively free from anthropogenic aerosol emissions (there are no industrial activities for at least 1000 km around) and it has no severe contributions from regional biomass burning emissions. The third sampling site is in the region of Alta Floresta, close to the border of the Mato Grosso, Pará and Amazonas states. This site is in the primary tropical forest, with land cleaning activities some kilometers around. This affects the aerosol concentration during the dry season, starting in June and finishing in the beginning of October. Aerosol measurements directly on the fires in the Amazon Basin were also done. In these measurements the particulate matter emitted directly from the combustion of cerrado and primary tropical forest burn were investigated. All these monitoring stations are in continuos operation. In this chapter the results from the years of continuous sampling will be presented and discussed.

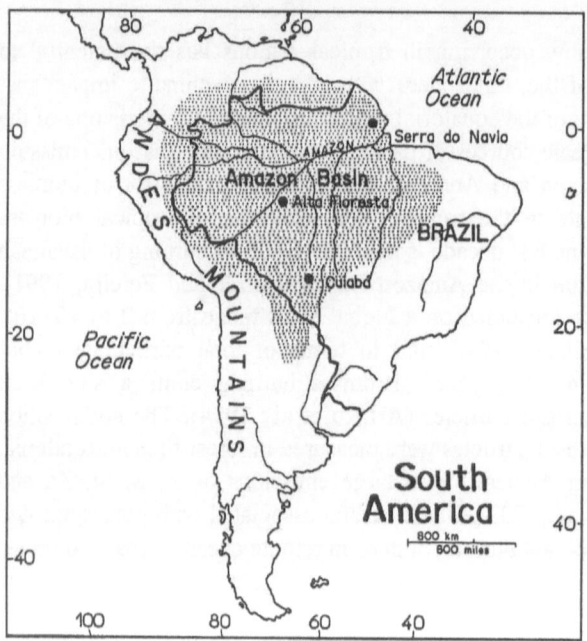

Fig. 1: Map of South America with the location of three monitoring stations in the Amazon Basin: Cuiabá, Alta Floresta and Serra do Navio

Sampling Procedure

Aerosols were sampled using stacked filter units (SFU). Coarse particles ($2<d_p<10$ µm) are separated on a 47-mm-diameter, 8-µm-pore-size Nuclepore filter, while a 0.4-µm-pore-size Nuclepore filter separates the fine particles ($d_p<2$ µm). The flow rate is typically 15 liters per minute resulting in a 50% cutoff diameter between fine and coarse aerosol fractions on about 2.0 µm. The SFU are fitted with a specially designed inlet which provided a 50% cutoff diameter of 10 µm so that only inhalable particles are sampled. SFU are loaded with

Nuclepore filters in a clean room at the University of São Paulo, transported in a sealed container, and hand-carried after the sampling.

Gravimetric Analysis

The fine and coarse fractions of the aerosol mass concentrations were obtained through gravimetric analysis of the Nuclepore filters. The filters were weighed before and after sampling in a Mettler M3 electronic microbalance with 1 μg sensitivity. Before weighting, filters were equilibrated for 24 hours at 50% relative humidity and 20°C temperature. Electrostatic charges were controlled by means of a ^{210}Po radioactive source. The detection limit for aerosol mass concentration is 0.3 μg m^{-3}. Precision was estimated to be about 15%.

Black Carbon Quantification

Black carbon concentrations were determined using an adaptation of the reflectance technique (Andreae et al., 1984). The black carbon concentration is proportional to the absorption of light originating from a source illuminating the sample. The equipment is calibrated by means of filters loaded with known amounts of black carbon originating from the combustion of acetylene. This method is not specific, and "natural" black carbon is certainly different from the soot carbon obtained by the combustion of acetylene. In this chapter, for simplicity, the value of the above measurement was termed "black carbon". The precision of this method is around 20%, and the detection limit for black carbon for the used sample volumes and filter areas is about 50 ng m^{-3}.

PIXE and PIGE analysis

Particle-induced X-ray emission (PIXE) (Johansson and Campbell, 1988) was used to measure concentrations of up to 28 elements (Mg, Al, Si, P, S, Cl, K, Ca, Sc, Ti, V, Cr, Mn, Fe, Co, Ni, Cu, Zn, Ga, Ge, Ar, Se, Br, Rb, Sr, Au, Hg, Pb). The Na concentration was measured using particle-induced gamma-ray emission (PIGE) (Asking et al., 1987). The samples were irradiated by a 2.4 MeV proton beam, supplied by a dedicated 5SDH Tandem Pelletron accelerator facility, at the LAMFI (Laboratório de Análise de Materiais por Feixes Iônicos) of the University of São Paulo, Brazil. Detection limits are typically 3 ng m^{-3} for elements in the range $12 < Z < 23$ and 0.2 ng m^{-3} for elements with $Z > 22$. These detection limits were calculated based on a sampling flow rate of 15 L per minute, sampling time of 72 hours and irradiation time of 600 s. The precision of the elemental concentration measurements is typically better than 10%. The PIGE system was used to quantify Na in the Serra do Navio samples. The PIGE detection limit for Na quantification was 25 ng m^{-3}, when the 440 keV gamma line (from the $^{3}Na_{(p,p\gamma')}$,^{23}Na reaction) is measured.

Ion Chromatography Analysis

Ion chromatography was used to determine the water-soluble concentrations of acetate, acetic acid, formate, formic acid, methyl sulfonic acid, H^+, Na^+, NH_4^{2+}, K^+, Mg^{2+}, Ca^{2+}, Cl^-, NO_3^-, SO_4^{2-}, $C_2O_4^{2-}$, PO_4^{3-}, in the Serra do Navio aerosol fine fraction. Water-soluble ionic components are only related to the fine fraction of the aerosol comprising most of the aerosol emissions and includes particles with transport and optical properties of interest. The ion chromatography measurements were performed by Dr. G. Ayers, at CSIRO, Australia, and Drs. A.H. Miguel and A.G.Allen, at the Chemistry Institute of the University of São Paulo, Brazil.

Absolute Principal Factor Analysis

To separate the different components present in the samples using elemental composition, absolute principal factor analysis (APFA) was used (Thurston and Spengler, 1985; Hopke, 1985). Using APFA it is possible to find the quantitative elemental profile instead of only a qualitative factor loading matrix as in traditional applications of factor analysis. The absolute elemental source profiles enable the identification of the factors and can be used to compare the factor composition with the assumed aerosol sources.

In principal factor analysis a model of the variability of the trace element concentrations is constructed so that the set of intercorrelated variables is transformed into a set of independent, uncorrelated ones. The APFA procedure obtains the elemental mass contribution of each identified component by calculating the absolute principal factor scores (APFS) for each sample (Artaxo et al., 1988; 1990). The elemental concentrations are subsequently regressed on the APFS to obtain the contribution of each element for each component. The source profiles thus obtained can be compared with values from the literature to gain information on enrichment and atmospheric chemistry processes (Hopke, 1985). The measured aerosol mass concentration can also be regressed on the APFS to get the aerosol total mass source apportionment.

Results and Discussion

In two different vegetation formations - cerrado and primary forest - aerosol samples were collected directly over fires. For the cerrado fires, sampling was performed in Brasília ecological reserve of IBGE (Instituto Brasileiro de Geografia e Estatística). These fires were planned and prescribed. Primary tropical forest direct emissions were sampled in Rondônia State, near Ariquemes village. stacked filter units were used in the sampling procedure and PIXE and ion chromatography were used in the aerosol characterization. Table 1 shows the fine fraction average concentration for the measured species for each ecosystem and fire type. In order to be able to compare concentrations with large spatial variability, the elemental and the ionic concentrations were normalized to the fine fraction mass concentration. The comparison of our measurements with similar data collected by other authors is shown in Table 2. The emission factors for Cl, SO_4^{2-} and Ca were normalized to K emissions. It is possible to observe that emissions of sulfates, reported by several authors, vary by a factor 8. Crutzen et al. (1979) and Bingemer et al. (1992) have studied the sulfur cycle over a tropical forest ecosystem and found large emissions of sulfur compounds in both natural release and biomass burning emissions. Table 3 shows the factor analysis calculations for the elements and ionic species measured over the fires in the cerrado and tropical forest. Potassium and chlorine are always closely associated with zinc. Magnesium and calcium are closely related to both cerrado and forest fire emissions. The organic acids (formic and acetic) are not directly associated with any other element or ion. Some factors are similar for cerrado and forest, like fine particulate matter (FPM) and acetate, indicating high organic influence in the FPM concentration.

Table 1: Ionic and elemental composition of aerosol direct emissions from biomass burning in the Amazon basin. Samples collected in savannah and tropical forest for different phases of fire combustion. Values expressed as ratios to fine particle mass in percentages

	Savannah (% of fine mass)		Tropical forest (% of fine mass)	
	Flaming phase	Smoldering phase	Flaming phase	Smoldering phase
Ac⁻	0.28±0.11 (56)	0.25±0.11 (32)	0.23±0.10 (21)	0.27±0.08 (39)
Fo⁻	0.024±0.012 (55)	0.023±0.013 (32)	0.019±0.008 (13)	0.025±0.014 (29)
NO_3^-	0.61±0.36 (54)	0.36±0.32 (31)	0.11±0.08 (15)	0.12±0.07 (25)
SO_4^{2-}	0.74±0.36 (56)	0.35±0.33 (32)	0.87±0.64 (21)	0.36±0.31 (39)
$C_2O_4^{2-}$	0.08±0.05 (56)	0.06±0.05 (30)	0.06±0.06 (19)	0.04±0.04 (34)
Na^+	0.022±0.016 (39)	0.024±0.020 (20)	0.016±0.011 (17)	0.020±0.037 (25)
NH_4^+	0.10±0.08 (28)	0.047±0.037 (11)	0.09±0.06 (14)	0.046±0.039 (23)
K^+	2.9 ± 2.0 (56)	1.3 ± 1.7 (32)	0.7 ± 0.6 (21)	0.43±0.25 (39)
Mg^{2+}	0.035±0.024 (36)	0.032±0.019 (20)	0.024±0.021 (18)	0.026±0.028 (20)
Ca^{2+}	0.10±0.09 (42)	0.09±0.09 (24)	0.08±0.03 (20)	0.06±0.04 (37)
Al	2.3 ± 6.0 (14)	0.7 ± 1.2 (11)	0.4 ± 0.4 (8)	0.5 ± 0.5 (18)
Si	4.2 ± 1.6 (17)	2.3 ± 1.3 (10)	-	-
P	0.049±0.019 (9)	0.043±0.012 (16)	0.048±0.011 (11)	0.033±0.010 (20)
Cl	1.5 ± 1.4 (53)	0.7 ± 1.0 (27)	0.12±0.14 (8)	0.08±0.05 (16)
Fe	0.15±0.26 (13)	0.09±0.15 (12)	0.031 (1)	0.048±0.027 (8)
Cu	0.006±0.004 (54)	0.004±0.002 (26)	0.004±0.001 (20)	0.003±0.002 (36)
Zn	0.019±0.015 (54)	0.012±0.015 (12)	0.006±0.004 (20)	0.004±0.002 (25)
Br	0.06±0.04 (35)	0.04±0.02 (8)	0.05±0.02 (8)	0.05±0.06 (9)
Soot	13 ± 7 (56)	8 ± 7 (32)	8 ± 7 (21)	4.4 ± 2.6 (39)

Mean, Standard Deviation and number of samples in which the element or ion was measured above analytical detection limits. Means were calculated using only values above the detection limit.

Table 2: Ratios for K of the average values for emissions of Cl, SO_4^{-2} and Ca observed in direct emissions of biomass burning plumes. Samples from savannah and tropical forest. Also values observed by other authors are shown

Type of Vegetation	Cl	SO_4^{-2}	Ca
Savannah (Brasília) (flaming)[a]	0.52	0.26	0.03
Savannah (Brasília) (smoldering)[a]	0.54	0.27	0.07
Tropical forest (Rondônia) (flaming)[a]	0.17	1.24	0.11
Tropical forest (Rondônia) (smoldering)[a]	0.19	0.84	0.14
Savannah (Brasília) (flaming)[b]	0.33	0.1	0.1
Tropical forest (Rondônia) (flaming)[b]	0.48	0.54	0.16
Tropical forest (Rondônia) (smoldering)[b]	0.25	0.38	0.21
Savannah[c]	0.17	0.95	0.05
Savannah[d]	0.59	0.18	0.18
Savannah[e]	0.17	0.17	0.018

Notes: a) This work; b) Ward et al., 1992 (BASE-B, Brasília, Marabá); c) Artaxo et al., 1993 (Brushfire/80, Brasília); d) Lacaux et al, 1993 (Decafe, Ivory Coast, Africa); e) This work (Brasília/90)

Table 3a: Principal factor analysis results for biomass burning direct emissions in the Amazon basin savannah, Brasília

Variable	Factor 1	Factor 2	Factor 3	Factor 4	Factor 5	Factor 6	Factor 7	Communality
K	**0.97**	-0.03	0.006	0.08	-0.005	0.08	-0.02	0.96
Cl	**0.93**	-0.09	-0.03	-0.009	-0.05	0.15	-0.012	0.90
Zn	**0.55**	-0.30	0.26	-0.25	-0.01	0.23	0.42	0.75
Cu	0.16	0.07	-0.04	0.23	0.18	-0.04	**0.87**	0.87
Mg	0.006	0.05	0.16	0.11	**0.82**	-0.08	0.13	0.73
Br	0.18	0.04	0.24	0.16	-0.16	**0.71**	0.08	0.65
Ca	-0.07	0.06	-0.20	-0.15	**0.79**	-0.12	0.02	0.71
Soot	0.04	-0.04	-0.11	**0.85**	-0.06	0.09	0.29	0.84
FPM	**-0.50**	**0.74**	0.09	-0.25	-0.002	-0.018	0.25	0.93
Ac	-0.007	**0.94**	0.10	0.06	0.10	-0.05	-0.06	0.92
Fo	-0.10	0.12	**0.86**	-0.08	0.006	-0.06	-0.003	0.78
NO_3^-	**0.53**	-0.21	0.44	-0.003	-0.21	0.16	0.24	0.65
SO_4^-	**0.78**	-0.007	-0.005	0.18	-0.06	0.06	0.24	0.71
$C_2O_4^{2-}$	**0.62**	**0.55**	-0.08	0.35	0.11	-0.25	-0.06	0.90
Na^+	-0.24	-0.11	**-0.55**	**-0.58**	-0.04	-0.19	0.19	0.78
NH_4^+	0.06	-0.14	-0.17	0.006	-0.06	**0.83**	-0.07	0.75

Table 3b: Principal factor analysis results for biomass burning direct emissions in tropical rain forest in Rondônia

Variable	Factor 1	Factor 2	Factor 3	Factor 4	Factor 5	Factor 6	Factor 7	Communality
K	**0.83**	-0.31	-0.04	0.22	0.10	0.10	-0.10	0.87
Cl	**0.75**	0.03	0.19	0.14	-0.20	0.21	-0.12	0.72
Zn	**0.79**	-0.8	0.02	-0.25	0.20	-0.12	0.25	0.82
Cu	0.19	0.07	-0.14	0.012	-0.08	0.03	**0.87**	0.80
Mg	0.19	0.27	**0.80**	-0.26	0.16	0.10	-0.24	0.91
Br	**0.85**	-0.16	0.35	0.21	0.09	-0.03	0.14	0.92
Ca	0.16	-0.10	**0.91**	0.21	0.06	0.03	-0.01	0.92
Soot	0.36	**-0.56**	-0.35	-0.02	0.30	0.30	-0.02	0.76
FPM	-0.18	**0.86**	-0.19	0.09	0.08	0.06	0.18	0.86
Ac	0.002	**0.86**	0.12	0.26	0.14	0.14	-0.07	0.87
Fo	-0.07	0.27	0.06	**0.92**	0.08	-0.02	0.04	0.94
NO_3^-	-0.07	-0.02	0.07	0.16	-0.04	**0.70**	0.49	0.76
SO_4^-	0.19	**-0.69**	-0.15	-0.04	0.58	0.07	0.06	0.89
$C_2O_4^{2-}$	0.21	0.21	0.24	0.30	**0.70**	-0.20	-0.14	0.79
Na^+	-0.12	-0.07	-0.03	0.10	0.04	**-0.82**	0.11	0.71
NH_4^+	-0.44	-0.11	0.21	-0.47	**0.58**	0.04	-0.08	0.81

A large number of samples have been collected by the sampling monitoring stations up to now: 313 in Alta Floresta, 185 in Serra do Navio and 324 in Cuiabá. The samplings cover several years for the three sampling sites. This allows detailed analysis of seasonal variations in the aerosol concentration and composition. Biomass burning occurs mainly at the end of the dry season (August to beginning of October) at the central Amazon Basin. Figure 2 shows the time series for the fine and coarse particulate matter (FPM and CPM) concentrations in all the samples collected in the monitoring stations. For the Alta Floresta and the Cuiabá sites, the biomass burning contribution appears as a strong increase in the fine fraction concentrations during the dry season. In Cuiabá, from a typical inhalable particle matter (IPM = FPM+CPM) concentration of 10 - 20 $\mu g\ m^{-3}$ during the wet season, the concentration goes as high as 100 $\mu g\ m^{-3}$ during the burning season. For Alta Floresta the increase in the concentrations is still more dramatic. The IPM concentration increases from 30 - 40 $\mu g\ m^{-3}$ to concentrations up to 500 $\mu g\ m^{-3}$.

Aircraft measurements during the burning season show high concentrations of IPM and black carbon in large areas of the Amazon Basin (Pereira et al., 1996). Remote sensing measurements also show these high concentrations for areas as large as 2 million km^2 in this period. This situation is illustrated in Fig. 3. Soil dust contributes to the increment in the coarse fraction aerosol during the dry season, being suppressed when the rains start, usually in the middle of October. In January and February, with heavy rains, the coarse fraction aerosol concentration drops to concentrations bellow 10 $\mu g\ m^{-3}$ in Cuiabá and 20 $\mu g\ m^{-3}$ in Alta Floresta.

Alta Floresta Aerosol Concentration

Aerosol Samples from 1990 to 1997

Cuiabá Aerosol Mass Concentration

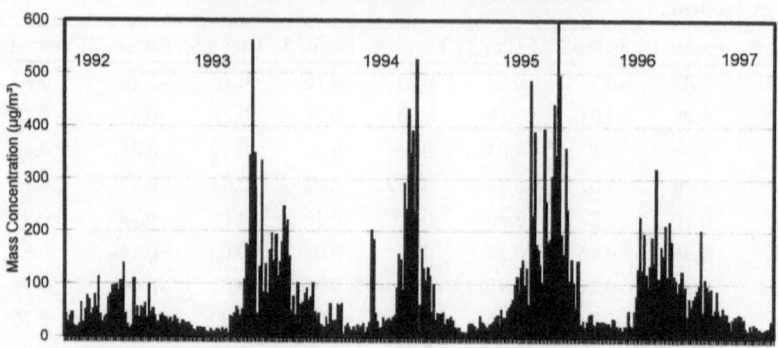

Samples from 1990 to 1996

Serra do Navio Aerosol Concentration

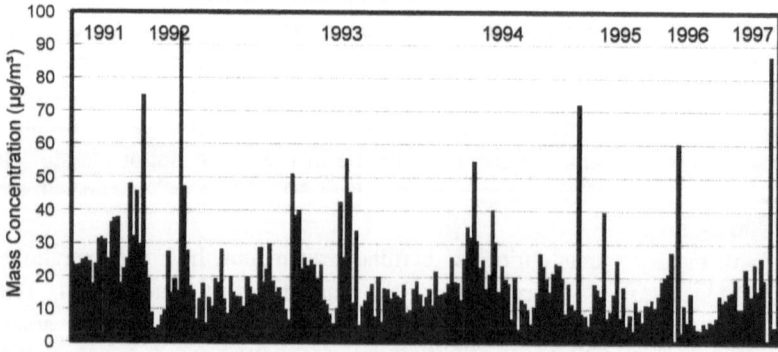

Aerosol Samples from 1991 to 1997

Fig. 2: Fine and coarse aerosol fraction time series from three atmospheric aerosol monitoring stations in Amazon Basin

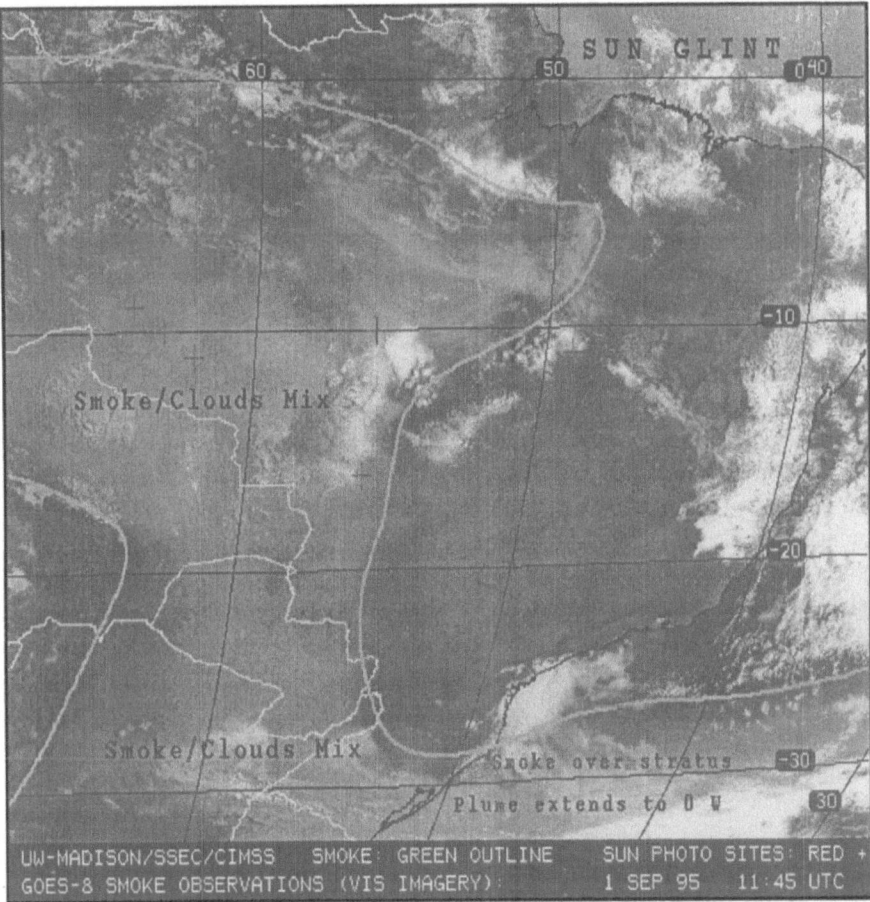

Fig. 3: GOES Satellite image in the visible channel in September 1, 1995

In Alta Floresta the FPM concentration drops earlier than the CPM concentration, because the farmers stop burning the primary forest as soon as the rain starts. The time series for black carbon concentrations measured both in Alta Floresta and in Cuiabá, used as a biomass burning tracer, presents the same behavior as the FPM time series.

The Serra do Navio sampling site is not as strongly affected by biomass burning as Alta Floresta and Cuiabá. Some increase in the FPM concentration during the biomass burning season can be detected, but far less pronounced than it is in Alta Floresta and Cuiabá. Serra do Navio also presents a different rain pattern when compared with the central Amazon. The most rainy period occurs between March and June, with an average precipitation of approximately 2,700 mm. In Serra do Navio there is not a well defined dry season, with a monthly precipitation of 80 mm month^{-1} during September and October, the driest period. Thus, much lower inhalable particulate matter concentration was observed in Serra do Navio. A background value of 10 - 20 μg m^{-3} was measured. On the other hand, the coarse particle mass concentration presents almost constant values during the year, showing clear episodes of higher concentrations (40 - 60 μg m^{-3}) during December and February, possibly an effect of the long range transport of Saharan dust (Swap et al., 1996). During this time of the year, the position of the ITCZ (ntertropical convergence zone) allows the intrusion of Saharan dust into the Amazon Basin. The Amazon importation of African aerosol has been reported and broadly discussed in the literature. (Prospero et al, 1981; Swap et al, 1992, Swap et al, 1996).

As samples are being constantly collected, only some of the filters have already been analyzed by PIXE and the other complementary analytical techniques. Table 4 presents the average concentrations for the fine fraction. Some other aerosol compound concentrations are also showed.

A comparison between the three sampling stations' fine fraction aerosol concentrations shows that Alta Floresta presents much larger concentrations than the other stations. The FPM concentration in Alta Floresta is four times greater than it is in Cuiabá and almost seven times greater than in Serra do Navio. The same behavior can be observed for black carbon concentrations. The average concentrations also indicate a greater influence of soil dust in Alta Floresta when compared to the other sampling stations.

Cuiabá and Serra do Navio present concentrations of the same order of magnitude, but Cuiabá has a stronger influence of biomass burning, as was expected. Cuiabá presented three times more black carbon and twice as much potassium as Serra do Navio. Black carbon and potassium are largely emitted in biomass burning processes. Although sulfur and chlorine are also emitted in biomass burnings, the average concentrations for these elements are similar in these two sampling stations. This indicates a marine contribution of these elements in Serra do Navio.

Table 5 presents the average concentrations of the coarse fraction aerosol. The highest concentrations were again obtained in Alta Floresta. Cuiabá and Alta Floresta present a significant contribution of iron, aluminum and silicon, indicating that soil dust aerosol particles contribute during the dry season. Coarse fraction aerosol particles from Serra do Navio present very high concentrations of sodium, chlorine and magnesium, clearly from the sea salt aerosol contribution.

Absolute principal factor analysis applied to the Serra do Navio concentration time series was able to identify four factors in the fine fraction aerosol and three factors in the coarse fraction. Table 6 presents the VARIMAX rotated factor loading matrix obtained from the APFA technique.

Table 4: Average concentrations, in ng m^{-3}, for the fine fraction aerosol particles ($d_p < 2\mu m$) collected in three sampling sites in the Amazon Basin. FPM is the fine particulate mass concentration expressed in $\mu g\ m^{-3}$

	Serra do Navio			Cuiabá			Alta Floresta		
	Mean	Mean std. dev.	(n)	Mean	Mean std. dev.	(n)	Mean	Mean std. dev.	(n)
Na	194	13	(125)	-	-	-	-	-	-
Mg	109.0	5.8	(125)	-	-	-	366	60	(80)
Al	129	14	(125)	91.5	7.9	(136)	420	49	(251)
Si	298	33	(125)	134	13	(109)	728	899	(172)
P	1.870	0.069	(6)	10.0	1.2	(72)	48.1	6.3	(130)
S	348	26	(125)	389	33	(136)	847	63	(295)
Cl	13.7	3.3	(9)	10.40	0.89	(136)	51.0	9.9	(53)
K	176.2	14	(125)	326	28	(136)	915	102	(296)
Ca	37	3.2	(125)	29.1	2.5	(136)	50.6	5.6	(285)
Sc	1.220	0.067	(125)	-	-	-	4.51	0.76	(79)
Ti	10.8	1.2	(125)	6.92	0.59	(136)	22.5	2.3	(263)
V	0.46	0.11	(19)	0.86	0.18	(25)	1.34	0.75	(3)
Cr	1.72	0.58	(5)	3.42	0.42	(66)	29.3	16.5	(10)
Mn	3.63	0.33	(125)	3.60	0.31	(136)	3.36	0.32	(156)
Fe	81.9	8.3	(125)	175	15	(136)	303	47	(278)
Co	0.560	0.050	(125)	-	-	-	5.5	1.0	(195)
Ni	0.170	0.026	(13)	1.09	0.39	(8)	3.65	0.60	(126)
Cu	15.8	-	(1)	1.55	0.19	(69)	5.44	0.68	(152)
Zn	1.80	0.12	(125)	5.81	0.50	(136)	8.52	0.81	(292)
Ga	0.130	0.014	(54)	-	-	-	0.66	0.17	(45)
Ge	0.098	0.011	(45)	-	-	-	0.47	0.12	(24)
As	0.373	0.034	(80)	-	-	-	0.62	0.10	(39)
Se	0.255	0.018	(125)	-	-	-	0.511	0.077	(74)
Br	3.79	0.33	(125)	5.61	0.60	(87)	18.5	4.6	(32)
Rb	0.78	0.10	(44)	1.32	0.21	(41)	6.7	1.7	(97)
Sr	0.548	0.055	(125)	0.70	0.10	(49)	1.20	0.19	(43)
Zr	-	-	-	1.22	0.13	(95)	2.44	0.51	(13)
Au	0.185	0.022	(51)	-	-	-	1.37	0.17	(71)
Hg	0.321	0.036	(54)	-	-	-	2.10	0.25	(103)
Pb	0.739	0.050	(125)	1.68	0.18	(87)	3.20	0.42	(208)
H$^+$	1.96	0.23	(125)	-	-	-	-	-	-
Na$^+$	210	15	(125)	-	-	-	-	-	-
NH$_4^+$	138	16	(125)	-	-	-	-	-	-
K$^+$	150	14	(125)	-	-	-	-	-	-
Mg^{2+}	22.6	1.3	(125)	-	-	-	-	-	-
Ca^{2+}	27.1	2.1	(125)	-	-	-	-	-	-
Cl$^-$	6.09	0.65	(125)	-	-	-	-	-	-
NO$_3^-$	20.6	3.1	(85)	-	-	-	-	-	-
SO$_4^{2-}$	996	73	(125)	-	-	-	-	-	-
C$_2$O$_4^{2-}$	79.6	5.5	(125)	-	-	-	-	-	-
PO$_4^{3-}$	6.50	0.65	(35)	-	-	-	-	-	-
acetate	6.56	0.72	(56)	-	-	-	-	-	-
acetic ac.	12.9	1.6	(62)	-	-	-	-	-	-
formate	5.84	0.71	(125)	-	-	-	-	-	-
formic ac.	6.44	0.79	(125)	-	-	-	-	-	-
MSA	8.02	0.53	(125)	-	-	-	-	-	-
MPF *	6.82	0.57	(125)	10.70	0.92	(136)	47.1	4.9	(305)
BC	746	60	(125)	2075	180	(136)	4720	340	(302)

Table 5: Average concentrations, in ng m^{-3}, for the coarse fraction aerosol particles ($2<d_p<10\mu m$) collected in three sampling sites in the Amazon Basin. CPM is the fine particulate mass concentration expressed in $\mu g\ m^{-3}$

	Serra do Navio			Cuiabá			Alta Floresta		
	Mean	Mean std. dev.	(n)	Mean	Mean std. dev.	(n)	Mean	Mean std. dev.	(n)
Na(PG)	2060	110	(142)	-	-	-	-	-	-
Mg	269	12	(142)	-	-	-	487	75	(128)
Al	404	42	(142)	830	63	(135)	4600	890	(302)
Si	770	91	(142)	2120	160	(135)	3850	370	(300)
P	7.58	1,2	(38)	18.7	1.0	(135)	107.9	7.9	(240)
S	137.0	6.3	(142)	135	12	(135)	237	15	(249)
Cl	1059	51	(142)	17.1	1.1	(135)	64.0	5.2	(193)
K	135	10	(142)	430	37	(135)	468	29	(302)
Ca	161	17	(142)	316	24	(135)	390	44	(302)
Sc	3.19	0.25	(142)	-	-	-	9.1	4.1	(6)
Ti	46.2	4.2	(142)	79.8	6.4	(135)	214	19	(300)
V	2.36	0.25	(99)	-	-	-	8.59	0.83	(114)
Cr	1.59	0.14	(54)	-	-	-	29	26	(3)
Mn	59.2	0.62	(142)	8.18	0.77	(135)	17.9	1.7	(277)
Fe	472	42	(142)	1291	85	(135)	2219	200	(303)
Co	1.50	1.31	(142)	-	-	-	22.0	2.9	(260)
Ni	0.820	0.075	(88)	-	-	-	8.1	1.4	(114)
Cu	10.3	3.7	(142)	-	-	-	15.4	1.9	(154)
Zn	7.35	2.2	(142)	-	-	-	8.02	0.71	(269)
Ga	0.673	0.077	(36)	-	-	-	2.6	1.2	(69)
Ge	0.504	0.072	(17)	-	-	-	1.47	0.38	(0.23)
As	1.46	0.14	(89)	-	-	-	1.74	0.36	(41)
Se	0.459	0.027	(76)	-	-	-	0.94	0.12	(100)
Br	1.721	0.098	(77)	-	-	-	1.91	0.46	(28)
Rb	1.54	0.28	(34)	-	-	-	2.71	0.33	(135)
Sr	2.22	0.17	(142)	-	-	-	5.08	0.91	(194)
Zr	-	-	-	-	-	-	14.0	2.0	(67)
Au	0.680	0.080	(24)	-	-	-	2.21	0.26	(80)
Hg	1.49	0.23	(14)	-	-	-	5.08	0.68	(99)
Pb	2.83	0.56	(55)	5.01	0.35	(135)	5.43	0.51	(182)
CPM*	14.62	0.93	(142)	23.0	1.9	(135)	56.1	5.3	(304)

Mean and mean standard deviation are shown. Numbers in parentheses are the quantity of samples in which the concentrations of the element were above the detection limit.

Table 6: Component Loading Matrix after VARIMAX rotation based on the results from Serra do Navio aerosols. Explained communality is also shown

Variable	Fine mode					Coarse mode			
	Biogenic	Soil dust	Sea salt	Mn mining	Commu- nality (%)	Soil dust, biogenic	Sea salt	Mn mining	Commu- nality (%)
CPM	-	-	-	-	-	**0.87**	0.38	0.19	94
BC	**0.97**	0.08	0.06	0.02	94	-	-	-	-
FPM	**0.89**	0.33	0.08	-0.05	91	-	-	-	-
Na	**0.60**	-0.20	**0.70**	0.05	88	0.04	**0.95**	0.13	93
Mg	-	-	-	-	-	0.37	**0.76**	-0.14	76
Al	0.01	**0.98**	-0.07	0.06	97	**0.95**	0.09	0.23	97
Si	0.03	**0.98**	-0.06	0.02	97	**0.96**	0.09	0.02	93
S	**0.96**	-0.04	0.17	0.09	96	0.30	**0.89**	0.10	90
Cl	-	-	-	-	-	0.10	**0.96**	0.03	94
K	**0.94**	0.17	0.13	0.10	94	**0.88**	0.32	-0.01	87
Ca	0.14	**0.95**	0.17	0.03	95	**0.89**	0.07	-0.23	85
Sc	0.33	**0.79**	0.14	-0.20	79	**0.81**	0.13	-0.25	76
Ti	-0.04	**0.97**	-0.07	0.09	96	**0.90**	0.10	0.36	95
Mn	0.00	0.19	0.04	**0.91**	86	-0.01	0.05	**0.86**	74
Fe	-0.02	**0.98**	-0.04	0.13	98	**0.76**	0.12	**0.56**	91
Co	0.08	**0.98**	-0.00	0.07	96	**0.72**	0.18	**0.45**	94
Zn	**0.92**	0.18	0.19	0.05	92	0.05	0.04	0.01	99
Se	**0.77**	0.22	0.29	-0.12	74	-	-	-	-
Br	**0.72**	0.32	0.28	-0.21	74	-	-	-	-
Sr	0.12	**0.93**	0.17	-0.03	90	**0.91**	0.24	-0.06	89
Pb	**0.74**	**0.44**	0.15	-0.20	80	-	-	-	-
H$^+$	**0.97**	-0.05	0.01	-0.07	95	-	-	-	-
Mg^{2+}	**0.57**	0.01	**0.76**	0.08	91	-	-	-	-
Ca^{2+}	0.29	**0.90**	0.22	-0.03	95	-	-	-	-
Cl$^-$	0.10	0.28	**0.68**	-0.12	56	-	-	-	-
NH$_4^{2-}$	**0.95**	-0.02	-0.12	0.02	92	-	-	-	-
SO$_4^{2-}$	**0.95**	-0.06	0.19	0.10	96	-	-	-	-
C$_2$O$_4^{2-}$	**0.92**	0.09	0.27	-0.01	92	-	-	-	-
Formic Ac.	**0.84**	0.24	0.12	-0.25	84	-	-	-	-
MSA	**0.78**	-0.23	0.36	0.29	88	-	-	-	-

For the fine fraction, the first factor is associated with natural biogenic aerosol source. This source has strong contributions of the ionic compounds, K, S, and FPM. Black carbon associated to this factor can be explained as a small influence of biomass burning. The elemental compositions of biomass burning and natural biogenic aerosol are similar, making difficult their separation by APFS technique. The second and the third factors are clearly associated with soil dust and sea salt aerosol particles, repectively, with high component loadings for Al, Si, Ca, Ti, Fe and Co to the second factor and Na, Cl and Mg to the third factor. A fourth factor, associated only with Mn, represents the local contribution from Mn mining activities in Serra do Navio. Serra do Navio has the biggest Brazilian manganese mining area. The APFA procedure allows obtaining absolute source profiles in concentration units. Fig. 4 shows the absolute elemental profiles for the identified aerosol sources in the fine (Fig. 4a) and in the coarse (Fig. 4b) aerosol fractions.

Table 7: Component Loading Matrix after VARIMAX rotation based on the results from Cuiabá aerosols. Explained communality is also shown

Variable	Fine fraction				Coarse fraction		
	Soil dust	Biom.burning/ biogenic	Antropo- genic	Communa- lity (%)	Soil dust	Biom. burning/ biogenic	Commu- nality (%)
CPM	-	-	-	-	**0.81**	**0.53**	0.94
FPM	0.37	**0.90**	0.08	0.96	-	-	-
BC	**0.48**	**0.81**	0.21	0.92	-	-	-
Al	**0.94**	0.31	0.08	0.98	**0.95**	0.25	0.97
Si	**0.84**	0.35	0.27	0.90	**0.95**	0.26	0.98
P	-	-	-	-	0.11	**0.95**	0.92
S	0.20	**0.87**	0.25	0.85	**0.70**	**0.62**	0.88
Cl	-	-	-	-	**0.61**	**0.66**	0.81
K	**0.43**	**0.88**	0.11	0.97	**0.83**	**0.52**	0.97
Ca	**0.65**	**0.60**	0.27	0.86	**0.86**	0.37	0.88
Ti	**0.90**	0.37	0.15	0.97	**0.94**	0.30	0.97
Mn	**0.67**	**0.59**	0.65	0.92	**0.80**	**0.54**	0.94
Fe	**0.89**	0.34	0.22	0.95	**0.96**	0.19	0.96
Cu	0.19	0.19	**0.94**	0.95	-	-	-
Zn	**0.46**	**0.74**	**0.43**	0.95	-	-	-
Pb	**0.58**	0.45	**0.51**	0.81	**0.61**	0.22	0.42

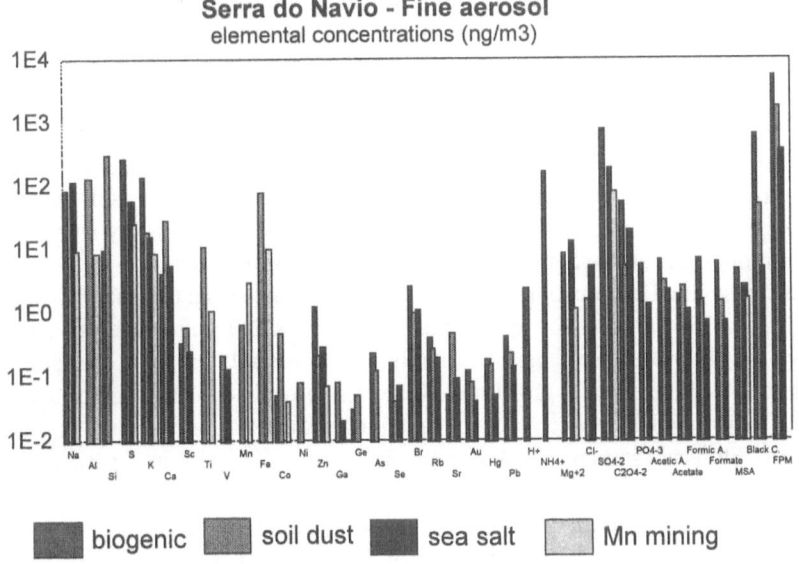

Fig. 4a: Fine mode aerosol elemental profiles of Serra do Navio

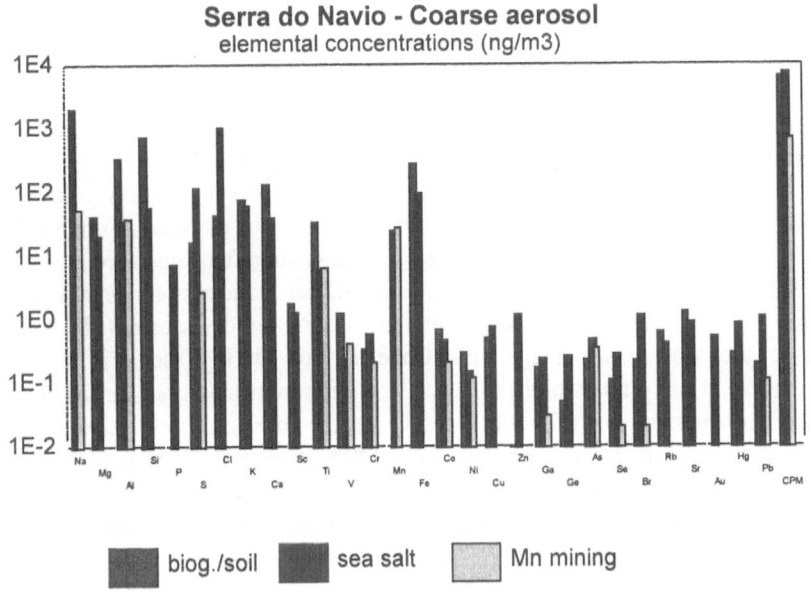

Fig. 4b: Coarse mode aerosol elemental profiles of Serra do Navio

The APFA of the elemental concentration in the Cuiabá sample is shown in Table 7. Three factors were identified in the fine fraction and two in the coarse fraction aerosol. The first factor in the fine fraction is from soil dust. FPM is related to the second factor, associated with biomass burning and biogenic emissions (high factor loadings for FPM, black carbon, S and K). Black carbon associated with the second factor shows the influence of the biomass burning emissions in the composition of the Cuiabá atmospheric fine aerosol. A third component, with significant contributions of Cu, Zn and Pb, is possibly associated with anthropogenic aerosol emissions from Cuiabá urban activities.

The coarse fraction aerosol present in Cuiabá originates from two main sources: soil dust, which explains the major portion of the CPM concentration, and a mixture of biomass burning contribution and natural biogenic emissions. Similarities in the elemental profiles of these two aerosol sources always makes it difficult to distinguish between biogenic and biomass burning emissions.

In Alta Floresta samples APFA presented the same difficulty in separating the natural biogenic aerosol emission from the biomass burning contribution. In order to separate these important aerosol sources the samples were divided into two groups. The first contains samples collected at the end of the dry season. These samples have a major influence of biomass burning contribution. The second group is formed by samples collected during the wet season or the beginning of the dry season. These samples have much less or no biomass burning influence, but, on the other hand, have an important natural biogenic aerosol contribution. The separation of two groups was based on the absolute concentration of black carbon. At the beginning of the Alta Floresta burning season the black carbon absolute concentration rises sharply, and after the first rains it goes down to background concentration values again. Using the Alta Floresta black carbon concentration as a biomass burning activities indicator, July 19 to October 8 was defined as the typical annual biomass burning period in Alta Floresta. All samples collected during this period belong to the first group (here called dry season group) with biomass burning influence. The other samples belong to the second sample (here called wet season group), characterized by mainly natural biogenic aerosol.

Figure 5 shows the absolute elemental profiles for the identified aerosol sources in the Alta Floresta fine particulate fractions during the dry season (Fig. 5a) and during the wet season (Fig. 5b). Concentration during the dry season is approximately five times higher than the concentration measured during the wet season. It is possible to verify that natural biogenic emissions dominate the aerosol mass concentration for the fine fraction (FPM) during the wet season. In this period the soil dust contribution is only 24% of the FPM concentration. During the dry season the soil dust contribution to the FPM concentration is still lower (13% of the FPM concentration). During the dry season almost all the fine aerosol is from biomass burning emissions.

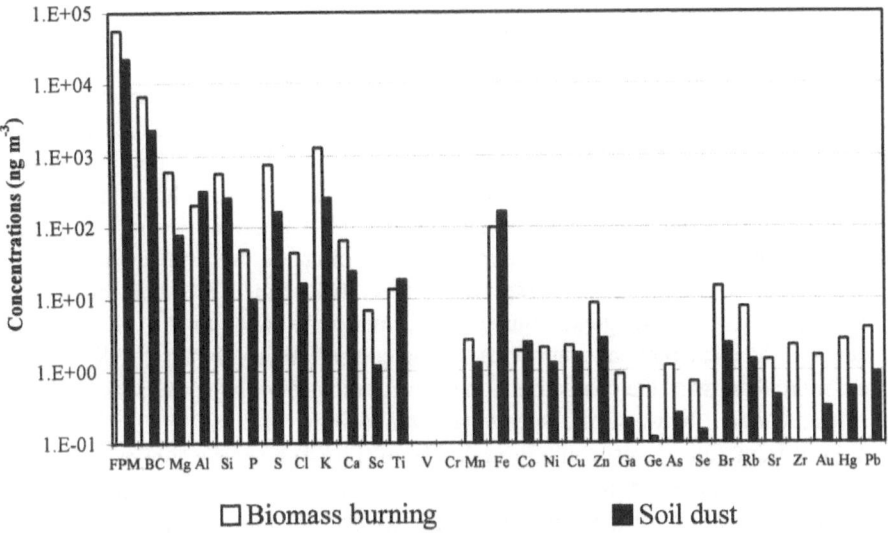

Fig. 5a: Fine fraction aerosol elemental profiles of Alta Floresta during the dry season

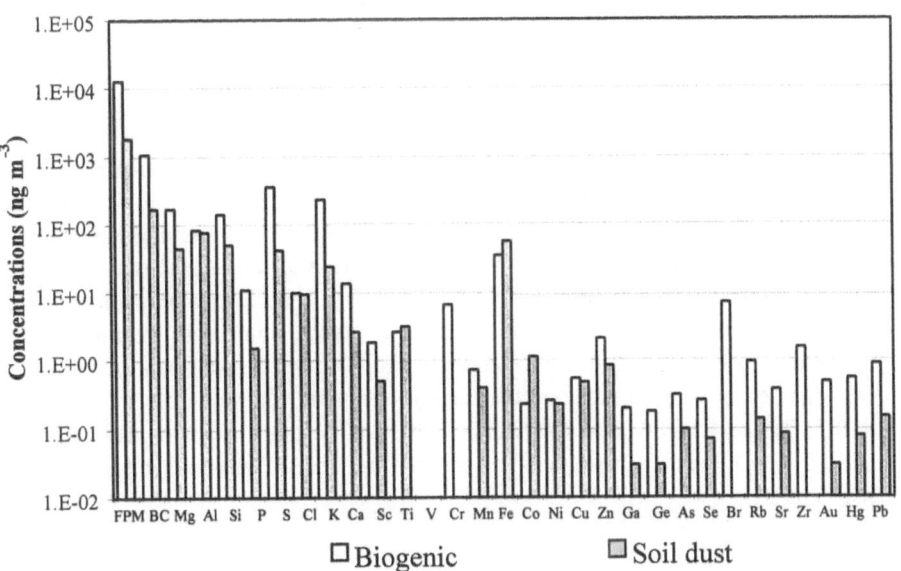

Fig. 5b: Fine fraction aerosol elemental profiles of Alta Floresta during the wet season

The APFA for the Alta Floresta coarse fraction aerosol is presented in Fig. 6. The APFA of the Alta Floresta coarse fraction aerosol during the dry season (Fig. 6a) was able to distinguish three main aerosol sources: soil dust, biomass burning and natural biogenic aerosols. During this period soil dust aerosol is the major source, but biomass burning and natural biogenic aerosol contributions are significant. In this case, the distinction between biomass burning and biogenic aerosol during the dry season was possible because for the coarse fraction aerosol the PIXE technique was able to quantify phosphorus. Phosphorus is largely emitted in biogenic activities and can be used as a tracer of natural biogenic aerosol. During the wet season the major source of coarse fraction aerosol in Alta Floresta is from biogenic processes. More than 70% of the CPM concentration is associated with the biogenic source in this period. Similar to the fine fraction aerosol, the elemental concentrations for the coarse fraction during the dry season are more than twice the coarse fraction concentrations during the wet season.

Comparison between the elemental composition profiles for the biogenic aerosol from Alta Floresta and Serra do Navio (Fig. 4) reveals a similarity in their composition. In terms of percentage of the particulate mass attributed to the biogenic source, S, K, Ca, Zn, As, Se, Br, Rb and Pb have an agreement better than a factor of two for fine fraction aerosol.

Since 1992 INPE (Instituto Nacional de Pesquisas Espaciais - Brazilian space research institute) and NASA (National Aeronautics and Space Administration - United States) have been operating a sun photometer (SP) network in the Amazon Basin (Holben et al., 1995). A SP was installed in Alta Floresta. Each SP is able to measure the aerosol optical thickness (AOT) (Horvath, 1993) for six different solar radiation wave lengths: 339, 380, 441, 672, 873 and 1022 nm. The SP can also estimate the amount of water vapour in the atmosphere. The AOT is related to light scattering and light absorbing in the atmosphere by suspended particles (Horvath, 1993). As shown in Fig. 3, the presence of suspended matter in the atmosphere can change some atmospheric optical properties. Figure 7 shows a strong increment in the AOT measured in Alta Floresta for all the measured wave lengths during the dry season. The SP measures the AOT during the day every 15 minutes, when there is a clear sky (without clouds). Alta Floresta aerosol samples were measured day and night, for a few days. Hence, some data treatment was performed to make compatible SP and the Alta Floresta aerosol data in order to put together the AOT results and the aerosol elemental concentrations in the same APFA. Tables 8 and 9 show the results of the conjugated analysis of AOT results and 60 Alta Floresta aerosol samples for the fine and the coarse fraction aerosol respectively.

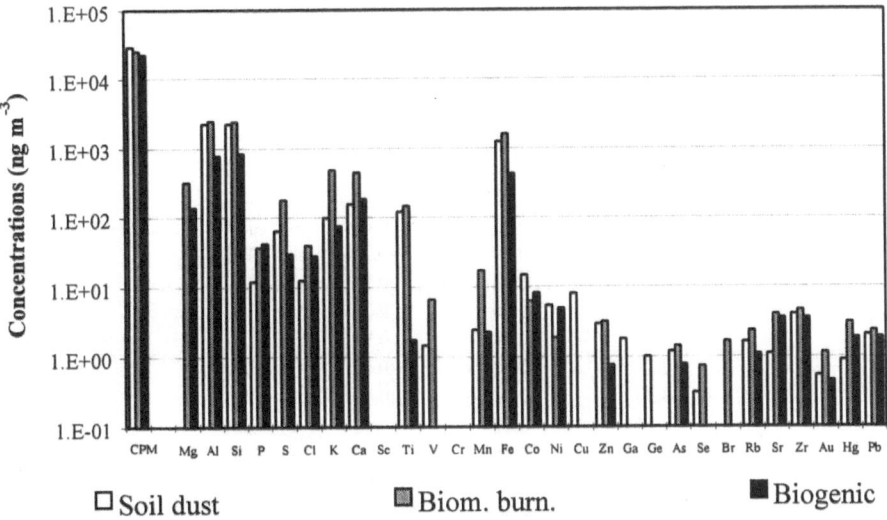

Fig. 6a: Coarse fraction aerosol elemental profiles of Alta Floresta during the dry season

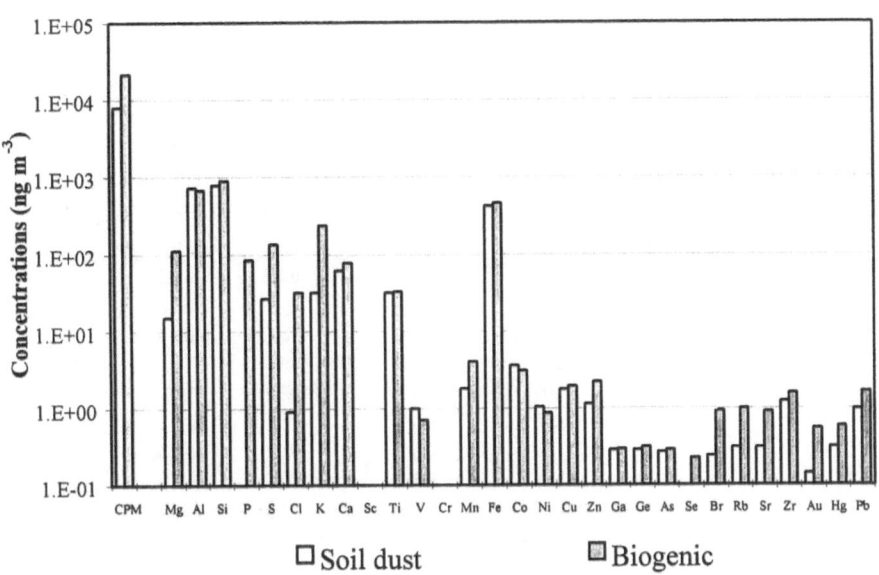

Fig. 6b: Coarse fraction aerosol elemental profiles of Alta Floresta during the wet season

Table 8: Component Loading Matrix after VARIMAX rotation for AOT and the fine fraction aerosol elemental concentration data

Variable	Factor 1 AOT	Factor 2 biomass burning	Factor 3 soil dust	Factor 4 water vapour	Communality (%)
AOT 339 nm	**0.93**	0.27	0.06	-0.07	95
AOT 380 nm	**0.94**	0.28	0.09	-0.05	98
AOT 441 nm	**0.94**	0.29	0.14	-0.06	99
AOT 672 nm	**0.93**	**0.31**	0.15	0.05	99
AOT 873 nm	**0.91**	**0.31**	0.07	0.05	98
AOT 1022 nm	**0.89**	0.28	0.02	0.27	95
water vapor	0.11	0.09	**-0.44**	**0.89**	97
FPM	**0.97**	**0.58**	0.18	-0.12	84
BC	**0.41**	**0.86**	0.22	-0.03	97
Al	0.08	0.21	**0.95**	-0.14	98
S	**0.31**	**0.91**	-0.02	0.08	94
K	**0.30**	**0.92**	0.16	-0.06	96
Ca	0.24	**0.80**	**0.42**	-0.01	88
Ti	0.10	0.28	**0.94**	-0.12	99
Fe	0.14	0.12	**0.97**	-0.12	98
Zn	**0.39**	**0.88**	0.13	0.04	95
Pb	0.24	**0.83**	0.23	0.16	82

Table 9: Component Loading Matrix after VARIMAX rotation for AOT and the coarse fraction aerosol elemental concentration data

Variable	Factor 1 AOT	Factor 2 soil dust	Factor 3 soil dust / biomass burn.	Factor 4 biogenic	Fator 5 water vapor	Communality (%)
AOT 339 nm	**0.95**	0.13	0.16	0.08	-0.08	96
AOT 380 nm	**0.96**	0.15	0.18	0.10	-0.06	98
AOT 441 nm	**0.94**	0.19	0.23	0.09	-0.07	99
AOT 672 nm	**0.94**	0.20	0.24	0.10	0.04	99
AOT 873 nm	**0.93**	0.17	0.18	0.13	0.19	98
AOT 1022 nm	**0.91**	0.14	0.12	0.13	0.27	95
water vapor	0.16	-0.29	-0.11	**0.10**	**0.91**	96
CPM	**0.32**	**0.67**	0.16	**0.53**	-0.03	86
Al	0.14	**0.76**	**0.57**	-0.09	-0.23	99
Si	0.16	**0.73**	**0.62**	-0.07	-0.20	99
P	0.22	-0.22	0.13	**0.91**	0.12	95
S	**0.34**	0.05	**0.67**	**0.37**	0.17	73
K	0.25	**0.38**	**0.75**	**0.41**	-0.13	96
Ca	0.25	0.27	**0.90**	0.04	-0.06	94
Ti	0.17	**0.78**	**0.54**	-0.08	-0.19	98
Mn	0.20	**0.40**	**0.86**	0.00	-0.08	94
Fe	0.18	**0.81**	**0.49**	-0.10	-0.16	97
Co	0.21	**0.88**	0.06	-0.08	-0.06	84
Zn	**0.34**	**0.57**	**0.55**	0.25	-0.21	85

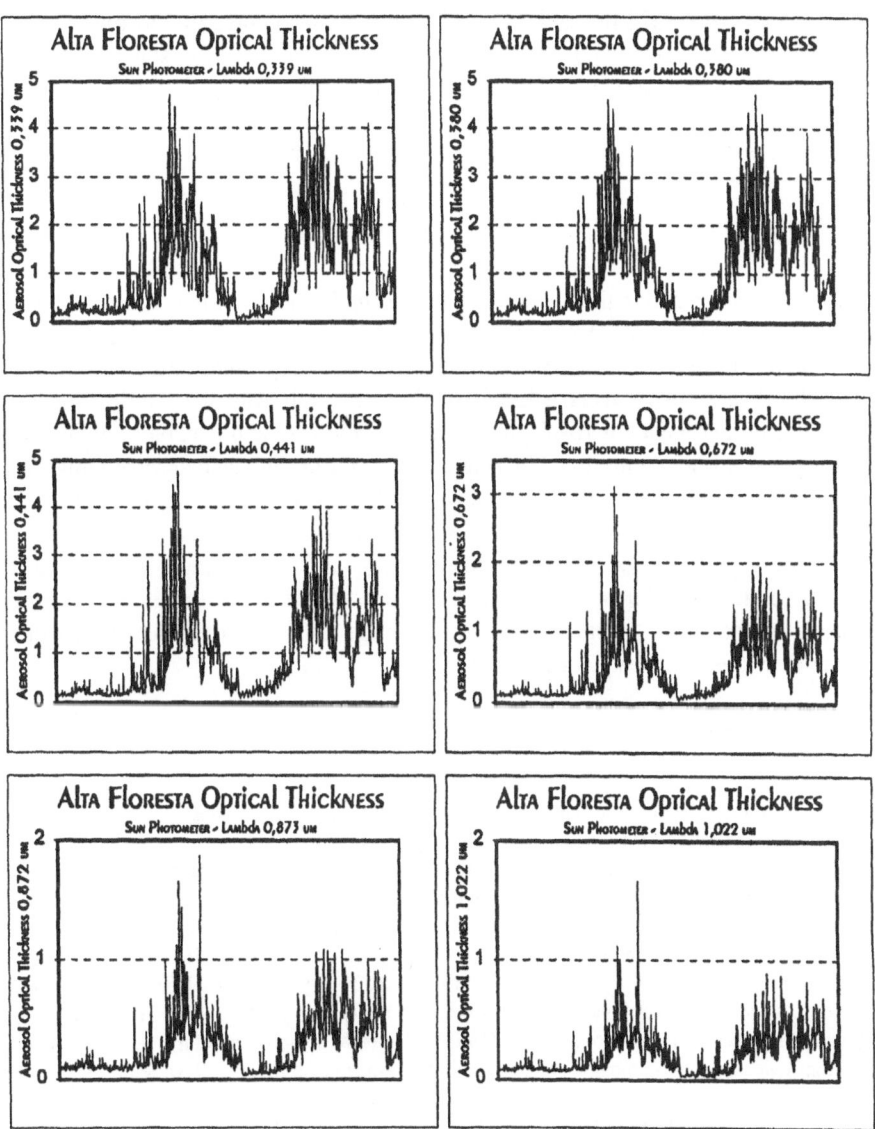

Fig. 7: Aerosol optical thickness for six wave lengths measured in Alta Floresta from June 20, 1993 to November 9, 1994

In the fine fraction aerosol analysis four factors were retained, the AOT, biomass burning, soil dust and water vapor respectively. It is possible to verify some interference between the AOT and the biomass burning factors. Significant component loadings for FPM, black carbon, S, K and Zn were obtained in the

AOT factor. AOT-672 nm and AOT-873 nm were also correlated to the biomass burning factor. Water vapor has a negative correlation with the soil dust factor. This behavior was expected since the suspended soil dust particle concentration increases with decreasing humidity. Table 9 shows that there is almost no influence of the coarse particulate aerosol in the AOT results.

Conclusion

The results presented in this work show the strong influence of the aerosol biomass burning emissions in the Amazon atmospheric composition during the dry season. The inhalable aerosol mass concentration reaches very high values during this period, when smoke covers a large fraction of South America. During the dry season, fine particulate aerosol present in Alta Floresta and Cuiabá is dominated by biomass burning emissions. During the wet season biogenic aerosol emissions dominate. The Eastern Amazon atmosphere has an important marine influence, and the Serra do Navio results point to Saharan dust aerosol import phenomena.

A conjugated analysis of the aerosol elemental concentration and the AOT measured in Alta Floresta shows an influence of the biomass burning aerosol present in the lower troposphere in the average atmospheric optical properties, with yet unknown consequences.

References

Andreae, M.O. (1983). Soot carbon and excess fine potassium: long-range transport of combustion derived aerosols. *Science*, 220:1148-1151.

Andreae, M.O.; T. W. Andreae; R. J. Ferek; H. Raemdonck (1984). Long-range transport of soot carbon in the marine atmosphere, *Sci. Total Environ.*, 36:73-80.

Artaxo, P., Storms, H., Bruynseels, F., Van Grieken, R., Maenhaut, W. (1988). Composition and sources of aerosols from the Amazon Basin. *J. Geophys. Res.*, 93:1605-1615.

Artaxo, P., Maenhaut, W., Storms, H. e Van Grieken, R. (1990). Aerosol characteristics and sources for the Amazon Basin during the wet season. *J.Geophys. Res.*, 95:1671-16985.

Artaxo, P., Gerab, F. e Rabello, M. L. C. (1993). Elemental composition of the aerosol particles from two atmospheric monitoring stations in the Amazon Basin. *Nucl. Inst. Meth.*, B75:277-281.

Artaxo, P.; Yamasoe, M.M.; Martins, J.V.; Kocinas, S.,;Carvalho,S.; Maenhaut,W. (1993). Case study of atmospheric measurements in Brazil: Aerosol emissions in the Amazon Basin biomass burning. In: *Fire in the environment: The ecological, atmospheric and climatic importance of vegetation fires;* Crutzen, P.J. and Goldammer, J-G. (Eds.). Dahlem Konferenzen ES13. John Wiley and Sons, Chichester. Pp:139-158.

Artaxo, P., Gerab, F., Yamasoe, M. A. and Martins, J. V. (1994). Fine mode aerosol composition at three long term atmospheric monitoring sites in the Amazon Basin. *J. Geophys. Res.*, 99D:22857-22868.

Artaxo, P. e Hansson, H-C. (1995). Size distribution of biogenic aerosol particles from the Amazon Basin. *Atmospheric Environment*, 29:393-402.

Asking, L., Swietlick, E., Garg, M. L. (1987). PIGE analysis for sodium in thin aerosol samples. *Nucl. Inst. Meth.*, B22:368-371.

Bingemer, H. G., Andreae, M. O., Andreae, T. W., Artaxo, P., Helas, G., Jacob, D. J., Mihalopoulos, N. and Nguyen, B. C. (1992). Sulfur gases and aerosols in and above the Equatorial African rain forest. *J. Geophys. Res.*, 97:6207-6217.

Cashier, H.; Buat-Menard, P.; Fontugne, M.; Rancher, J. (1985). Source terms and source strengths of the carbonaceous aerosols in the tropics. *J.Atmos.Chem.*, 3:469-489.

Crozat, G.; Domerge, J.L.; Baudet, J.; Bogui, V. (1978). Influence des feux de brousse sur la composition chimique des aérosols atmosphériques en Afrique de l'Ouest. *Atmos.Environ.*, 12:1917-1920.

Crutzen, P.; Andreae, M.O. (1990). Biomass burning in the tropics: Impact on atmospheric chemistry and biogeochemical cycles. *Science*, 250:1669-1678.

Crutzen,P.J.; Heidt, L.E.; Krasnec, J.P.; Pollock,W.H., Seiler, W. (1979). Biomass burning as a source of the atmospheric gases CO, H_2, N_2O, NO, CH_3Cl and COS *Nature*, 282:253-256.

Dingle, A. M., (1966). Pollen as condensation nuclei, *J. Rech. Atmosph.*, 2, 231-237.

Garstang, M.; Scala, J.; Greco, S., Harriss, R.; Beck, S.; Browell, E.; Sachse, G.; Gregory, G.; Hill, G.; Simpson, J.; Tao, W.K.; Torres, A. (1988). Trace gas exchange and convective transport over the amazonian rain forest. *J.Geophys. Res.*, 93:1528-1550.

Greco, S., Swap, R., Garstang, M., Ulanski, S., Shipman, M., Harriss, R.,Talbot, R., Andreae, M. O. and Artaxo, P. (1990). Rainfall and surface kinematic conditions over Central Amazonia during ABLE-2B. *J. Geophys. Res.*, 95D:17001-17014.

Holben, B. N., Setzer, A. W., Eck, T. F., Pereira, E. B. and Slusker, I. (1996). Effect of dry season biomass burning on Amazon basin aerosol concentrations and optical properties, 1992-1994. *J.Geophys. Res.*, 101(D14):19465-19481.

Hopke, P.K. (1991). *Receptor modelling for air quality management.* Elsevier Science Pub., Amsterdam.

Horvath, H. (1993). Atmospheric light absorption: a review. *Atmosph. Env.*, 27A(3):293-317.

Johansson, S. A. E. and Campbell, J. L. (1988). *PIXE: A novel technique for elemental analysis.* John Wiley and Sons, Chichester.

Lacaux, J.-P.; Cachier, H.; Delmas, H. (1993). Biomass burning in Africa. An overview of its impact on atmospheric chemistry. In: *Fire in the environment: The ecological, atmospheric and climatic importance of vegetation fires.* Crutzen, P.J. and Goldammer, J-G. (Eds.). Dahlem Konferenzen ES13. John Wiley and Sons, Chichester. Pp.:159-191.

Maki, L. R. and Willoughby, K. J. (1978). Bacteria as a source of freezing nuclei. *J. Applied Meteorology*, 17:1049-1053.

Parkin, D. W., Phillips, D. R., Sullivan, R. A. L. and Johnson, L. R. (1972). Airborn dust collections down the Atlantic. *Quart. J. Roy. Meteor. Soc.*,98:798-808.

Pereira, E.B.; Setzer, A.W.; Gerab, F.; Artaxo, P.; Pereira, M.C.; Monroe, G. (1996). Airborne measurements of biomass burning aerosols in Brazil related to the Trace-A experiment. *J.Geophys.Res.*, 101:23983-23992.

Prospero, J. M., Nees, R. T. and Uematsu, M. (1987). Deposition rate of particulate and dissolved aluminium derived from Saharan dust in precipitation at Miami. *J. Geophys. Res.*, 92:14723-14731.

Prospero, J. M., Glaccum, R. A. and Nees, R. T. (1981). Atmospheric transport of soil dust from Africa to South America. *Nature*, 289:570-572.

Salati, E. and Vose, P.B. (1984). Amazon Basin: A system in equilibrium. *Science*, 225:129-138.

Savoie, D. and Prospero, J.M. (1977). Aerosol concentration statistics for the northern tropical Atlantic. *J. Geophys. Res.*, 82:5954-5964.

Schnell, R. C. (1982). Kenia leaf litter: a source of ice nuclei. *Tellus*, 34:92-95.

Setzer, A.; Pereira, E.B. (1991). Amazon biomass burning in 1987 and estimate of their tropospheric emissions. *Ambio*, 20:19-22.

Simoneit, B. R. T. (1989). Organic matter of the troposphere - V: Application of molecular marker analysis to biogenic emissions into the troposphere for source reconciliations. *J. Atmosph. Chem.*, 8:251-275.

Swap, R.; Garstang, M.; Greco, S.; Talbot, R.; Kallberg, P. (1992). Saharan dust in the Amazon basin. *Tellus*, 44B:133-149.

Swap, R.; Ulanski, S.;, Cobbett, M.; Garstang, M. (1996). Temporal and spatial characteristics of Saharan dust outbreaks. *J.Geophys.Res.*, 101:4205-4220.

Talbot, R. W., Andreae, M. O., Berrescheim, H., Artaxo, P., Garstang, M., Harriss, R. C., Beecher, K.M. and Li, S. M. (1990). Aerosol chemistry during the wet season in Central America: Thr influence of long term range transport. *J. Geophys. Res.*, 95:16955-16969.

Thurston, S. R. and Spengler, J. D. (1985). A quantitative assessment of source contributions to inhalable particulate matter pollution in metropolitan Boston. *Atm. Env.*, 19:9-25.

Vali, G., Christensen, M., Fresh, R. W., Galyvan, E. L., Maki, R. R. e Schnell, R. C. (1976). Biogenic ice nuclei. Part II: Bacterial sources. *J.Atmosph. Sci.*, 33:1565-1570.

Vitousek, P.M.; Sanford, R.L. (1986). Nutrient cycling in moist tropical forest. *Ann.Rev.Ecol. Syst.*, 17:137-167.

Ward, D.E.; Hardy, C.C. (1991). Smoke emissions from wildland fires. *Environ. Internat.*, 17:117-134.

19 Are Tropical Estuaries Environmental Sinks or Sources? [*]

Egbert K. Duursma

Res. les Marguerites, app. 15, 1305, chemin des Revoirs, La Turbie, F-06320, France

Abstract

Even when an estuary is at steady state, it shows the typical features of a sink for contaminants, although input from rivers into the estuary may result in equal output from the estuary to the sea, thus in such estuaries accumulation of material is practically insignificant. The essential properties which determine the processes and transfer of contaminants to particulate matter are the chemical properties of the contaminant, the specific surface of particulate matter, its composition and its specific surface exchange capacities. A correlation exists between these factors. It is very essential to understand the complete picture of how compounds are sorbed to particulate matter, and not only part of it. For estuaries with a long freshwater-seawater mixing section and a turbidity maximum, the question can be posed whether the distribution of the dissolved and particulate contaminants along the estuary is conservative, meaning that the total inflow of contaminants (particulate and dissolved) from the river equals the discharge from the estuary into the sea, while there is no loss to or gain from the bottom. Model calculations have been made to solve this problem for different adsorption and complexing processes. Whether tropical estuaries are sinks or sources has to be answered in a time context. A tropical estuary newly contaminated will be a sink for a very long period, although equilibrium between input and output may be sooner obtained than in temperate climates. It may afterwards become a source, where the bottom sediments act as a buffer with very long turnover times, up to 1000 years.

Introduction

The major terrestrial freshwater and fluvial sediment runoff to the world oceans takes place between 30° N and 30° S (Milliman, 1991; Al-Gharib, 1992). These two discharges occur at different velocities, which has their consequence for contaminant transport. Since most contaminants are "strongly" adsorbed to particulate matter, the conclusion is often drawn that estuaries and coastal regions, where fluvial sediments are deposited, are sinks of contaminants.

---

[*] Also published in *Tecnologia Ambiental*, CETEM No. 6, 1995, 36 pp.

Though this conclusion seems in the first instance correct, it nevertheless requires a thorough investigation of the involved sorption and transport processes in order to be fully acceptable. Even when an estuary is at steady state, it shows the typical features of a sink for contaminants, although input from rivers into the estuary may equal output from estuary to the sea, thus in such estuaries accumulation of material is practically insignificant. A basic representation of this phenomenon must be given in order to distinguish between "natural" and man-made sources of contaminants (Mee, 1978; Mee and Osuna-Lopez, 1991).

Principles

In order to understand changes of concentrations between compartments and transfer of molecules from one compartment to another, it is necessary to define properly those compartments and the processes involved.

1. Compartment or phase: Unit of space in which molecules have a defined state of freedom (or non-freedom) of movement and reaction.
2. Processes: Reactions between molecules and transfer of molecules inside and between compartments.
3. Equilibria: Reactions and transfer processes counterbalanced by feedback processes.

Particulate Matter and Sediments

In marine and freshwater systems there is practically always some suspended material present, either settling or floating due to turbulence or having a similar density as sea water. The amount (concentration) of particulate matter in the oceans, coastal seas and estuaries ranges from a few μg PM (particulate matter) L^{-1} in open oceans to tenths of mg L^{-1} in estuaries. Some rivers may even have loads of a few g L^{-1}. Characteristic is that the PM concentrations can be locally and temporarily different, which means that it is essential, for purposes of studying processes in which PM is involved, to determine the PM concentrations as a function of time and space. In the next sections we will focus on the essential properties which determine the processes and transfer of contaminants between different conpartments.

Specific Surface of Particulate Matter

Particulate matter suspended in sea or freshwater is generally low grain-sized and has, due to these low grain sizes, a relative large specific surface (SpS).

The grain size/specific surface relation is given by formula (1), valid for spheres:

$$SpS = \frac{(4\pi r^2)}{\frac{3}{4}\pi r^3 \sigma} \quad \frac{3}{r\sigma} \quad cm^2\, g^{-1} \tag{1}$$

where r = radius of grain in cm and σ = specific weight in g cm^{-3}. Using grain sizes ($d = 2r$) in μm, the specific surface is:

$$SpS = \frac{6\times10^4}{d\times\sigma} \quad cm^2\, g^{-1} \tag{2}$$

For example, grains with $d = 1\,\mu m$ ($r = \frac{1}{2} \times 10^{-4}$ cm) and a σ of 2.0 g cm^{-3}, the SpS = 6×10^4 cm^2 g^{-1}.

Composition

In principle the composition of the surface and crystal structures of PM can differ extensively, but there are a number of common features which can be taken into account with respect to the sorptive and desorptive properties. With deviation from spheres, which is the case for aggregates, the SpS is usually larger than given in formulas (1) or (2). Through adsorption of water molecules (intermediate layer) on PM a similarity of charge conditions is created, which depends solely on the polarity-charge distribution. Organic matter, with surface-active properties (depending again on polarity), is or can be adsorbed onto the PM surfaces, thus producing sites for specific reactions between dissolved substances and these organic molecules.

Hence, in spite of PM having different origins like biotic debris of living matter, including faeces, organic aggregates (from dissolved organic compounds), atmospheric dust (fine desert materials), and minerals (from terrestrial sources), a number of similarities may be expected, at least with respect to surface properties.

Determination of Surface-Active Properties

There is no complete theory to define the surface-active properties of PM that determines its sorption capacity. The best empirical methods are taken over from soil science, which encounters the same problems. In fact three methods are applied, determining either:

1. The SpS (specific surface) by adsorption of ethylenoglycol (Dyal and Hendricks, 1950; Duursma and Eisma, 1973).
2. The base exchange capacity (BEC), with as a principle a first saturation of all exchange sites with Na$^+$, and subsequent replacement of these with NH$_4^+$. The determination of Na$^+$ in the NH$_4$Cl$^-$ percolate gives the base exchange capacity in meq 100g^{-1} (Duursma and Eisma, 1973).

3. The specific exchange capacity based on the properties of the clay minerals themselves. Each clay mineral has a specific exchange capacity, and by multiplying this value with its relative content in a sediment, the specific exchange capacity Q' can be calculated by:

$$Q' = \overline{a} \left(\frac{n}{10} \right) \times \left(\frac{P}{10} \right) \times k \tag{3}$$

Where k is 10, 25, 25 and 100 meq 100g^{-1} for kaolinite, illite, chlorite and montmorillonite, respectively, n = size fraction and P is % of mineral in this size fraction (Grim, 1953).

As is demonstrated for 45 ocean and coastal-sea sediments (Fig. 1), the correlation between the SpS, Q' and the BEC is not too bad (from original data of Duursma and Eisma, 1973), although for the SpS this is not a linear one.

Strength of Sorption

The term sorption covers two processes, (i) the rapid (matter of hours) sorption of substances to the surface of particulate matter, and (ii) that of absorption which is the binding of substances inter-layerly in clay minerals and inside crystal lattices. The latter process is considered slow (matter of days-weeks-months). The reverse reaction of desorption involves both surface desorption, inter-layer and crystal lattice releases, each having different speeds.

In agriculture the availability of soil-attached trace metals is of essence for plant growth and the BEC is an indicative factor for the availability of soil trace metals to the roots of plants. The actual amounts available can be determined by weak leaching techniques. These can also be applied for aquatic soils and particulate matter as shown in Table 1.

Table 1: Leaching of metals from marine sediments was carried out by applying two media (in excess) of acetic acid/ammonia acetate and acetic acid, creating pHs of 5.4 and 2.3, respectively. Radiotracers of metals had been previously sorbed by Mediterranean fine-grained clay sediment for 7 months (Duursma et al., 1975)

	Mn	Fe	Co	Zn	Sr	Ru	Ag	Cd	Cs	Ce	Pb
				% Metal leached after 7 months sorption							
pH = 5.4	50	0	33	4-33	100	10	0	100	2	50	50
pH = 2.3	100	75	100	10-80	100	25	0	100	30	90	50

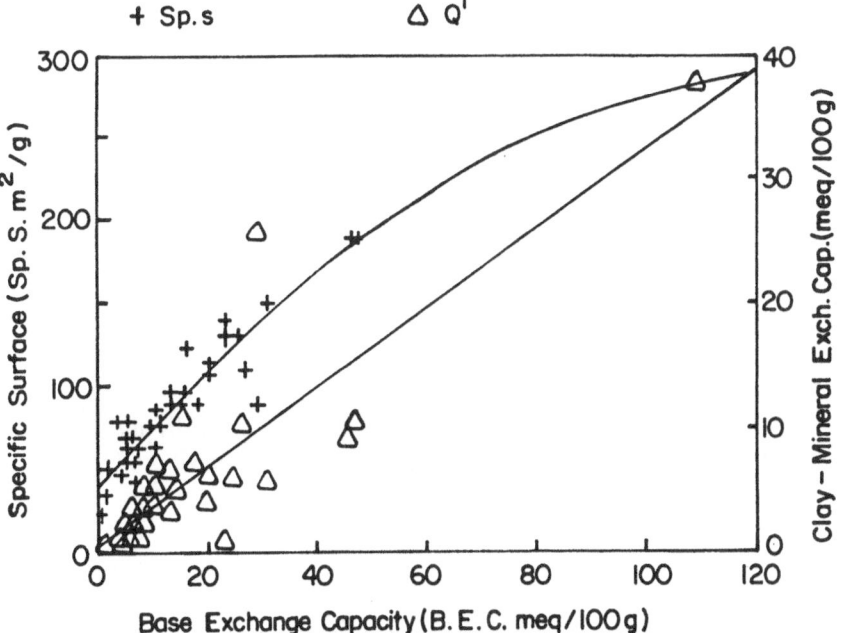

Fig. 1: Correlation of the base exchange capacity (BEC) with the specific surface (SpS) and the calculated clay-mineral exchange capacity (Q') of 45 different ocean and coastal sediments. Graph prepared from the original data of Duursma and Eisma (1973)

It is supposed that the treatment at pH 5.4 leaches the adsorbed part, and leaching at pH 2.3 removes part of the exchangeable absorbed fraction.

Most peculiar is that Cd and Sr seem to be only adsorbed, while the alkaline metal Cs is rather firmly bound. This is due to the fact that Cs is inter-layerly bound in a clay mineral like illite, its ionic radius matching the interlayer space.

Conclusion

It is essential to understand the complete picture of how compounds are sorbed to PM, and not only part of it. It is common pratice to correlate the adsorption of some metals (and also organochlorines) to the amount of clay, aluminium or organic matter of PM, and to correlate metal adsorption to only the clay or aluminium content. However, the causality of these correlations needs to be proven, since organic matter (POM, particulate organic matter) can be proportional to the clay content. This was determined for sediments of three estuaries together with a correlation coefficient r of 0.88 (Duursma et al., 1989).

Exchange Processes and Equilibria

In determining the distribution of natural and anthropogenic substances between the compartments water, PM, sediments and organisms, many errors are made by either approaching the involved processes too theoretically or by neglecting major competitive reactions that occur. A number of essential points will be treated.

Sea water contains a variety of cations, anions and uncharged molecules, and any reaction between two of them in the dissolved phase will have to compete with identical reactions of other (competitive) molecules. This has to be understood from the start, and should never be neglected.

Basic Exchange Reactions

The basic exchange reaction of for example a metal is given by the following reaction, from which the distribution coefficient k_d can be defined:

$$Me^{n+}Cl_n + nNaPM \underset{k_2}{\overset{k_1}{\leftrightarrow}} MePM + nNaCl \tag{4}$$

in which Me = a metal or a radionuclide in concentrations <<< than Na; k_1 et k_2 are two reaction speed constants, defined as:

$$k_1 = f_1 (Me^{n+}Cl_n) (NaPM)^n$$

$$k_2 = f_2 (MePM) (NaCl)^n \tag{5}$$

in which f_1 and f_2 are constant factors.

At equilibrium $k_1 = k_2$ and we obtain from (5) the following equation:

$$\frac{f_1}{f_2} = \frac{(MePM) (Me^{n+} Cl_n)}{Me^{n+}Cl_n} \times \frac{(NaCl)^n (Na\,PM)^n}{(Na\,PM)^n} \tag{6}$$

The second term is quasi constant because the dissolved Na and particulate Na are in excess to the dissolved Me and particulate Me, respectively. In consequence changes in dissolved Me to particulate Me will not significantly change the ratio dissolved Na to particulate Na. Hence the second ratio remains constant. Since f_1/f_2 is a constant, the first ratio $(MePM)/(Me^{n+}Cl_n)$ will be also a constant factor. This is called the distribution coefficient k_d.

Hence:

$$k_d = \frac{\mu g\ Me / g\ PM}{mg\ Me / g\ water} \quad \text{(dimensionless g g}^{-1}) \tag{7}$$

Since, however, reaction (4) is only semi-thermodynamic in nature, the k_d's have to be understood as apparent distribution or partitioning constants. However, as long as their values are known in space and time, processes of transfer and equilibrium can be studied with help of this factor.

So far the k_d was defined as a dimensionless factor, which was dimensionless on the basis of g g^{-1}. k_d values thus obtained can be used in determining the % distribution of metals between water and particulate matter. Knowing the k_d (as g g^{-1}), the suspended load, the water transport and the PM transport (sometimes because of settling and resuspension slower than the water), the effective transport can be calculated. It is only necessary to know for each spot and time what percentage of the metal is in solution and what percentage is particulate. k_d can also be expressed (dimensionless) in mL mL^{-1}, which is required for diffusion calculations [see formula (14)].

Suppose for the metal cadmium (Cd) there is P µg Cd L^{-1} (total Cd in solution particulate) and S mg PM L^{-1}. We take 1 L water as 1000 g and neglect the S mg PM in suspension. When the percentage Cd in solution is given as X % Cd, X can be found using the definition of k_d (7):

$$k_d = \frac{(\dfrac{100-X0}{100}\,P)\Big/(\dfrac{S}{1000})}{(\dfrac{X}{100}\,P)\Big/1000} \qquad (8)$$

From this equation the percentage X can be derived:

$$X = \frac{10^8}{Sk_d + 10^8} \text{ \% dissolved cadmium} \qquad (9)$$

This formula (9) can be graphically plotted, with the result that for rapidly exchanging metals on PM, the percentage dissolved can easily be read (Fig. 2 of Duursma and Bewers, 1986). For a k_d of 104 and S of 1 mg L^{-1}, X is 99,0 %. For the same k_d but with an S of respectively 10 and 100 mg L^{-1}, X is 90 and 50 %. This can have consequences at long distance, as is demonstrated by an example of plutonium. This nuclide, originating from the effluents of Sellafield on the Irish Sea, was detected in sediments close to Spitsbergen (Svalbard, 78° N and 20° W) at 2800 km distance. Plutonium is "strongly" adsorbed to PM and bottom sediments, with a k_d between 10^4 and 10^5. Surely the Irish Sea sediments have not been transported so far, but the transport must have occurred in dissolved form by the Gulf Stream, since S in the oceans is $\ll 1$ mg L^{-1}.

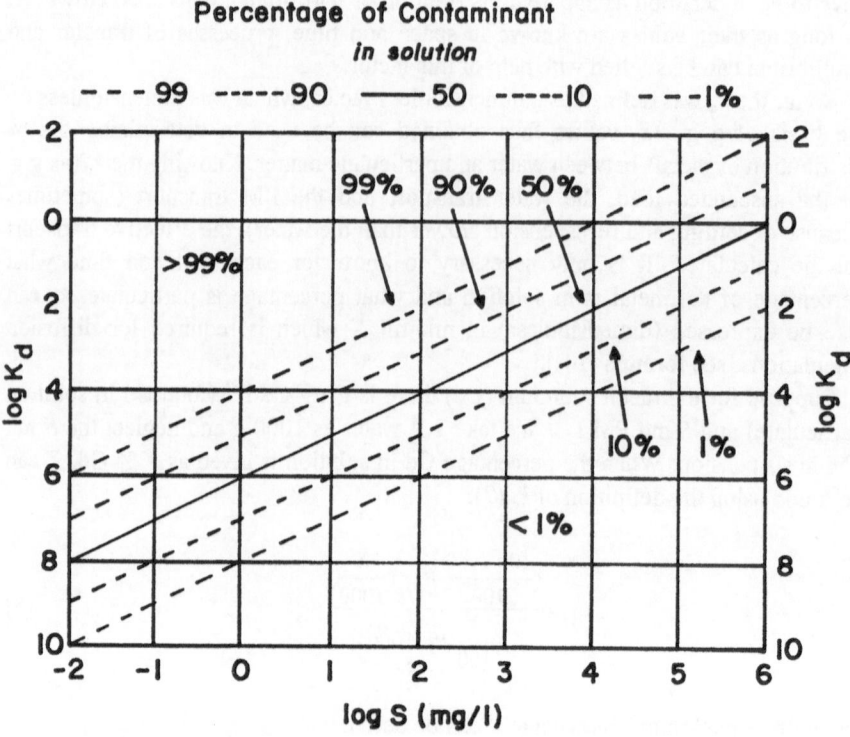

Fig. 2: Percentage of contaminant in solution as dependent on k_d and particulate matter (PM = S) load, according to formula (9)

Competition Between Complexation and Sorption

Most natural waters, and in particular freshwaters, contain in addition to suspended matter dissolved organic (or inorganic) matter that is capable of complexing trace metals, thus favouring them to be in dissolved form. For this three-compartment system a theoretical approach is possible. The reactions involved are the following:

$$Me^{++} + L^- \leftrightarrow MeL^+ \qquad\qquad k_{st} = \frac{MeL^+}{(Me^{++})\,(L^-)}$$

$$Me^{++} + MP^- \leftrightarrow MeMP \qquad\qquad k_d = \frac{(MePM)}{(Me^{++})} \qquad (10)$$

in which k_{st} is the stability constant of the above complexing reaction. This competition effect between complexation and sorption can be made clear with the folowing example. To a sea water suspension of PM and containing radioactive ^{60}Co, leucine is added. Sampling of 5 ml suspension is carried out after and PM on a filter. These are analysed for radioactivity mL^{-1} filtrate and radioactivity g^{-1} PM.

The apparent distribution coefficient k_{da} becomes, according to (7), the ratio of total particulate metal concentration to total dissolved metal concentration, which is:

$$k_{da} = \frac{(MePM)}{(Me^{++}) + (MeL^+)} \qquad (11)$$

from which, according to (10):

$$(MeL^+) = (Me^{++})(L^-) \times k_{st} \qquad (12)$$

From (10), (11) and (12) we can derive that:

$$\frac{1}{k_{da}} = \frac{1}{k_d}(1 + k_{st}\, L^-) \qquad (13)$$

Since k_d is the distribution coefficient without addition of ligand and k_{da} with addition of ligand, this latter formula can serve for determining stability constants (k_{st}) of complexing organic matter with metals. Just a simple experiment is required, such as adding progressive amounts of particulate matter to a solution containing metal and organic matter, or by adding progressive amounts of organic matter to a resuspension containing metal(s). But for natural waters the situation is more complicated, since the results are affected by competing reactions of other metals and H^+ ions with the same ligand. This is illustrated in Fig. 3 (Duursma, 1970). This figure shows that the determined $1/k_{da}$ values for ^{60}Co and [65]Zn seem to match the determined $1/k_{da}$ only when the competition of the major sea water ions H^+, Ca^{++} and Mg^{++} are taken into account. For these calculations the formulas (10) to (13) have been extended for these cations.

Conclusions

Any reaction in natural systems between compartments is multi-competitive.

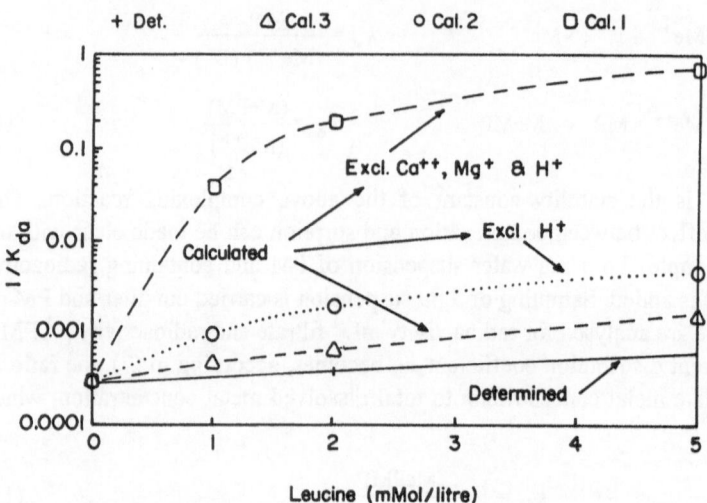

Fig. 3: Determined and calculated reciprocals of the apparent distribution coefficient k_{da} of ^{60}Co in relation with leucine additions to sea water. From Duursma (1970)

Correlation of k_d with Sediment and Metal Properties

There have been always intensive discussions on how exchange reactions occur between metals in solution and available sites on PM. Usually people believe that only the ionic metal species take part in the adsorption-desorption reactions.

Particulate matter has all kind of possibilities for adsorption reactions, like ion exchange, chemical binding and precipitation-dissolution (the latter are more or less independent of the surface properties). Taking this into account, many scientists believe that threshold phenomena may also play a role, so that exchange sites may become saturated. For ion exchange, however, this threshold approach does not seem to hold, since the principle of ion exchange is equilibrium, and concentrations in the water compartment would have to be extremely high to have 100% exchange sites occupied (in exchange forsomething else, which will then also become high in concentration in the water). Neither does it hold for precipitation, since precipitation may continue in many layers on the surface. For binding the threshold model might hold, although binding also has exchange reactions up to a certain equilibrium.

Results of k_d with Sediment and Radionuclide Properties

During 1965 and 1973 intensive studies were carried out to try to understand the major reactions of fallout radionuclides with various ocean sediments. The studies comprised the cooperative efforts of 20 national laboratories together with the marine laboratory of the International Atomic Energy Agency (IAEA), in Monaco.

Radionuclide sorption experiments were carried out with 35 characteristic ocean and coastal sediments which were sent to Monaco. Table 2, after Duursma and Eisma (1973), demonstrates clearly that for those 35 ocean sediments the range of k_d/BEC and that of the half-time of sorption ($t_{1/2}$) is large, but the order of magnitude is characteristic for each radionuclide.

Table 2: Average k_d/BEC (1/(meq/100g)) ratios and their range for 35 marine sediments and 14 radionuclides (Duursma and Eisma, 1973). The half-time of sorption ($t_{1/2}$) is indicative of the kind of sorption, where ion exchange is fastest and precipitation on surfaces slowest

Radionuclide	k_d/BEC Average	k_d/BEC Range	$t_{1/2}$ of sorption Days
^{45}Ca	45	?	<1
^{54}Mn	1000	200-2000	6.1±4.6
^{59}Fe	1250	400-2500	18.0±6.5
^{60}Co	?	100-4000	8.2±5.1
^{65}Zn	150	100-400	1.9±1.3
^{86}Rb	620	?	<0.5
^{90}Sr/Y	1	0.5-1.5	2.3±1.7
^{95}Zn/Nb	1000	200-1600	5.7±2.3
^{106}Ru	2850	1400-5000	8.9±3.3
^{137}Cs	50	25-90	0.7±0.4
^{144}Ce	4000	2000-8000	10.5±4.4
^{147}Pm	3300	2800-6700	11.4±3.5
^{210}Pb	2700	?	?
^{239}Pu	52		?

Conclusion

Although correlations may be found, extrapolation to other sediments is speculative.

Determined and Calculated Diffusion Coefficients of Ten Radionuclides

Since diffusion of radionuclides in sediments is a very slow process, the determination of diffusion coefficients is a tedious process. Therefore it has been made possible by calculation, using the k_d and applying formula (14) developed by Duursma and Hoede (1976):

$$D_{\text{metal}} = \frac{D_{\text{molecular}}}{1 + k_d} \tag{14}$$

In order to confirm the validity of this calculation, it was compared with the direct method of determining the diffusion coefficient (Table 3).

The results given in Table 3 demonstrate that a fairly rough agreement exists, at least in the order of magnitude. This is rather astonishing, taking into regard that the k_ds were determined during a four-week experiment, and k_d may be higher when sorption occurs over the period of a a year as shown by Duursma and Bewers (1970) for ^{65}Zn.

Conclusion

Although it is determined that k_d are only empirical factors which can depend on many circumstances, at least in one sediment it was demonstrated that the order of magnitude of experimentally determined diffusion matched that of the calculated ones.

Table 3: The diffusion technique with instantaneous source (Duursma and Bosh, 1970) was applied for the direct determination of D_{metal}, having as sources ten radioisotopes. The time of the experiment for allowing diffusion was 1 yr. The sediment was Mediterranean clay sediment. With the same sediment and radionuclide k_d was determined. By applying formula (14) the 'calculated' D_{metal} was determined

Radionuclide	Diffusion Coefficient (D_{metal}) $\times 10^{-10}$ cm^2/sec (average values)	
	Calculated	Determined
^{90}Sr	17,500	9,000
^{137}Cs	60	?
^{136}Ru	22	95
^{59}Fe	4.5	82
^{65}Zn	12.2	10
^{60}Co	25	15
^{147}Pm	2	7
^{54}Mn	4	4
^{95}Zr	2.8	33
^{144}Ce	2	13

Residence Times of Metals in an Estuary

In many estuaries of the world, bottom sediments are contaminated with heavy metals. Dissolved metals are both sorbed to PM, settled to the bottom and migrating by diffusion into the bottom.

A case study on process, comparable to what actually happens at Sepetiba Bay west of Rio de Janeiro (Lacerda and Barcellos, 1993), has been made in a Dutch estuary, the Haringvliet, where for a number of years the dissolved and sediment-attached metals were analyzed. From these analyses Van de Vrie and Duursma (1986) calculated the turn-over time (residence time) and 50% adaptation time of three metals, Zn, Cd and Pb, with a 2-box-model (Table 4).

Table 4: Turn-over and 50% adaptation times for Zn, Cd and Pb for water and a 10 cm top layer of the sediment of the Haringvliet (Netherlands)

| | Turn-over time (yr) | | | | 50% adaptation time (yr) | | | |
| | water | | sediment | | water | | sediment | |
	Box 1	Box 2	Box 1	Box 2	Box 1	Box 2	Box 1	Box 2
Zn	42	86	975	1230	18	37	24	45
Cd	42	86	975	1230	34	68	35	69
Pb	39	86	975	1230	28	59	36	70

Conclusion

From these data it can be concluded that it will take at most 1230 yr to contaminate a bottom layer of 10 cm (Box 2) or to have it completely decontaminated, supposing the water over the estuary becomes clean. For 50% of the contamination level this will happen in 45 yr for Zn, 69 yr for Cd and 70 yr for Pb, respectively, again for Box 2.

Estuaries, Sinks or Sources?

For estuaries with a long freshwater-seawater mixing section and a turbidity maximum, the question can be posed whether the distribution of the dissolved and particulate contaminants along the estuary is conservative, meaning that the total inflow of contaminants (particulate and dissolved) from the river equals the discharge from the estuary into the sea, while there is no loss to or gain from the bottom. Model circulations have been made to solve this problem for different adsorption and complexing processes (Duursma and Ruardij, 1989).

In Fig. 4, results from such a model calculation are presented with as boundary conditions $k_d = 10^5$ (k_d = (µg contaminant g^{-1} particulate matter)/µg contaminant/ mL water) and $k_{st} = 10^2$ (k_{st} = (Me$_{compl}$.) / (Me) × (Organic matter)), called a stability coefficient of a metal with organic matter. For a given particulated matter distribution with turbidity maximum, the amount of total metal (T_{MET} = total dissolved + total particulated metal, expressed in µg L^{-1} water) also shows a maximum at the point of the residence times of water (which are constant) and particulate matter (which is depend on the PM concentration).

Once these boundary conditions for the processes involved have been established, it is possible to investigate whether real sources or sinks exist in the system. These have to be found from the difference of the actually determined concentrations as compared to the "mixing" concentrations.

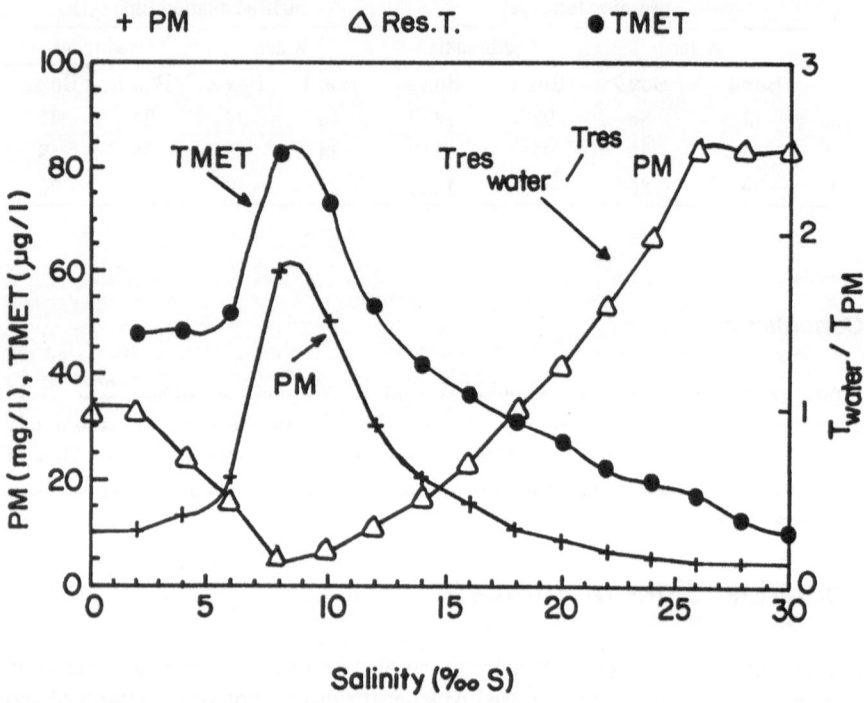

Fig. 4: Calculation of the total amount of contaminant in solution and in particulate matter for a model estuary with steady state contaminant input (from the river) and output (to the sea), where the distance is a linear function of the salinity. There is no loss to or gain from the bottom, which means a full conservative situation. Run 3 of the model given by Duursma and Ruarddij (1989). The residence time of particulate matter is the largest at the turbidity maximum

With these calculations in mind it is possible to determine the contamination from effluents into the estuary. The total transport per cross-section should differ along the estuary. In that case the exchange between water and bottom sediments plays a buffering role with turn-orver times as mentioned in Table 4.

Organochlorines

For organochlorine pesticides similar calculations can be made, although time-dependent transformation processes are involved. In particular in the tropics such processes of transformation are more rapid than in temperate water (Duursma, 1976), and are additionally accelerated in estuaries by the presence of microbiota attached to particulate matter. This goes also for the new generation of organophosphorus pesticides (Readman et al., 1992) which have their impact in tropical estuarine and lagoon systems.

In 1962 the first edition of the "Silent Spring" by Rachel Carson (1987) was published, which brought a large change in public and scientific opinion on the use of persistent pesticides. DDT and its metabolites were found as residues in the lipids of very remote living animals and birds, while the decrease in number of top predators in the food chain was very clear with birds. The idea of accumulation through the food chain was launched and is still quoted today as a major environmental principle.

Since DDT was banned from the early 1970s, although for anti-malaria programmes production may still be going on, the DDT concentrations have decreased largely in environmental samples. Attention is now focused on PCBs (polychlorinated biphenyls), which product has been produced and applied worldwide during the last 40 years, in particular as a useful nonflammable grease or oil. Beside other applications it was used as isolation liquid in electrical transformers and grease in mines.

However, PCBs are even more persistent than DDT and have similar and for some of its congeners higher toxicity values, such as in vivo ED-50 inhibition body weight gain of 750 and 1120 μmol kg^{-1} (234 and 350 mg kg^{-1}) for the PCB congeners 105 and 118 respectively (Goldstein and Safe, 1989). For about ten years, there has been a ban on PCBs production and use. However, the world is left with about 1.2 million tons of PCB (Tanabe, 1988), of which 65% is in storage or in use, 31% in the environment (three-fifths of it in the marine environment), while only < 3.2% is degraded or burned. The option to burn PCB has not yet been applied extensively, due to the very high temperatures required, and the possible production of toxic oxines.

PCB Behaviour in Estuarine Compartments of Water, PM and Sediments

PCB congeners are non-polar compounds and have therefore a low solubility in water. They adsorb to solid surfaces (being less polar than water) and "dissolve" easily in lipids and fats. Their degree of non-polarity is defined by the partition ratio between water and octanol, giving a partition coefficient k_{ow} which is defined as:

$$k_{ow} = \frac{(PCB)_{octanol}}{(PCB)_{water}} \tag{15}$$

Each PCB congener has a specific k_{ow} value, which almost corresponds proportionally with its IUPAC number (Brownawell and Farrington, 1985; and Duursma et al., 1986).

The principle of a partition coefficient k_d, as defined by formula (7), is adopted for studying the partition of PCB congeners between the dissolved and particulate compartment:

$$k_d = \frac{\mu g \ PCB \ g^{-1}(dry)PM}{\mu g \ PCB \ g^{-1}water} \tag{16}$$

A study was made to determine this k_d in an estuarine system where a large variety of salinities, PM concentrations and PCB concentrations occurred. As shown in Fig. 5 (after Duursma et al., 1965), log k_d (which is the left bar) is at the level of about 5 (k_d about 10^5), independent of the salinity (right bar) of the concentration of PCB in solution (middle bar).

PCB Accumulation in Organisms

Distribution Coefficients (k_d)

The thesis whether accumulation through the food chain is correct or incorrect is essential to understanding the impact of this pollutant on ecosystems and man. In 1982 a German scientist (Schneider, 1982) posed the thesis that organochlorines in marine organisms were in apparent equilibrium with the environmental concentrations of these compounds in the water compartment. This thesis implies that although there is uptake of organochlorines through the food chain, the eventual level of contamination is independent of this food chain. Calculated on a (lipid) fat basis, the level of DDT and PCB should be equal for those organisms occurring in the same region. figure 6 (after Duursma et al., 1986) presents k_d values for Σ PCBs (on a lipid basis) of a number of benthic organisms, which individually have different body lipid contents. The log k_d are, even for eggs of shrimp, all of the level of 6.5 + 0.2. The findings by Duursma et al. (1989) for mussels confirm these individual PCB congeners, since the standard error for k_d is very low for k_d determinations over a period of 1.5 yr. For some congeners there is even little difference between the stations.

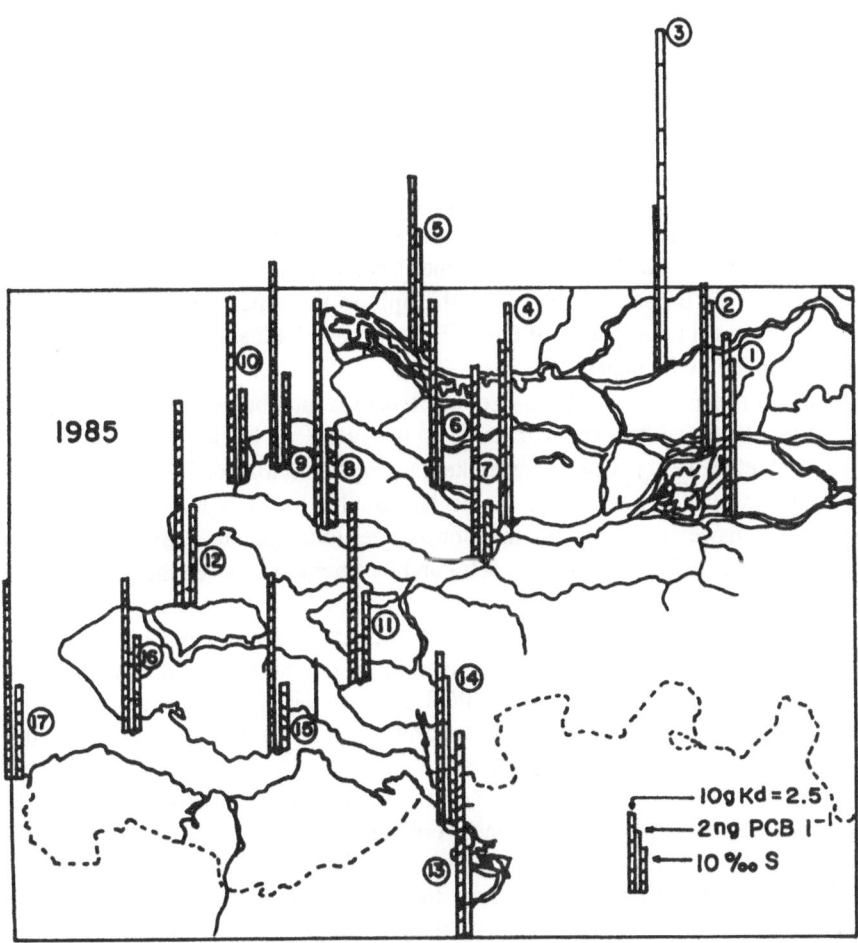

Fig. 5: Concentrations of dissolved PCB (*middle bars*), log k_d (*left bars*) of PCB partitioning between water and particulate matter and salinities (*right bars*). From Duursma et al. (1989)

The Eel case

For the European eel *Anguilla anguilla*, which reproduces in the open sea, the question was posed to what extent contamination by PCBs, accumulated in the lipid fraction of these organisms during their stay in contaminated freshwater systems, continues to be present during migration into the open sea. Although such specific partitioning of PCB congeners has not yet been demonstrated for eel, the principle of constant partitioning between water and lipids in living organisms suggests that loss of PCB congeners should occur when eel migrates from freshwater systems to the sea, where lower dissolved PCB concentrations occur. It is known that during migration to their ocean reproduction regions eels fast and have to live on their reserve of lipids and organs such as their alimentary tract. This may have a particular effect on their metabolism and as such also on the processes of PCB congener loss from the lipid to the water compartments. Shifts may also occur in the spectra of the PCB congeners, indicating specific metabolic activity on the PCB congeners.

A simulation experiment was set up where a number of female and male migration adult silver eels, captured in the Waal, a branch the river Rhine (Duursma et al. 1991). The eels were kept under starvation conditions for 91 days in running seawater. Eels were periodically sampled for PCB analysis, in order to obtain an answer to the above-mentioned questions and, in particular, to how starvation affects PCB congener spectra and partitioning between the lipid (in eel) and water compartment.

The results of this study showed that a constant partitioning coefficient k_d (defined as the ratio of PCB in lipid, given in $\mu g\ g^{-1}$ and dissolved PCB in $pg\ g^{-1}$ with time was not found for selected PCB congeners in eel. There was no loss of PCB from the eel to the water compartment, such as should be expected on the basis of the equal partitioning hypothesis.

The explanation given is that for aquatic organisms equal partitioning of PCB between water and lipids in organisms is only possible when the organisms have an active metabolism. This would match the results that for various marine organisms under normal *in situ* feeding conditions such an equilibrium is attained, while for starving eels with no metabolic activity the exchange between PCBs in lipids and the water compartment outside these eels has been blocked.

For other reasons, the exchange of PCBs with the environment is also "blocked" for terrestrial organisms, such as birds and mammals which feed on marine organisms. Therefore they can accumulate PCBs in their lipids to a much higher level than was found for aquatic organisms. This also goes for man.

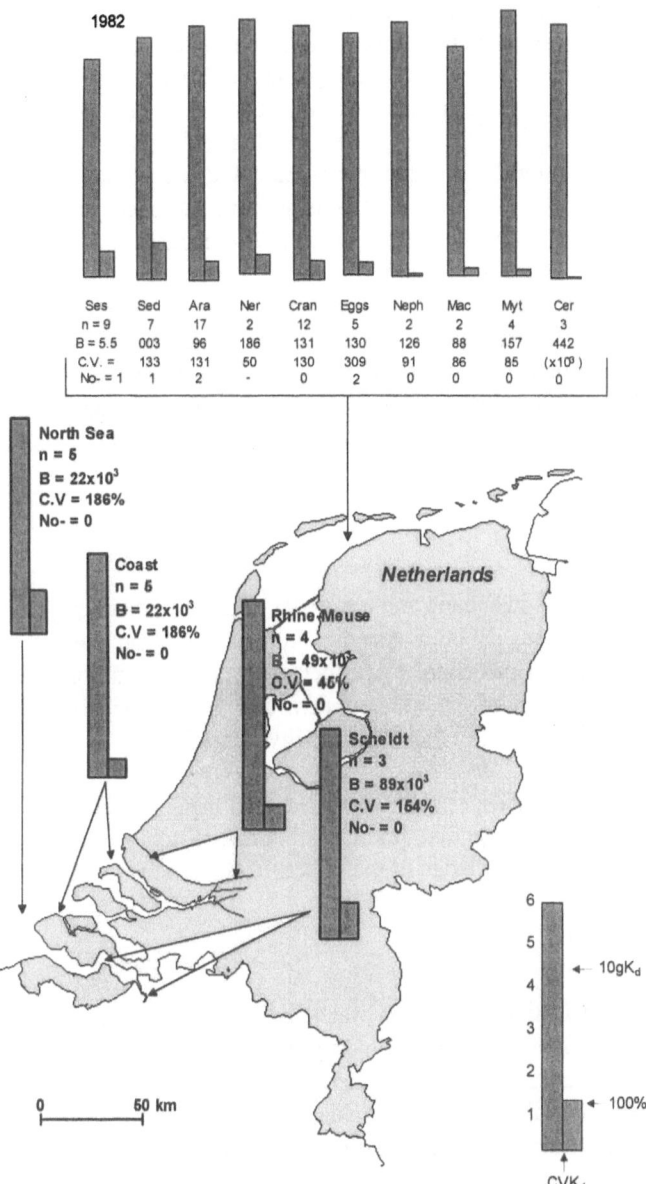

Fig. 6: Log k_d of partitioning of PCB between lipids in various organisms and sea water (*upper bars*) and of PCB between particulate matter and sea water (*lower four bars*). From Duursma et al. (1989). Ses. = seston or PM, Sed = sediment, Ara = *Arenicola marina,* Ner = *Nereis virens,* Cran = *Crangon crangon,* Eggs = eggs from shrimp, Neph = *Nephtys hombergii,* Mac = *Macoma balthica,* Myt = *Mytilus edulis* and Cer = *Cerastoderma edule*

The principle of accumulation in the food chain has therefore to be understood in another sense. Within the aquatic environment there is no accumulation through the food chain, at least when there is active feeding and metabolism involved. But outside the aquatic system, birds, mammals and other organisms can accumulate PCB to high levels in their lipids. Obviously these PCBs must have arrived there through food and drinking water. What is not true is that there is a gradual increase towards the top predator, since all of the organisms have the same chance of accumulating PCB in their lipids. In that sense it rather depends on the food consumed than on the place in the food chain.

The principle of accumulation in the food chain has also often been used for metals and radionuclides. Here there is even less reason for a general principle, since accumulation is different for each metal, and specific with respect to many processes involved. It is known, for example, that mercury accumulation is time-dependent, increasing with age in, for example, Mediterranean tuna fish (Bernhard, 1985, 1988).

Conclusion

The behaviour of contaminants in aquatic and terrestrial organisms, including man, that feed on aquatic organisms is both specific and complex and cannot be approached by too simplistic a principle. Nevertheless, exchange equilibria may exist between the contamination of the water system and the aquatic organisms only, of which use can be made to test the water quality by analysing for example shellfish.

Equilibria will certainly be faster established at elevated temperatures, such as in the tropics, which also indicates that findings obtained in temperate waters cannot be extrapolated without knowledge of the effects of temperature.

General Conclusions

1. Contaminants may be partitioned between water, sediment, particulate matter and organism compartments, for which partition coefficients (k_d) can be determined. These k_d are useful empirical factors to calculate the percentage distribution between dissolved and particulate matter, the diffusion of contaminants into the bottom and accumulation in organisms. Thus analysing sedimentary materials and organisms may give useful indications of the concentrations of the contaminants in the water compartment.

2. The question of whether tropical estuaries are sinks or sources has to be answered in a time context. A newly contaminated tropical estuary will be a sink for a very long period, although equilibrium between input and output may be obtained sooner than in temperate climates. It may afterwards become a source where the bottom sediments act as a buffer with very long turn-over times of up to 1000 years.

References

Al-Gharib, I. (1982). *Apport des isotopes à vie moyenne de l'Uranium et du Thorium,* ^{210}Pb *et* ^{10}Be *dans l'ètude de l'èrosion chimique et physique de deux grands bassins: Amazone et Congo.* Thesis, University of Nice-Sophia Antipolis, pp.284.

Bernhard, M. (1985). Mercury accumulation in the pelagic foodchain. In: *Environmental Inorganic Chemistry.* A.E. Matertell and K.J. Irgolic (eds.) Verlag Chemie, Weinheim, pp.:349-358.

Bernhard, M. (1988). *Mercury in the Mediterranean.* UNEP Regional Seas Reports and Studies, No. 89.

Brownawell, B.J. and J.W. Farrington (1985). Partitioning of PVB's in marine sediments. In: *Marine and Estuarine Geochemistry.* A. C. Sigle and A. Hattori (eds.), Lewis Publ. Chap. 7, 1-24.

Carson, R. (1987). *The Silent Spring.* 25 Anniversary Ed., Houghton Mifflin, Boston, USA, 448pp.

Duursma, E.K. (1970). Organic chelation of ^{60}Co and ^{65}Zn by leucine in relation to sorption by sediments. In: *Organic Matter in Natural Waters.* D.W.Hood (ed.), Inst. Mar. Sc. Univ. of Alaska Publ. no. 1:387-397.

Duursma, E.K. (1976). *Role of pollution and pesticides in brackish water aquaculture in Indonesia.* UNPD/FAO, W/H9164, FI:INS/72/003/4, May 1976, pp.42.

Duursma, E.K. and Bewers, J.M. (1986). Application of *k*d's in marine geochemistry and environmental assessment. In: *Application of Distribution Coefficients to Radiological Assessment Models.* T.H. Sibley and C. Myttenaere (eds.) Elsevier Appl. Sci. Publ., London. pp.:138-165.

Duursma, E.K. and C.J. Bosch (1970). Concerning diffusion in sediments and suspended particles in the sea. Part B. Methods and Experiments. *Neth. J. Sea Res.* 4:395-469.

Duursma, E.K.; Dawson, R. and J.Ros Vicent (1975). Competition and time of sorption for various radionuclides and trace metals by marine sediments and diatoms. *Thalass. Jugoslavia,* 11:47-51.

Duursma, E.K. and Eisma, D. (1973). Theoretical experimental and field studies of radioisotopes concerning molecular diffusion of radioisotopes with sediments and suspended particles in the sea. Part C: Applications to field studies. *Neth. J. Sea Res.,* 6:265-342.

Duursma, E.K. and C. Hoede (1967). Theoretical, experimental and field studies concerning molecular diffusion of radioisotopes in sediments and suspended solid particles of the sea. Part A: Theories and mathematical calculations. *Neth. J. Sea Res.,* 3:423-457.

Duursma, E.K., J. Nieuwenhuize and J.M. Van Liere (1986). Polychlorinated biphenyl equilibria in an estuarine system. *Sci. Total Environm.,* 79:141-155.

Duursma, E.K., J. Nieuwenhuize and J.M. Van Liere (1989). Partitioning of organochlorines between water, particulate matter and some organisms in estuarine and marine systems of the Netherlands. *Neth. J. Sea Res.,* 20:239-251.

Duursma, E.K., J. Nieuwenhuize, J.M. Van Liere, C.M. de Rooy, J.I.J. Witte and J. Van der Meer (1991). Possible loss of PCBs from migrating European silver eels; a 3-month simulation experiment. *Mar. Chem.,* 36:215-232.

Duursma, E.K. and P. Ruardij (1989). Conservative mixing in estuaries as affected by sorption, complexing and turbidity maximum: a simple model calculation. *Mar. Chem.,* 28:251-258.

Dyal, R.S. and S.B. Hendricks (1950). Total surface of clays in polar liquids as a characteristic index. *Soil Sci.,* 62:420-432.

Goldstein, J.A. and S. Safe (1989). Mechanisms of action and structure-activity relationships for the chlorinated dibenzo-p-dioxins and related compounds. In: *Halogenated Biphenyls,*

Triphenyls, Naphtalenes, Dibenzodioxins and Related Compounds. R.D. Kimbrayh and A.A. Jensen (eds.). Topics in Environmental Health 4, Chap. 9, 2nd. ed. Elsevier, Amsterdam, pp.:239-293.

Grim, R.E. (1953). *Clay Mineralogy.* Mc. Graw-Hill, London, 384pp.

Lacerda, L.D. and C. Barcelos (1993). Cadmium and zinc pathways differentiation in coastal environments. In: *Perspectives for Environmental Geochemistry in Tropical Countries.* J.J.Abrão, J.C. Wasserman and E.V. Silva Filho (eds.), Niterói, Brasil, 137-141.

Mee, L.D. (1978). Coastal Lagoons. In: *Chemical Oceanography*, J.P. Riley and R.Chester (eds.), 2nd ed., Academic Press, London, pp.:441-490.

Mee, L.D. and J.I. Osuna-Lopez (1991). The trace metal chemistry of a tropical hypersaline basin: evidence from field and laboratory experiments. *Ciências Marinas*, 17:19-45.

Milliman, J.D. (1991). Flux and fate of fluvial sediment and water in coastal seas. In: *Ocean Margin Processes in Global Change*. R.F.C. Mantoura, J.M. Martin and R. Wollast (eds.), John Wiley and Sons Ltd., Chichester, U.K., pp.:69-89.

Readman, J.W., L.Liong Wee Kwong, L.D.Mee, J. Bartocci, G. Nilve, J.A. Rodrigues-Solano and F. Gonzales-Farias (1992). Persistent organophosphorus pesticides in tropical marine environments. *Mar. Poll. Bull.*, 24:398-402.

Schneider, R. (1982). Polychlorinated biphenyls (PCBs) in cod tissues from the western Baltic: Significance of equilibrium partitioning and lipid composition in the bioaccumulation of lipophilic pollutants in gill-breathing animals. *Meere sf. Rep. Mar. Res.* (Ber. D. Wiss. Kom. Meeresf.), 29:69-79.

Tanabe, S. (1988). PCB problems in the future: foresight from current knowledge. *Environm. Pollut.*, 50:5-28.

Van De Vrie, E.M. and E.K. Duursma (1986). Residence times of contaminants in a lake-canal system: a delta case study in the SW Netherlands. *Mar. Chem.*, 18:171-188.

Index

Springer
and the
environment

At Springer we firmly believe that an
international science publisher has a
special obligation to the environment,
and our corporate policies consistently
reflect this conviction.
We also expect our business partners –
paper mills, printers, packaging
manufacturers, etc. – to commit
themselves to using materials and
production processes that do not harm
the environment. The paper in this
book is made from low- or no-chlorine
pulp and is acid free, in conformance
with international standards for paper
permanency.

Springer

Lecture Notes in Earth Sciences

Vol. 37: A. Armanini, G. Di Silvio (Eds.), Fluvial Hydraulics of Mountain Regions. X, 468 pages. 1991.

Vol. 38: W. Smykatz-Kloss, S. St. J. Warne, Thermal Analysis in the Geosciences. XII, 379 pages. 1991.

Vol. 39: S.-E. Hjelt, Pragmatic Inversion of Geophysical Data. IX, 262 pages. 1992.

Vol. 40: S. W. Petters, Regional Geology of Africa. XXIII, 722 pages. 1991.

Vol. 41: R. Pflug, J. W. Harbaugh (Eds.), Computer Graphics in Geology. XVII, 298 pages. 1992.

Vol. 42: A. Cendrero, G. Lüttig, F. Chr. Wolff (Eds.), Planning the Use of the Earth's Surface. IX, 556 pages. 1992.

Vol. 43: N. Clauer, S. Chaudhuri (Eds.), Isotopic Signatures and Sedimentary Records. VIII, 529 pages. 1992.

Vol. 44: D. A. Edwards, Turbidity Currents: Dynamics, Deposits and Reversals. XIII, 175 pages. 1993.

Vol. 45: A. G. Herrmann, B. Knipping, Waste Disposal and Evaporites. XII, 193 pages. 1993.

Vol. 46: G. Galli, Temporal and Spatial Patterns in Carbonate Platforms. IX, 325 pages. 1993.

Vol. 47: R. L. Littke, Deposition, Diagenesis and Weathering of Organic Matter-Rich Sediments. IX, 216 pages. 1993.

Vol. 48: B. R. Roberts, Water Management in Desert Environments. XVII, 337 pages. 1993.

Vol. 49: J. F. W. Negendank, B. Zolitschka (Eds.), Paleolimnology of European Maar Lakes. IX, 513 pages. 1993.

Vol. 50: R. Rummel, F. Sansò (Eds.), Satellite Altimetry in Geodesy and Oceanography. XII, 479 pages. 1993.

Vol. 51: W. Ricken, Sedimentation as a Three-Component System. XII, 211 pages. 1993.

Vol. 52: P. Ergenzinger, K.-H. Schmidt (Eds.), Dynamics and Geomorphology of Mountain Rivers. VIII, 326 pages. 1994.

Vol. 53: F. Scherbaum, Basic Concepts in Digital Signal Processing for Seismologists. X, 158 pages. 1994.

Vol. 54: J. J. P. Zijlstra, The Sedimentology of Chalk. IX, 194 pages. 1995.

Vol. 55: J. A. Scales, Theory of Seismic Imaging. XV, 291 pages. 1995.

Vol. 56: D. Müller, D. I. Groves, Potassic Igneous Rocks and Associated Gold-Copper Mineralization. 2nd updated and enlarged Edition. XIII, 238 pages. 1997.

Vol. 57: E. Lallier-Vergès, N.-P. Tribovillard, P. Bertrand (Eds.), Organic Matter Accumulation. VIII, 187 pages. 1995.

Vol. 58: G. Sarwar, G. M. Friedman, Post-Devonian Sediment Cover over New York State. VIII, 113 pages. 1995.

Vol. 59: A. C. Kibblewhite, C. Y. Wu, Wave Interactions As a Seismo-acoustic Source. XIX, 313 pages. 1996.

Vol. 60: A. Kleusberg, P. J. G. Teunissen (Eds.), GPS for Geodesy. VII, 407 pages. 1996.

Vol. 61: M. Breunig, Integration of Spatial Information for Geo-Information Systems. XI, 171 pages. 1996.

Vol. 62: H. V. Lyatsky, Continental-Crust Structures on the Continental Margin of Western North America. XIX, 352 pages. 1996.

Vol. 63: B. H. Jacobsen, K. Mosegaard, P. Sibani (Eds.), Inverse Methods. XVI, 341 pages, 1996.

Vol. 64: A. Armanini, M. Michiue (Eds.), Recent Developments on Debris Flows. X, 226 pages. 1997.

Vol. 65: F. Sansò, R. Rummel (Eds.), Geodetic Boundary Value Problems in View of the One Centimeter Geoid. XIX, 592 pages. 1997.

Vol. 66: H. Wilhelm, W. Zürn, H.-G. Wenzel (Eds.), Tidal Phenomena. VII, 398 pages. 1997.

Vol. 67: S. L. Webb, Silicate Melts. VIII. 74 pages. 1997.

Vol. 68: P. Stille, G. Shields, Radiogenetic Isotope Geochemistry of Sedimentary and Aquatic Systems. XI, 217 pages. 1997.

Vol. 69: S. P. Singal (Ed.), Acoustic Remote Sensing Applications. XIII, 585 pages. 1997.

Vol. 70: R. H. Charlier, C. P. De Meyer, Coastal Erosion – Response and Management. XVI, 343 pages. 1998.

Vol. 71: T. M. Will, Phase Equilibria in Metamorphic Rocks. XIV, 315 pages. 1998.

Vol. 72: J. C. Wasserman, E. V. Silva-Filho, R. Villas-Boas (Eds.), Environmental Geochemistry in the Tropics. XIV, 305 pages. 1998.